Vegetation Dynamics of Mongolia

D1920213

To: our friends Bernd and Antje from Russian colleagues! With best wishes!

Peter Gunin

[illegible]

12.01.2003.
Berlin

Geobotany 26

Series Editor

M.J.A. WERGER

The titles published in this series are listed at the end of this volume.

Vegetation Dynamics of Mongolia

edited by

Peter D. Gunin
Elizabeth A. Vostokova
Nadezhda I. Dorofeyuk
Laboratory of Ecology of Arid Areas,
Severtsov Institute of Ecology & Evolution,
Russia

Pavel E. Tarasov
Department of Geography,
Moscow State University,
Russia

and

Clanton C. Black
Department of Biochemistry & Molecular Biology,
The University of Georgia,
U.S.A.

Kluwer Academic Publishers
Dordrecht / Boston / London

A C.I.P. Catalogue record for this book is available from the Library of Congress.

ISBN 0-7923-5582-2

Published by Kluwer Academic Publishers,
P.O. Box 17, 3300 AA Dordrecht, The Netherlands.

Sold and distributed in North, Central and South America
by Kluwer Academic Publishers,
101 Philip Drive, Norwell, MA 02061, U.S.A.

In all other countries, sold and distributed
by Kluwer Academic Publishers,
P.O. Box 322, 3300 AH Dordrecht, The Netherlands.

Printed on acid-free paper

All Rights Reserved
©1999 Kluwer Academic Publishers
No part of the material protected by this copyright notice may be reproduced or
utilized in any form or by any means, electronic or mechanical,
including photocopying, recording or by any information storage and
retrieval system, without written permission from the copyright owner

Printed in the Netherlands.

CONTENTS

INTRODUCTION TO STUDIES ON THE VEGETATION OF MONGOLIA 1

NATURAL AND ANTHROPOGENIC FACTORS AND THE DYNAMICS OF VEGETATION DISTRIBUTION IN MONGOLIA 7

1.1 Introduction 7

1.2 Natural features of Mongolia 8

1.3 Landscape-ecological regions 14

1.4 Landscape and ecological factors of vegetation dynamics 29

1.5 Conclusion 43

LATE QUATERNARY VEGETATION HISTORY OF MONGOLIA 45

2.1 Introduction 45

2.2 An overview of previous studies 46
Studies on the vegetation history of Mongolia 46
Modern pollen spectra: problems of interpretation 48
Pollen preservation 48
Long distant transport and re-deposition of pollen 48

2.3 Data used in this study 49
Modern data 49
Fossil data 50

2.4 Regional pollen records from individual sites 54
Hoton-Nur Lake 54
Achit-Nur Lake 57
Dood-Nur Lake 59
Daba-Nur Lake 60
Yamant-Nur Lake 62
Gun-Nur Lake 64

2.5 Holocene changes in the distribution of tree and shrub taxa in Mongolia validated by plant macrofossil records 65
Results 65
Interpretation and discussion 67

2.6 Spatial reconstruction and mapping of Mongolian vegetation during the last 15,000 years using a «biomization» method 68
Summary of the method 68
Implementation for Mongolia 69
Validation of the method: present-day pollen-derived biome reconstruction 72
Application to the fossil pollen data 73

2.7 General discussion and conclusions 76

ASSESSING PRESENT-DAY PLANT COVER DYNAMICS 79

3.1 Introduction. Modern Methods for Studying and Monitoring Plant Cover 79
3.2 Mountain Plant Community Dynamics 81
3.3 Plant Community Dynamics in Plains and Rocky Areas 96
3.4 Dynamics of Water-Associated Vegetation 118
3.5 Conclusions 128

ANALYSIS OF PRESENT-DAY VEGETATION DYNAMICS 131

4.1 Basic changes in vegetation 131
4.2 Regressive plant community successions 134
4.3 Progressive plant community regeneration 149
4.4 Mapping vegetation dynamics 157
4.5 Conclusions 163

STRATEGIES FOR NATURE MANAGEMENT AND VEGETATION CONSERVATION 165

5.1 Introduction. Methods for vegetation conservation 165
5.2 Restoration and conservation of botanical successions 173
5.3 Systems for the conservation of botanical diversity 188
5.4 Conclusions 202

SUMMARY CONCLUSIONS AND RECOMMENDATIONS 203

References 205
Appendix 1 217
Appendix 2 227
Index 233

PREFACE

Mongolia is situated where contrasting geological structures intersect, within the zone of interaction of different systems of global atmospheric circulation. Here are situated the watershed between the basins of the Arctic and Pacific Oceans and the closed depressions of Central Asia. Here, at the junction of the Siberian taiga forests, Dahurian steppes, and Gobi desert is a crossroads of plant and animal distribution. The territory of Mongolia as a whole is characterized by great diversity and a particularly complex spatial structure of soil and vegetation cover.

Low human population density and preservation to a great extent of a traditional economy are the prerequisites of large-scale nature management in the region. The predominant ecosystems have not drastically changed and frequently function in a manner close to natural regimes. Some of the Earth's heavily transformed biomes (such as steppes, which are destroyed by plowing almost everywhere) are well-conserved in Mongolia and the adjacent areas of China. Mongolia is one of those regions which can make a critical contribution to the ecological health of our planet due to its unique characteristics and undamaged state.

However, we must bear in mind that anthropogenic pressure on the natural complexes of Mongolia, most importantly the vegetation cover, is increasing rapidly as well. More and more areas and a much broader set of ecosystems are being converted into agricultural lands. Due to the high vulnerability characteristic of arid and semi-arid regions, even relatively weak anthropogenic impacts can seriously damage the natural balance and trigger processes of progressive degradation and desertification. Long-term economic pressure on the vegetation cover often leads to considerable changes in the natural dynamics of development of the latter, including changes in its direction. For example, despite the fact that the dry steppes of Mongolia have experienced increasing moisture over the last two or three centuries, long-term degradation of grazing lands had led to the growth of xerophytic features in the grass cover.

Various areas of Mongolia are faced with the occurrence of such negative phenomena as local deflation, sheet and linear erosion. As a result, the dynamic processes of vegetation cover change intensify. Grazing lands experience continuous degradation, and forests are decreasing in area due to uncontrolled cutting and fires. The continued existence of rare plant and animal species, as well as ecosystems, is still threatened. Further development of land resources leads to a decrease in diversity of the original biota and simultaneously encourages the spread of broadly distributed and even cosmopolitan weed species.

At the same time, the strategy for conservation of plant communities and ecosystems as a whole in Mongolia is not still sufficiently detailed to meet the requirements of current ecological and economic conditions. Many aspects of the strategy are still being tested – new approaches and solutions are indicated. Moreover, tactical solutions to problems of nature protection in Mongolia are affected not only by the high dynamic rate of anthropogenically induced processes in the contemporary vegetation cover, but also by inherited paleogeographical features of its development. Up until now the latter have been weakly investigated, and are virtually absent from any publications, especially in English.

The main goal of this monograph is to evaluate the results of related studies and to reveal characteristic features of the dynamics of the main types of the vegetation cover in Mongolia in the Holocene Era and the present day based on ecological and geobotanical approaches. Measures of a general character aimed at conservation of certain ecosystems and the cenotic diversity of some parts of the country are also discussed. Special attention is paid to the status and development of the network of strictly protected areas.

We hope this overview of experience to date in the given field for Mongolia will be useful for other areas and countries, especially in Central Asia, which has similar natural conditions. Above all it examines the methodology of assessment of the contemporary status and degree of disturbance of the plant cover, and estimates dynamic trends in vegetation development under severe and contrasting ecological conditions, taking into account the character and degree of anthropogenic influence on natural complexes.

This book is the result of personal field studies carried out over ten years in Mongolia, and also based on published materials by participants in the Joint Russian-Mongolian Complex Biological Expedition of the Russian and Mongolian Academies of Sciences during the last quarter of this century.

In reporting the results of our studies on vegetation dynamics in Mongolia I wish to acknowledge the help of colleagues. I am especially grateful for insights gained from V.Pyankov, A.Prishchepa, B.Dashnyam, B.Choizhamts, and Ch.Dugarzhav.

I am very grateful to E.Meteltseva and V.Sokolovskaya, who provided unpublished data on modern and fossil pollen spectra necessary for this study. Thanks are also expressed to N.Gorban and T.Nakagawa for their help in figure drawing, to V.Milgram for computer assistance, and to T.Webb III for important comments on Chapter 2 of this manuscript.

The Latin names of plants are those used by I.A.Gubanov, Conspectus of Flora of Outer Mongolia (Vascular plants), Valang Publishers, Moscow, 1996.

Peter D. Gunin

INTRODUCTION TO STUDIES ON THE VEGETATION OF MONGOLIA

Mongolia is at the center of the Asian continent and for a long time it has attracted attention from European scientists. Russian scientists and explorers also were keenly interested in this part of Central Asia. Their investigations were both pragmatic and scientific in character and were aimed at understanding this country which occupied the «crown» of Asia. Studies on the vegetation cover of Mongolia carried out by Russian scientists and explorers can be roughly subdivided into three periods:

– Brief route observations done alongside the main tasks of an expedition;

– Mainly floristic investigations aimed at collecting a comprehensive inventory of the region's flora;

– Geobotanical investigations accompanied by studies on the distribution of plant communities and their mapping, as well as by studies on the physiological features of plants and their ecology.

The leading role of Russian scientists in studies on the plant cover of Mongolia is recognized worldwide. The first information on plants of North-Eastern Mongolia was obtained at the beginning of the 18th century by D.G.Messerschmidt, the pioneer in Siberian flora investigations. The herbarium he had collected by 1724 was admired by prominent Russian botanists, but unfortunately it did not survive (Grubov, 1955).

At the first part of the 19th century trade between Russia and China became more

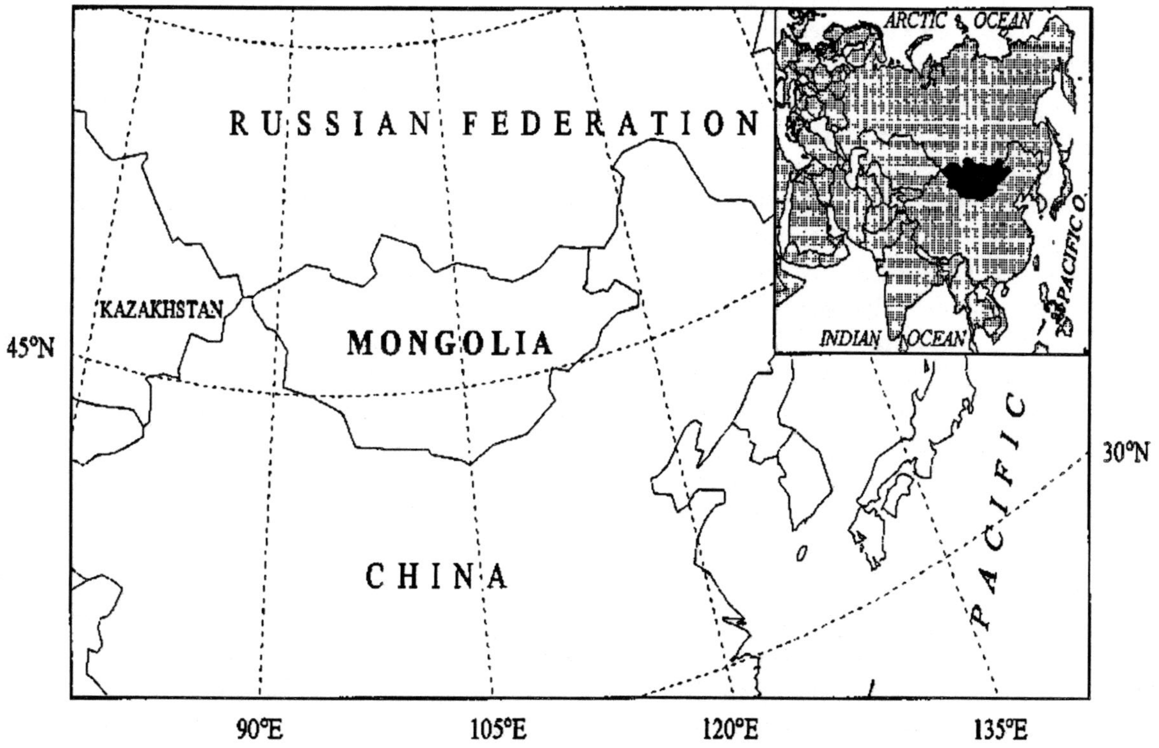

Figure 1. Position of Mongolia in the center of Asia

intense with the itineraries of Russian trade, religious and diplomatic missions crossing Mongolia. Very often persons interested in botany were participants of these missions. In the 19th century the first published works on new plant species and genera from Mongolia appeared in the 1830's (Turczaninov, 1832; Bunge, 1835). The first systematic plant list (484 species) created by Maximovicz (1859) served as an important contribution for studies on Mongolian flora. The list covered botanical collections of Russian travelers in Mongolia from 1830 to 1847. E.R.Trautvetter (1872) added several new taxons, thus broadening the list of Mongolian flora up to 529 species (Maximovicz, 1889).

The first fundamental contributions on the vegetation cover of the country was made by expeditions of the Russian Geographical Society at the end of 19th – beginning of 20th century under the leadership of N.M.Przhevalsky, G.N.Potanin, M.B.Pevtsov and P.K.Kozlov. Processing the huge herbarium collected by these expeditions enabled K.I.Maximovicz, the distinguished expert in Eastern Asia flora, to describe numerous new plant taxonomic units and to begin assembling the «Flora of Mongolia». Unfortunately only the first part of the new list was published in 1889. Simultaneously with expeditions of the Russian Geographical Society, the Mongolian flora at the end of 19th beginning of 20th century was studied by other Russian scientists including: D.A. and E.N. Klements, P.N. Krylov, I.V. Palibin, V.V. Sapozhnikov, B.A. Fedchenko, G.E. Grum-Grzhimailo, and others.

At the beginning of the 20th century the idea to finish the «Flora of Mongolia» was picked up by the prominent Russian botanist V.L.Komarov. His classic work «The Introduction to Flora of China and Mongolia» appeared to be an introduction to work that was not completed (Gubanov & Kamelin, 1988a).

When the Mongolian People's Republic was founded as an independent state in 1921, Russian botanists had accumulated a large amount of materials on the vegetation cover of the country. And the majority of this herbarium i.e. about 40 thousand herbarium sheets (Grubov, 1955), was concentrated in Russia. But these collections of materials and botanical investigations were irregular and poorly organized. After the creation of the Mongolian People's Republic, and the establishment of close friendly relations with the Soviet Republic, floristic investigations not only broadened but they obtained a much more organized character. They were facilitated as well by establishment, in the 20th century, of the Mongolian Commission of the USSR Academy of Sciences, headed by V.L.Komarov, that in cooperation with the Scientific Committee of Mongolia planned and carried out extensive field work in Mongolia. They published their most important results of geological, geographical, soil, hydrological, zoological and botanical investigations in the «Proceedings of the Mongolian Commission of the USSR Academy of Sciences».

Floristic investigations had an important role in the Third Expedition of the Russian Geographical Society, headed by P.K.Kozlov in 1923–1926. The Expedition included such florists as N.V.Pavlov, N.P.Ikonnokov-Galitski, Ya.I.Prokhanov. They gathered and processed a huge collection of Mongolian plants, now stored in Moscow and in the Botanical Institute of St.Petersburg. Based on these collections, N.V.Pavlov published, in English, a synopsis of the flora of Northern and Central Mongolia with 950 species of vascular plants including 11 new species for science (Pavlov, 1929a). His other work (1929b) provided the first definitive description of the specific Khangai botanical-geographical province.

Within expeditions of the Mongolian Commission of the USSR Academy of Sciences in the 1930's, serious floristic investigations were carried out by V.I.Baranov, N.L.Desyatkin, Ye.G.Pobedimova and others in different regions of Mongolia. Detailed descriptions of the routes of expeditions and their contributions are given in the works of

A.A.Yunatov (1946, 1950) and V.I.Grubov (1955).

The greatest contribution to studies on the vegetation cover and forage resources of Mongolia belongs to A.A.Yunatov, who worked in the country continuously for more than 30 years beginning in 1940 (Yunatov, 1950, 1954, 1974; Tsatsenkin & Yunatov, 1951). His contributions to our understanding of the flora of Mongolia are enormous. For example, he collected and processed the huge herbarium that included more than 16 thousands sheets (now stored in the V.L.Komarov's Botanical Institute of the Russian Academy of Sciences and in the Institute of Biology of the Mongolian Academy of Sciences). In addition, A.A.Yunatov also considered the problems of the origin and history of Mongolian flora; he developed the first botanical-geographical zoning of the country which, after more precise definitions, (Grubov & Yunatov, 1952) is still used (Gubanov, 1996); and, in collaboration with his Mongolian student B.Dashnyam, he assembled the special contents of the first «Map of Mongolian Vegetation» published in 1979.

The next set of studies on the vegetation of Mongolia by Soviet botanists started in 1947 and was connected with the Mongolian Agricultural Expedition of the Academy of Sciences of the USSR (Lavrenko & Shulzhenko, 1962). Its botanical studies on the flora and vegetation of Mongolia were performed by E.M.Lavrenko and V.I.Grubov, two outstanding experts on Mongolian and Central Asian flora. Their contributions also cannot be overestimated. A.A.Yunatov continued to work in this Expedition and A.V.Kalinina (1974) actively participated in the geobotanical investigations. Floristic investigations were carried out mainly by V.I.Grubov, who during the 1947–1948 field seasons collected and brought to the Botanical Institute of the USSR Academy of Sciences about 10 thousand herbarium sheets. Afterwards, V.I.Grubov fulfilled a hugh task when he reworked the Mongolian herbarium collections accumulated in the Botanical Institute during the previous century from 1830 to 1951 (about 100 thousands sheets), identified plants, revised taxonomic units, and created a list of flora. From this work the first complete floristic review for Mongolia «Synopsis of Flora of Mongolia» appeared (Grubov, 1955), that included 1875 species of vascular plants, from 552 genera, related to 97 families. This annotated list of species also contained important sections on the history of floristic studies in Mongolia, the origin of its flora, a botanical-geographical zoning of the country, and an analysis of its flora. This work of V.I.Grubov became for years (till 1982) the main manual for plant identification (although it did not contain an identification key) and geography of Mongolian plants (Gubanov & Kamelin, 1988b).

An important floral contribution for the «Flora of Mongolia», was published by the Botanical Institute under the leadership of V.I.Grubov entitled, «Plants of Central Asia» (1963–1994). These 11 volumes contained new facts, new concepts on many taxons, and monographic descriptions of some genera and groups of Mongolian plants done mainly by Russian scientists.

In 1970 the Joint Soviet (Russian)-Mongolian Complex Biological Expedition of the Academies of Sciences of the USSR and Mongolia, headed by Academician E.M.Lavrenko, started investigations across the whole territory of Mongolia. From the very first year the Expedition was not only the greatest in the world on complex biological studies, but a large geographical expedition as well. From 200 to 300 Russian and Mongolian scientists participated in the Expedition annually with specialists in botany, zoology, climatology, geomorphology, soil sciences, paleogeography, etc. Nowadays we believe that the natural features of Mongolia were never studied so intensively and comprehensively as during the years of the Expedition activity (1970–1996). This lucky combination of field and station investigations across the entire territory allow us to consider

Mongolia as one of the world's most completely investigated countries despite its geographical isolation.

This brief history of studies on the vegetation cover of Mongolia does not allow characterization of the volume of investigations carried out by each Expedition specialist. The list of works published until 1997 contains more than 3500 titles including 40 large monographs published as Proceedings of the Expedition, «Biological Resources and Natural Conditions of Mongolia».

Botanical investigations were carried out mainly by specialists from the Botanical Institute of the USSR Academy of Sciences, the Forest Institute of the Siberian Section of the USSR Academy of Sciences, the Institute of Animal Evolution Morphology and Ecology of the USSR Academy of Sciences, Moscow State University, and the Institute of Botany of the Academy of Sciences of Mongolia. These investigations were implemented according to a comprehensive program and included: floristic and systematic investigations, geobotanical surveys of the territory, mapping of vegetation cover, studies on plant biology, searches for valuable objects from an economic point of view, chemical and resource assessments of forage, medicine, tanning and eatable plants, etc.

At station sites located in all of the natural zones and subzones of the country Expedition investigations on fundamental botanical studies were conducted according to the general program as well as in combination with specialists from other fields of biology, geographers, climatologists, and soil scientists. Botanists studied communities of the most important types of vegetation, identified their composition and structure, carried out detailed investigations on the biology and morphology of plants, anatomy, ecophysiology, and studied productivity plus seasonal and annual dynamics of plants in each ecosystem.

The results of these botanical studies are in a large number of publications (Meteltseva, 1986; Hilbig, 1981, 1991; Gubanov & Hilbig, 1993). The most important are:

1. Floristic studies until the middle 1970's, headed by V.I.Grubov gave many new floristic findings that made the «Synopsis of Flora of Mongolia» (Grubov, 1955) more complete. The «Determination of Vascular plants of Mongolia» (Grubov, 1982) contained 2239 species of 599 genera and 113 families. Further floristic studies of I.A.Gubanov and R.V.Kamelin, together with their Mongolian colleagues, produced new data on the flora composition and the distribution of some taxons over the territory. By early 1996, 2823 species of vascular plants, of 662 genera, and 128 families (Gubanov, 1996) had been identified in the flora of Mongolia.

2. The results of botanical-geographical investigations headed by E.M.Lavrenko and carried out together with geographers and soil scientists, supported the formulation of a new set of fundamental conclusions that changed our geographical concepts about the vegetation cover of Mongolia. As a result large works were published such as the «Map of Vegetation of Mongolia» (Lavrenko, 1979) and a series of vegetation maps, the «National Atlas of Mongolia» (Tsegmid & Vorobiev, 1990). These revealed considerable differences in the Mongolian steppes from more western territories, as well as regional differences of steppes in different sectors of the country. These also supported and described steppized deserts and the extreme arid Central-Asian deserts, that have no analogues over the world. And in mountains at the south of Mongolia, a specific type of altitude belt pattern was described.

3. At the forest stations in Khangai and Khentii, and during the field routes of the Forest team of the Expedition, works were completed on forest type definitions, on their geographical distribution, on the ecosystem structure of larch stands, on studies of regeneration processes with the main forest types, and on growth processes in planted forests. Original materials were obtained on the interrelationship of forests and steppes in different regions of Mongolia and on the water regulating and soil protecting role of forests in

Mongolia. These works resulted in four volumes of «Forests of Mongolia» published in 1978, 1980, 1983, and 1988, a «Map of the Forests of Mongolia» (Lavrenko, 1983) containing an original classification of forests, and a zoning scheme according to forest type distribution and forest hydrology.

4. A considered analysis of huge amounts of diverse materials from investigations accumulated by the Expedition during its long-term studies on the natural environments of Mongolia demonstrated the necessity of conducting investigations at new ecosystem levels, as well as the need to assess contemporary ecosystems of Mongolia. In 1989 a new stage in Expedition's activity started with the development of criteria for assessment, a cartographic representation of current ecosystems, and assessing the degrees of anthropogenic disturbance. In 1990 these criteria were developed along with methods for assessment and mapping of ecosystems. Then a first version of the map «Ecosystems of Mongolia» (Gunin & Vostokova, 1995b) was compiled. The final version of the map was published in 1995 simultaneously with a monograph written by a team of authors, «Ecosystems of Mongolia: Distribution and Contemporary State». These authors considered abiotic and biota conditions of contemporary Mongolia ecosystems, their distribution, changes under the impact of anthropogenic factors, and problems of natural zoning.

Alongside the Expedition of the Academy of Sciences, the Joint Soviet-Mongolian Hubsugul expedition (initiated mainly by scientists and students of Irkutsk and Mongolian State Universities) began to work with scientists from Czechoslovakia, Poland and Germany. This Expedition studied terrestrial and aquatic ecosystems of Hubsugul, the Darkhat depression, and part of the Selenga river basin. Botanists of the expedition obtained new information on the flora of the Northern parts of Mongolia and on the spatial distribution of vegetation at the junction of taiga and steppe regions. The results were published as Proceedings, «Natural Conditions and Resources of Hubsugul Region» (1972–1986). The «Atlas of Hubsugul Lake (MPR)» with maps of vegetation, forage meadows, geobotanical zones, diagrams of the floristic composition of grazing lands, and their productivity were the major contributions of this Expedition.

In addition Mongolian vegetation cover has been studied intensively by botanists from Germany, Poland, Czechoslovakia, and Hungary who together with Mongolian colleagues, conducted a series of floristic and botanical-geographical investigations and discovered numerous plant species formerly unknown in Mongolia and in the world (Blazkova, 1985; Hilbig, 1984, 1987, 1995; Hilbig & Mirkin, 1983; Pacyna, 1984; Vasak, 1971 and others).

Naturally, during the botanical investigations, which were accompanied by mapping the vegetation cover, the scientists paid attention to changes in plant cover. These changes were related mainly to the influences of exogenic factors both natural and anthropogenic. Thus, numerous field observations were made on exogenic dynamics of vegetation. The successional changes in ecosystems, defined as the result of ecological relationships i.e. the dynamics of vegetation, had not been previously documented.

Recently major changes have occurred in the vegetation of Mongolia connected either with the direct use of vegetation resources, or with the impact of different types of economic activities. These include both industrial and municipal constructions, use of land resources for agricultural purposes, mining works, and the construction of transport routes. In other words, different types of human-associated influences are increasingly important factors of vegetation cover dynamics. The dynamics of vegetation communities is considered here as ecosystem changes under the influences of different natural, natural-anthropogenic and anthropogenic processes.

CHAPTER 1

NATURAL AND ANTHROPOGENIC FACTORS AND THE DYNAMICS OF VEGETATION DISTRIBUTION IN MONGOLIA

1.1 INTRODUCTION

The natural features of Mongolia are defined by its position in the center of the vast Asian continent and a natural separation, or more precisely semi-separation, from adjacent territories. The inner regions are surrounded by mountains to the North, West, and the South-West. Even at the East and at the South-East, where such distinct barriers are absent, the steppe plains are locked by the long Great Khyangan highlands (Figure 1.1) and the Gobi Desert to the South seals the country.

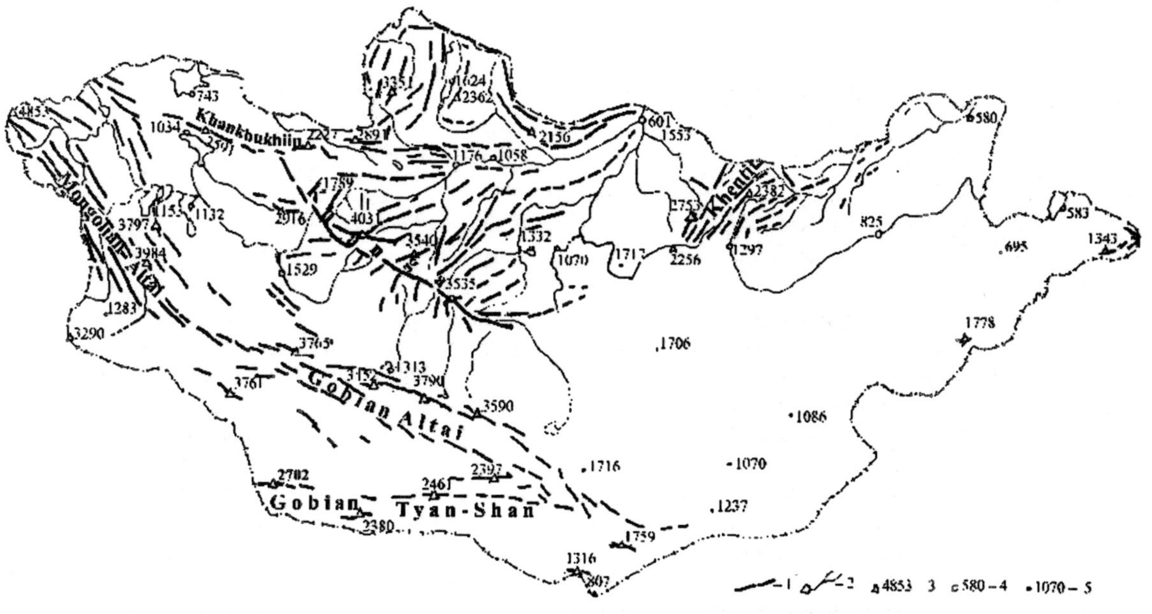

Figure 1.1. Topography

1 – stretch directions of main ridges; 2 – rivers and lakes; 3 – 5 – altitudes of: 3 – mountain summits, 4 – banks and shores of rivers and lakes; 5 – low knolls and plains

1.2 NATURAL FEATURES OF MONGOLIA

The central continental position of Mongolia, far from oceanic influences, defines its climate. Under such conditions the spectrum of natural climate zones becomes simple, but with extreme contrasts. The character of contrast in combinations of phenomena and processes is the dominant ecological feature of Mongolia. It can be seen both in time (diurnal, seasonal aspects) and in space. Various combinations of extreme ecological conditions in the territory of Mongolia are shown in Figure 1.2.

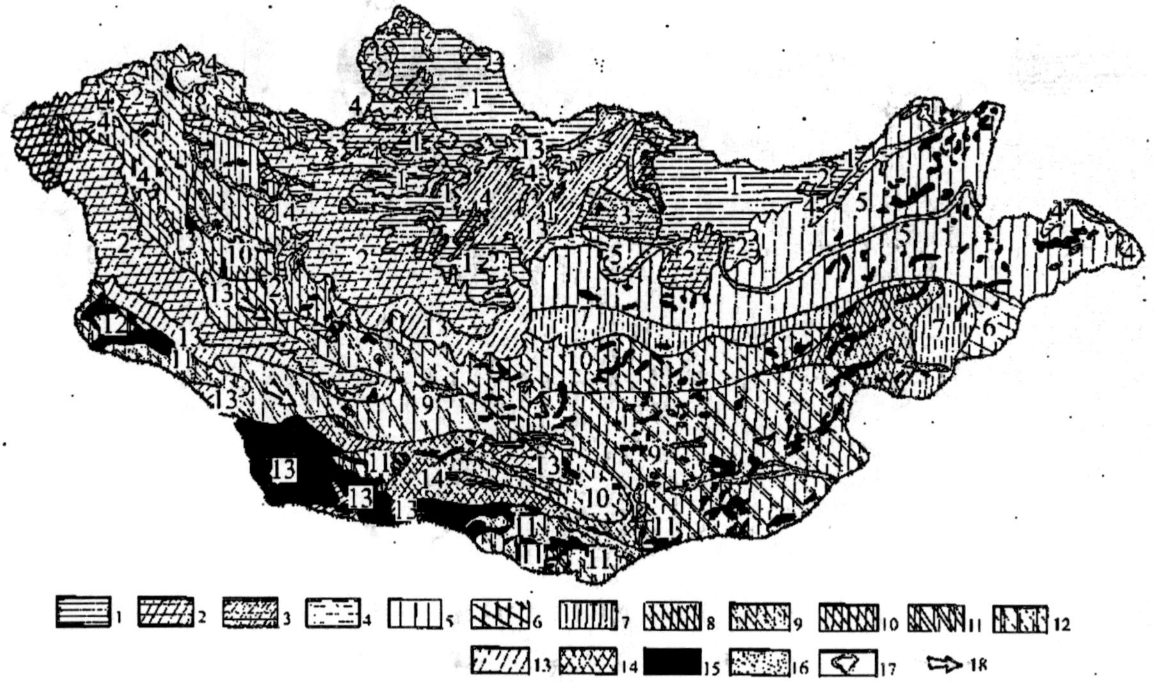

Figure 1.2. Combinations of natural extreme conditions

1–3 – Territories with permafrost and seasonally-frost rocks: 1 – subject to weak erosion; 2 – eroded and with potential erosion danger; 3 – eroded, with local salinization; 4 – territories with seasonally frost rocks in river valleys; weakly eroded; 5 – 6 – territories with salinized sites covering less than 50% of total area: 5 – mainly with sodium carbonate salinization, atmospheric precipitation rate up to 300 mm/year and solar radiation up to 1,400 kWh/m^2; 6 – with sodium carbonate and partially with chloride-sulfate salinization, atmospheric precipitation rate up to 250 mm/year and solar radiation more than 1,400 kWh/m^2, with few dust storms with wind velocity above 15 m/sec; 7 – 10 – territories more than 50% covered by salinized areas with mixed sulfate-chloride and chloride-sulfate salinization: 7 – with precipitation rate up to 200 mm/year, solar radiation values not more than 1,400 kWh/m^2, with high degree of deflation danger, with massifs of blown sands, with dust storms up to 30 days a year; 8 – the same as 7 but with more than 30 dust storm days a year; 9 – with precipitation rate less than 150 mm/year, solar radiation more than 1,500 kWh/m^2, with very high degree of deflation danger, with massifs of blown sands, more than 30 dust storm days a year, up to 30 days with wind velocities exceeding 15 m/sec; 10 – the same as 9, but with less than 30 dust storm days a year; 11 –12 – territories for more than 60% covered by salinized areas with mainly gypsum salinization: 11 – with precipitation rate less than 50 mm/year solar radiation more than 1,500 kWh/m^2, very higher degree of deflation danger, with massifs of blown sands, up to 50 dust storm days a year, more than 30 days with wind velocities more than 15 m/sec; 12 – the same as 11 but with less frequent dust storms; 13 – 14 – territories subject to linear erosion, eroded and with erosion danger, with infrequent shower rains: 13 – with precipitation rate more than 250 mm/year; 14 – the same as 13, with considerable participation of deflation dangerous sites sometimes with thick mantle of blown sands; 15. Local sites with predomination of salinized lands (more than 75%). 16. Lone large massifs of blown barkhan-ridge sands. 17. Salt lakes. 18. Dominant directions of wind transportation.

The mountainous character of much of the territory gives a zonal altitude differentiation of vegetation that separates and sharpens their contrast. On adjacent slopes of opposite exposure, ecosystems of absolutely different character face each other, e.g. dry steppes versus mountain Siberian forests. Alternating imposing mountains with intermountain depressions cause sharp zones with deeply penetrating wedge- and tongue-shaped ledges containing islands of community vegetation. These intermountain depressions also are characterized by altitude belts (Murzaev, 1952).

Mongolia, as a whole, lies rather high from a topographical point of view. The greatest part of the territory is at 1500 m or more above sea level. Even plains are within the interval of 800 – 1500 m. The highest peaks of more than 4000 m are concentrated in the Mongolian Altai with Mt. Kuiten-Uul – 4374 m, Mt. Munkhe-Khairkhan – 4204 m, and Khangai Mt. Otgon-Tenger – 4021 m. Mostly medium high and low mountains predominate in the remainder of the mountain territory, and in Central and Southern Mongolia low knolls, hills, and high plains prevail. The summit parts of mountains often are not sharp and have smoothed shapes. The slopes of mountain ridges in a latitudinal direction usually have a distinct asymmetric character with the slopes of northern exposition being rather gentle, and those facing south, mainly, are steep and stony. In the Eastern Hubsugul area, in the Eastern Khangai, and at the South-Eastern part of the country lava plateaus occur with numerous small dead volcanoes, active in Quaternary (Florensov & Korzhuev, 1982). The highest ridges of Mongolian Altai and Khangai posses a deeply etched glaciers-exagerated relief (Shmidt, 1974).

The topography almost everywhere reflects the tectonic origin and structure of the territory. It is especially clear at places with tectonic disjunctions. The positions of large mountains separated at the periphery into semi-isolated masses and small knolls, are caused by mobile regions in the more stable areas of the Mongolian microplate surrounded by a belt of fuzzy seismicity (Sinitsyn, 1948; Florensov & Korzhuev, 1982; Zonenshain & Savostin, 1979).

The location of major mountain systems defines its climate (Figure 1.3). Mountain barriers at the West and North-West intercept atmospheric flows carrying moisture from the Atlantic side. The Pacific monsoon fades rapidly and only can be traced to 110-120° E. In addition Mongolia is practically completely open for the dry Central Asian desert winds from the South. As a result, Mongolia has a severe continental climate with large daily and seasonal temperature amplitudes, and with very intensive solar radiation values reaching 1500 KWt/hour/sq.m or more. Mongolia is characterized by frequent hurricane winds with velocities exceeding 15m/s. Hurricanes usually are in the South and are accompanied by dust storms as a rule.

A characteristic feature for Mongolia is the combination of low relative humidity values with rather low winter temperatures (from –4 to –24° C in January). This phenomenon is especially typical for mountain regions with a dominating anticyclonic circulation when extensive temperature inversions occur (Lebedev & Kopanev, 1975).

The region as a whole is characterized by a long and cold winter and a short but warm summer (Beresneva, 1992). The development of vegetation cover depends on the quantity plus the seasonal and geographical distribution of precipitation. The total amount of precipitation is rather low; it decreases from 600 mm/year at the Hubsugul and Khentii Mts. to 25 mm/year or less in the Gobi desert (Table 1.1). Maximum precipitation is observed during the second half of summer. And spring, as a rule, is dry and with strong winds. The northern part of Mongolia is a territory with permafrost present and seasonal frost rocks (Tsegmid & Vorobiev, 1990).

The complicated separated relief and severe continental climate create conditions for two different processes, aridization (desertification) and very low temperatures (cryogenesis) that

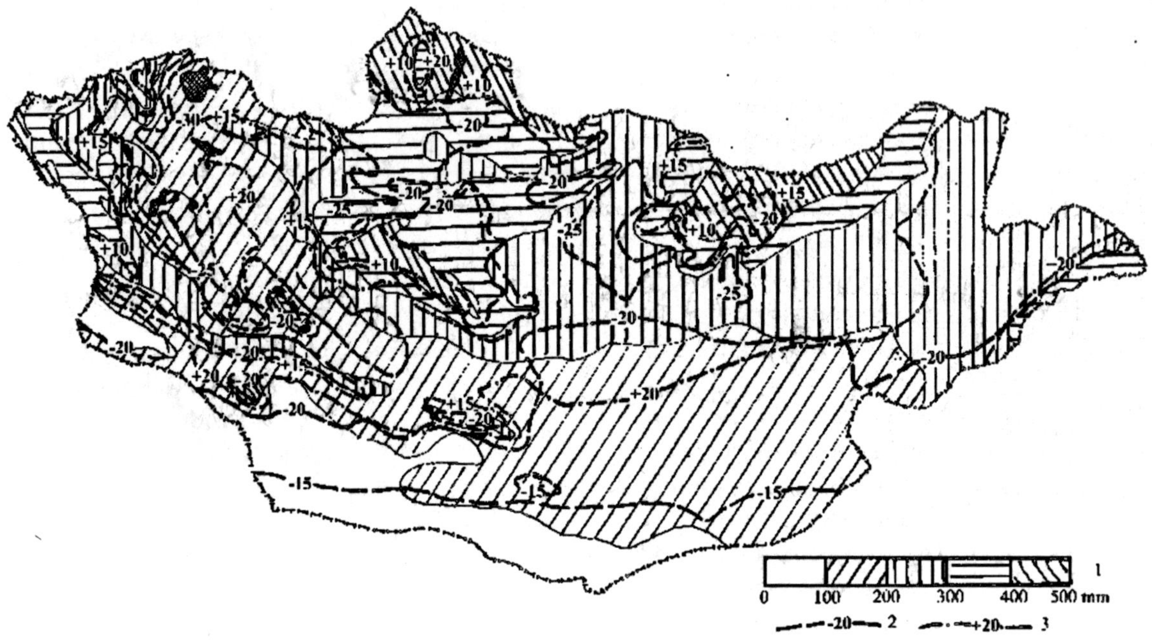

Figure 1.3. Climate scheme

1 – mean annual precipitation rate; 2 – January temperatures; 3 – July temperatures

strongly affect the distribution, structure, and dynamics of vegetation cover. Natural desertification factors – low water, soil erosion and salinization act mainly in the plains and foothills. Cryogenic processes develop mainly in mountains. Usually they are accompanied by moist soil movements by gravity, observed even in southern desert regions at sites with ground water outcrops.

In Mongolia's soil there is a broad distribution of detrital rock sediments that influence soil-forming processes. The prevailing soils are light loams and loamy sands, mainly not salinized, with considerable amounts of rock detritus and pebbles (Nogina, 1978, 1989; Nogina & Dorzhgotov, 1980; Evstifeev, 1980).

There is a well-known increase in aridity from north to south caused by an uneven development of river networks. Rivers are concentrated in the northern part of the country (Figure 1.4). The largest river system is formed by the river Selenga with tributaries Ider, Orkhon and Khara-Gol, that drains almost one third of the country. Watersheds of the Onon and Kerulen River occupy smaller territories. Almost two thirds of the area belong to closed basins. The greatest one is the basin of the Great Lake Depression. All of this part of Asia is crossed by a world waterdivide, i.e., dividing basins for rivers of the Arctic and Pacific oceans, and for basins of Central Asia. At the south the river network changes to a system of dry channels (sairas) often with a temporary stream in the upper reaches and well developed valleys with several terraces in lower reaches; these can be considered as remnants from previous pluvial epochs (Murzaev, 1952; Sinitsyn, 1962; Devyatkin, 1981, 1993; Dorofeyuk, 1994).

Another characteristic feature is the presence of several large lakes located mainly in mountains and in intermountain depressions. The largest fresh water lake, Hubsugul, occupies a rift valley and, along with Baikal Lake, is a unique natural phenomenon. A vast

Table 1.1. A. Annual atmospheric precipitation rate in Gobi, mm (Zolotokrylin & Gunin, 1986).

Station	Year 1975	1976	1977	1978	1979	1980	1981	1982	1983	1984
Erdene	138	68	70	118	162	–	–	43	48	–
Bayan-Under	44	80	59	75	131	108	103	87	99	10
Bayan-Tsagan	77	106	41	56	178	58	80	158	38	50
Bayan-Gobi	100	96	98	50	87	89	78	90	36	84
Shinedzhinst	143	125	101	109	168	95	109	147	55	86
Altai	16	25	6	7	42	7	41	14	16	43
Dzakhoi-Tooroi	38	38	55	48	103	33	40	57	16	85
Ekhiin-Gol	6	33	25	28	94	27	70	9	24	49
Gurvan-Tes	209	135	152	67	93	104	120	56	73	103

B. Annual atmospheric precipitation rate in Transaltai Gobi, mm (Slemnev et al., 1994).

Station	Years 1987	1988	1989	1990	1991	1992	1993
Bayan-Toroi	–	–	–	–	91.5	90.6	125.3
Ekhiin-Gol	2.0	15.8	4.5	12.2	24.6	99.5	164.6

Note: – not available

area is covered by waters of the salt lake Ubsu-Nur, and to the south – a system of lakes, Hara-Us-Nur, Hara-Nur and Hirgis-Nur, occupies the depression of the Great Lakes. In the central parts and at the south of the country the lakes often dry up during long (perennial) droughts.

Around lake depressions and within river floodplains aquatic ecosystems are formed, but their distribution mostly is in northern parts of the country. In southern parts the main sources of moisture for aquatic-type vegetation are aquifers for root systems, or water from deep aquifers discharging along tectonic disjunctions.

This central part of Asia is characterized by a specific high continental type of vegetation structure and of natural zones (Lavrenko, 1978, Lavrenko et al., 1991). Specifically these features are a reflection of the general climatic aridity, which influence all ecosystems to a certain degree. Thus, for mountain ridges of Khangai and Khentii, an exposed forest-steppe is rather typical. Mongolian Altai (where a forest belt is practically absent) has numerous intermountain depressions with an altitude belt vegetation pattern. Here the lowest central part of the depression is occupied mostly by desert vegetation communities; the middle strip is representative of that zones ecosystem and usually circles around the depression. The upper parts of these slopes are an altitude mountain vegetation belt. But the character of altitude vegetation belts changes considerably from north to south (Volkova, 1994; Karamysheva, 1988, Bannikova, 1983).

The geobotanical diversity of Mongolia is extremely high (Figure 1.5). Mongolia is 1200 km from the north to the south and one can find a spectrum of landscape as in the temperate belt of Eurasia, e.g., from typical forests to extra-arid deserts. The forest communities may have immediate contact, or can be at great distances, with steppes and sometimes one can see them very close to deserts. The contrast in these communities of such combination cause a compression of vegetation zones, and very high gradient changes are characteristic of the vegetation cover. But a wide spectrum of zone and inter-zonal transitions exists. Thus within the steppe zone there are subzones of moderately humid,

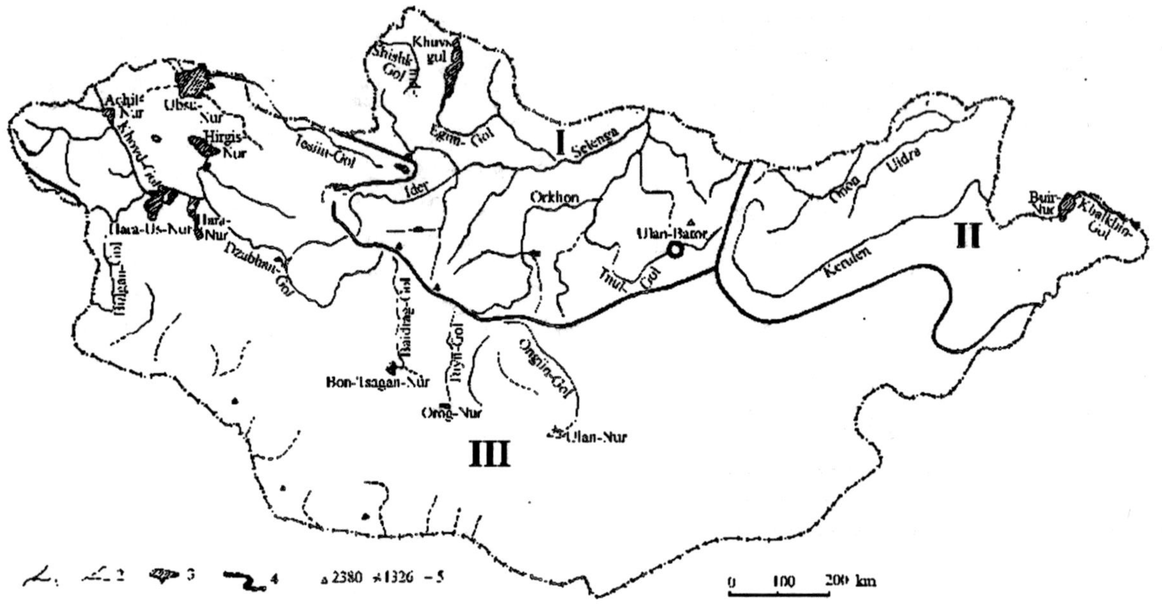

Figure 1.4. River basins and position of the world waterdivide

1 – rivers; 2 – intermittent temporary water streams (sairas); 3 – lakes; 4 – waterdivide line; 5 – altitudes.
I – III. Basins: I – Arctic ocean; II – Pacific ocean; III – endorheic basins of the Central Asia.

moderately dry, dry, and desertified steppes (Karamysheva, 1988), or within the desert zone these are subzones of northern (steppized), middle (true) southern, and extreme arid deserts (Rachkovskaya, 1977, 1993).

Zone and subzone boundaries in the central part of Mongolia often occupy latitudinal positions. They have relatively simple shapes except in the eastern plains where their directions are almost longitudinal. The complicated geography often changes the direction of zone boundaries. In the western part the directions also are changed to almost longitudinal. As a result, the penetration contours of many large vegetation communities reaches hundreds of kilometers. Especially consider the penetration, far to the north (up to 52° N) to the mountain slopes of Tannu-Ola and to the north-eastern slopes Mongolian Altai, of a desert-steppe and desert ecosystem dominated by *Nanophyton grubovii*.

The dominance of mountainous relief, from low knolls to highlands, leads to a close placement of latitudinal zones and altitude belt ecosystems (Gunin & Vostokova, 1995a, Lavrenko, 1979). Flat summit surfaces and the upper parts of slopes usually are occupied by short grass alpine meadows, mountain tundras, or thin forests with semi-bald mountain communities. In the north, if they occupy middle mountains, then in the center – in Khangai – they are at altitudes higher than 2800 m. The occurrence of high mountain vegetation decreases rapidly to the east from Tannu-Ola and Altai. Khangai and Hubsugul mountains posses considerable diversity of high mountain ecosystems, but in the mountains near Selenga and in Khentii only local poor vegetation fragments occurs.

The structure of the altitude vegetation belt patterns are very complicated. The highlands and middle mountains of Altai, Hubsugul and Khangai have up to six altitude belts: glacial, tundra, mountain-meadow, forest, forest-steppe, and steppe. To the south, in the mountains of Gobi Altai and Gobi Tyan-Shan

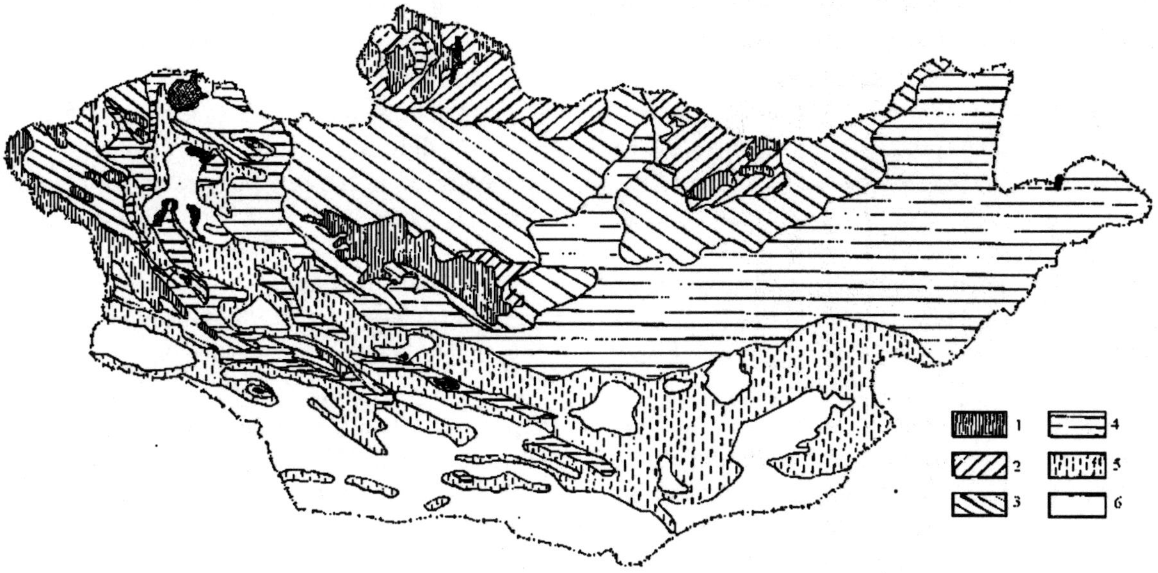

Figure 1.5. Scheme of zone-belt vegetation distribution.

1 – high mountain belt; 2 – forest belt; 3 – forest-steppe and forest-meadow belt; 4 – dry steppe zone; 5 – zone of desertified steppes; 6 – desert zone.

this structure is simplified, because the tundra, forest, and forest-steppe belts disappear.

The forest belt mainly contains larch forests sometimes mixed with Siberian pine and fir. In sandy sediments at the lower parts of slopes pine stands dominate and, together with larch, they form the forest-steppe belt. A forest belt is found only in the Hubsugul Mountains and North of the Selenga valley. To the south, forests occupy northern slopes as a rule, often in combination with steppe and meadow communities.

A vegetation zone consisting of a strip of forb-grass steppes is located at the North-East of the country. The south contains the broadest strip of dry steppes characterized not only by a grass dominance, but also a broad distribution of pea shrubs (*Caragana microphylla*, C. *stenophylla*, C. *pygmaea*) and sagebrushes (*Artemisia* spp.). These southern dry steppes change into desertified steppes further south where tall sod feather grass is replaced by short feather grass (*Stipa gobica*, *S. klementzii*). Further to the south the desertified steppes are replaced by semi-deserts which are called desert steppes (Lavrenko, 1978; Yunatov, 1950, 1974), and then by desert grass meadows (Rachkovskaya, 1989, 1993). The broad strip of steppe deserts is replaced at the south by the true Gobi desert, characterized by predominance of Central Asian semi-dwarf bushes, dwarf bushes, and bushes. The most widely spread plant communities over the Gobi are of *Haloxylon ammodendron*, *Anabasis brevifolia*, *Zygophyllum xanthoxylon*, *Reaumuria songarica* and others. E.I. Rachkovskaya was the first to characterize this zone of extra-arid deserts (Rachkovskaya, 1977; Rachkovskaya & Volkova, 1977; Volkova & Rachkovskaya, 1980). The vegetation communities are of an open type that only close at stream bottoms. Here the species commonly have a very well developed root system that can use the moisture of ground waters. Most often one finds *Haloxylon ammodendron*, *Nitraria sphaerocarpa*, and less often – *Tamarix ramosissima*.

The continental climate changes greatly across the territory of Mongolia due to the large distances leading to a clear differentiation of ecosystems. At least three longitudinal sectors can be defined. The western one includes Mongolian Altai, Great Lakes Depression, Dzungarian and Transaltai Gobi. The central sector comprises Khangai, Hubsugul region, Khentii, Khalkha plains, and partly the Gobian deserts. The eastern sector includes the Eastern Mongolian Plain, and outskirts of the Great Khyangan. Local peculiarities of natural vegetation communities are dictated to a great extent by combinations of ecological conditions controlling the water and mineral supply for plants. The composition of surface sediments and the availability of additional moisture from aquifers are of great importance. Rock growing plants are most wide spread in Gobi steppes and deserts with smaller areas occupied by sand-loving plants. Hydromorphic communities develop in river floodplains and river valleys; at the south they exist near springs discharging ground waters. A natural oasis type system is formed at such sites (Lavrenko & Yunatov, 1960; Gunin et al., 1980; Gunin & Vostokova, 1995a). The steppe drainage depressions are occupied by halophytic plant communities. In deserts, dominated by soils with a sulphate type of salinization (Nogina & Dorzhgotov, 1982), gypsophilous communities are widespread.

Hence parts of Mongolia posses various degrees of natural diversity caused by differences in isolation and influences of external factors. Their diversity is determined by a complex geography, their position in the world watershed, and a convergence of different systems of global atmospheric circulation. Here we see contacts between mountain-forest ecosystems of a Siberian type and Central-Asian ecosystems with inter-mountain depressions of desert, desert-steppe, and steppe types clearly experiencing an East-Asian influence. The true central position is occupied by the triad: forest-steppes – typical steppes – desert steppes. Siberian forests and true deserts are at their periphery like out-posts of vast territories with their corresponding vegetation; but out of Mongolia, i.e., Mongolia is at the south of Siberia and north of China's Inner Mongolian desert.

Vegetation landscapes that are most close to a definition of «Mongolian» are concentrated in the central part of the country. Here is Khangai with its forest-steppes, mountain base steppes and highlands, and the steppes of Khalkha with numerous rocky knobs and knolly high plains. The part of Khalkha covered by rocks and detritus is occupied by rock-loving plants, and the low salty depressions contain halophytic ecosystems (Vostokova, 1983; Vostokova & Kazantseva, 1995). To the south, in Gobi, the vegetation cover clearly exhibits features inherent to the deserts across Central Asia. For example, in Dzungarian Gobi, the Turan-Dzungarian influence is easily seen (Rachkovskaya, 1993). The Mountains of Hubsugul and Khentii are a continuation, to the south, of Siberia's mountain system and Mongolian Altai is part of the Altai Mountain range.

1.3 LANDSCAPE-ECOLOGICAL REGIONS

Mongolia's landscape-ecological zones were designated based on both natural and natural-anthropogenic landscape forming factors (Figure 1.2). Each zone was defined by taking into account different natural regions of the country (Murzaev, 1952; Yunatov, 1950, 1954; Lavrenko, 1970, 1978; Tsegmid, 1962; Neronov & Lushchekina, 1980; Preobrazhensky et al., 1984, Tsegmid & Vorobiev, 1990, etc.). The scheme of landscape-ecological zones shows six regions, comprising from one to six provinces (Figure 1.6). Provinces can be further divided into districts.

Generally landscape-ecological areas are characterized by relatively uniform landscape forming processes, similar macro- and microrelief, water availability, and a similar structure and composition of its soil and vegetation cover (Gunin & Vostokova, 1995a). Regions that exist under relatively uniform

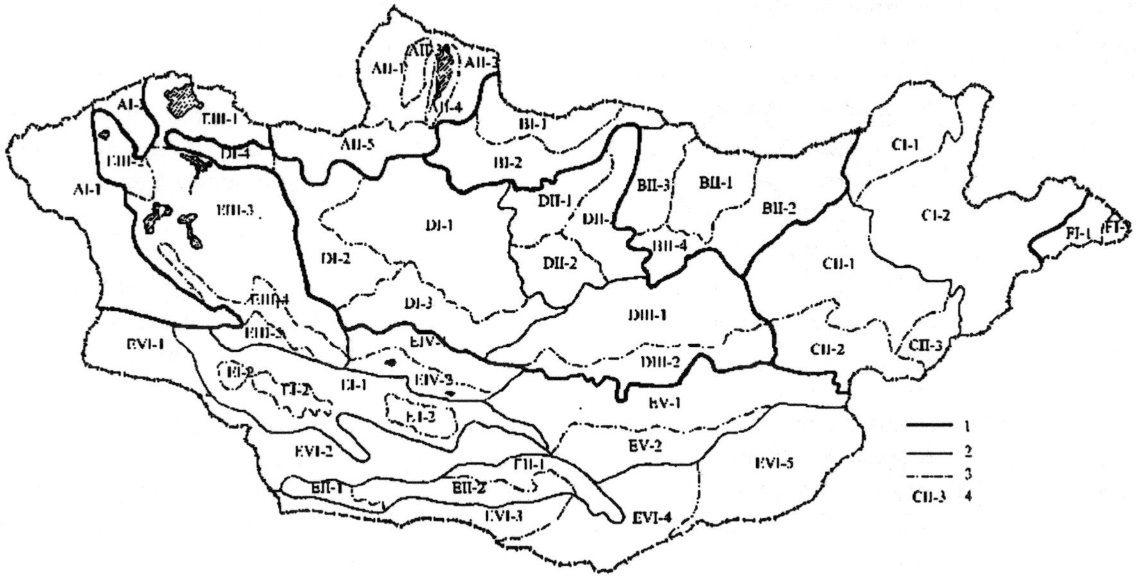

Figure 1.6. Scheme of landscape-ecological zones

1 – 3 – boundaries: 1 – regions, 2 – provinces, 3 – districts; 4 – indices of zoning units: **A – Altai-Sayan region, districts**: AI-1 – Mongolian Altai; AI-2 – Ureg-Nur; AII-1 Ulan-Taiga; AII-2 – East Hubsugul; AII-3 – Darkhat; AII-4 – Hubsugul; AII-5 – Sangilen. **B – Transbaikal region, districts**: BI-1 – Buteiliin; BI-2 – Bureg-Nur; BII-1 – Baga-Khentii; BII-2 – Tuul-Onon; BII-3 – West Khentii. **C – Daguria-East-Mongolian region, districts**: CI-1 – Uldzin; CI-2 – Menengiintal; CII-1 – Middle Kerulen; CII-2 – Dolongyn Gobi; CII-3 Dariganga. **D – Central Mongolian region, districts**: DI-1 – North-Khangai; DI-2 – West-Khangai; DI-3 – South-Khangai; DI-4 – Khankhukhiin; DI-5 – Tessiin; DII-1 – Orkhon; DII-2 – Burgetul; DII-3 – Darkhan; DIII-1 – Mandal Gobi; DIII-2 – North Gobi. **E – Central Asian region, districts**: EI-1 – mountainous; EI-2 – inter-mountain depressions; EII-1 – mountainous; EII-2 – piedmont; EIII-1 – Uvsunur; EIII-2 – Achitnur; EIII-3 – Lake district; EIII-4 – Middle-mountain; EIII-5 – Shargiin; EIV-1 – Khangai; EIV-2 – Bontsagan-Orognur; EV-1 Oigiingol-Saishand; EV-2 – Dalanzadgad; EVI-1 – Dzungarian Gobi; EVI-2 – Transaltai Gobi; EVI-3 – Alashan Gobi; EVI-4 – Gashun Gobi; EVI-5 – East Gobi. **F – Khyangan region, districts**: FI-1 – Middle Khalhingol; FI-2 – Modtoikhamar.

relief conditions with their geological structures form a landscape-ecological province. Based on these characteristics of landscape relationships, provinces are united into the unit of highest rank – the region.

The western and north-western parts of the country are occupied by the Altai-Sayan region, which includes the Mongolian-Altai and Hubsugul provinces. Within the Mongolian-Altai province two districts are defined. The first includes Mongolian Altai stretching from the north-west to the south-east for more than 1500 km. The second district comprises the Turgen Mountains and Lake Ureg-Nur.

The Mongolian Altai is the most extensive and largest mountain range of Mongolia. Here traces of ancient glaciation are found in addition to contemporary small glaciers and perennial snow patches. There are numerous glacier relief forms, e.g., excavated hollows and moraine ridges in the high mountains. Mongolian Altai is dissected by broad longitudinal valleys that separate several parallel mountain chains which, by erosion valleys and saddles, then are divided into separate ridges and peaks. The longitudinal valleys also are subdivided by dams into closed or semi-closed intermountain depressions. The Mongolian Altai high mountain ridges, that stretch from the north-west to the east-south-eastern direction, determine the distribution of vegetation cover. Thus, as a rule, north-eastern mountain slopes are not covered by forests, because mountains bar the flow of moist air from the west. One can find only isolated low productive larch stands on north facing slopes (Savin et al., 1978). But, on south-western exposure slopes in the upper reaches of the Black Irtysh River, there are small patches of larch stands mixed with Siberian stone pine (*Pinus sibirica*) and spruce (*Picea* spp.). In general there is an almost complete absence of a forest belt on the mountain ridges of Mongolian Altai. Here the meadow steppes are well developed, with cushion plant formations in combination with *Kobresia* and *Carex* formations. At the lower parts of slopes, dryer steppes are present that grade into desert steppes at the mountain base. The intermountain depressions are characterized by a vertical belt pattern of vegetation communities. One often finds small lakes and swampy-cold meadows at these depression bottoms. At the south of Mongolian Altai the mountain depression bottoms are occupied by halophilic plant communities (Figure 1.7).

In Mongolian Altai the vertical range of vegetation belts is between 500 to 800 m (Volkova, 1994). Deserts occupy the belt from 1800 to 2000 m on large southern slopes; but in the southern part of the district this belt lies within 2000 to 2400 m. The largest areas are occupied by mountain steppes, between the altitudes of 1800 to 2400 m. They differ greatly in vegetation composition and small spots are occupied by communities of *Juniperus sabina* (Volkova, 1994).

The large Mountain Turgen, with Lake Ureg-Nur, is situated at the upper north-west corner of Mongolian Altai. These mountains usually are characterized by vertical belt patterns of soil and vegetation cover, beginning with lichen, moss-lichen, dwarf-bush – *Kobresia*, tundras, and cold meadows at summit surfaces, ranging to forb – small sod – grass steppes at the lower slopes. Small fragments of larch stands with some *P. sibirica* are found on the wettest slopes. The salt Ureg-Nur Lake is surrounded by desert steppes that change to dry sod-grass and forb – grass mountain steppes up the surrounding slopes.

The Hubsugul province includes five districts characterized by extensively covered areas of high mountain tundras, sub-bald open woodlands and a clear belt of mountain forest. Dome-fault-block mountain ridges are at the west of the province. The heights of these mountains are often more than 3000 m. The highest summit of the ridge, mount Ulaan-Taiga, reaches 3351 m above sea level. Here high mountain tundra and mountain meadows occupy up to 50% of the area. The mountain slopes are occupied by a rather wide belt of sub-bald open woodlands with some tundra communities. These low-productive pine-larch forests, with well-developed lower level bush, form the upper forest boundary (Natural Conditions and Resources of Hubsugul Region, 1976). The mountain forest is composed of *Pinus sibirica* with some larch found mainly on north-western exposure slopes open to the Shishkhid-Gol and Busseiin-Gol river valleys. Fragments of the Siberian forest are on upper reaches of the Delger-Muren with green moss and cowberry – *Bergenia* -forests dominating.

In the east of the Hubsugul area dome - fault – block mountains with altitudes up to 2500 m are characterized by extensive granites and Neogene-Quaternary basalts. High mountain plant communities are located mainly at the north-eastern part of the highest ridges; these are moss, sedge-moss and bush tundras. Down the slopes the vegetation switches to larch-pine, yernik-pine, yernik – dwarf bush and moss open woodlands. But the largest territories are occupied by combinations of forests and forest-steppe vegetation at medium mountain heights and high plateaus. Here larch stands and derived birch stands dominate with occasional aspen. At burn sites, forests are replaced by cutting meadow and bush communities. The most diverse types of larch stands are on northern exposure slopes. Also here are found forb, shorthear, forb-cowberry and cowberry – green moss forest vegetations. In the valleys of Arig-Gol and Ure-Gol it is common to find forb larch stands,

Figure 1.7. Piedmont plains – in space image (scale 1:1 000 000).

and some cold hardy mountain forb meadow steppes.

To the East of Hubsugul a distinct district is formed in the Darkhat depression that occupys a rift valley. The Darkhat tectonic depression is filled with lacustrine-alluvial deposits from the Shishkhid-Gol River. The river current is weak, the river course heavily meanders, and the flood plain has numerous small lakes surrounded by grass-sedge swamps. The bottom of the depression is occupied mainly by *Kobresia*-fescue meadow steppes and yernik-*Kobresia* bushes. Spreading around the gentle slopes of the Darkhat depression from the east and west one finds larch stands with*Rhytidium* and forbs. They occupy the belt ranging from 1700 to 2000 m. The grass cover includes both forest meadow and tundra alpine species, the moss stratum is dominated by *Rhytidium rugosum* (Hedw.) Kindb. Typically forest communities are only fragments at granite outcrops on large northern mountain slopes.

The valley of upper Egiin-Gol and the rift depression of Hubsugul Lake stretching from the north to the south are characterized mostly by *Kobresia*-fescue and cryophyte-forb sod-grass steppes near the base of mountains. Up the slopes they are substituted by larch stands and sometimes with *Pinus sibirica.* The predominating forest types are forb-sedge and forb-cowberry. The summit surfaces above 2000 m are occupied by dryad tundras and sedge – *Kobresia* barrens.

Middle and low mountains on the southern outskirts of the Sengilen ridge are populated by a broad spread of mountain larch stands with forb grasses. The slopes of southern exposure and bases are occupied mainly by forb – small-sod grass steppes. The plains of the upper Tes River reaches, at altitudes from 1000 to 1600 m, are flat and swampy. But, nevertheless the water divide between drainage into the Central Asian basin and the Arctic Ocean basin pass there. The first one comprised of the Tes river, and the second the Selenga ridge tributaries. The Selenga's gently sloped plains are dominanted by forb – June grass – sod grass steppes, dry sod grass and wormwood – grass steppes. Plains near the Tes river sources are occupied by forb – sedge and *Kobresia* – sedge swamps.

The Transbaikal region comprises the Khentii and Middle Selenga Mountain ridges, which are a continuation of Transbaikal structures. This region includes two provinces – Middle Selenga and Khentii.

The Middle Selenga province comprises almost all of the Selenga river basin within Mongolia. The province includes dome – fault – block mountains of Buteiliin and Burengiin ridges laying in a north-eastern direction. These two medium height mountains with altitudes near 2171 m have quite different vegetation. The Buteiliin-Nuruu ridge is covered mainly by forests. Slopes of different orientations are covered by pine forests with larches or larch stands with ledum-green moss cover, cowberry – *Bergenia*, or cowberry – green moss cover. Smaller areas are occupied by larch stands with forbs-sedge cover. At burn sites and old forest cuttings, birch and aspen stands form a cover with forb-grass-sedge. The south-west slopes of Buteiliin-Nuruu, a sandstone elivium, is covered by pine stands with a sedge – brake fern –forb or a rhododendron-cowberry-forb cover. Forest burns and cuttings are overgrown by birch stands with a forb cover or by forb – fescue and forb – small sod grass steppes.

The ridge Burengiin Nuruu located to the south is characterized by an extensive development of steppes mainly on southern exposure slopes. Only the slopes of north-western exposure have forb-larch and pine stands.

The broad Selenga valley stretching from the west to the east is partly covered by swamps. The river channel forks into numerous branches. Low terraces are dominated by grass-forb meadows and willow thickets. The lower parts of slopes and high terraces are covered by forb-fescue steppes.

The mountains of Khentii stretch from the south-west to the north-east above 2500 m. The highest summits are: Asralt-Khairkhan – 2800 m, Altan-Bulgii – 2656 m, and Khiidiin-Sardag – 2665 m. High summit surfaces are occupied mainly by lichen and lichen-moss tundras. Below a narrow belt is formed by sub-bald Siberian pine – larch open woodlands with yernik regrowth and a continuous moss cover; then a broad belt of Siberian stone pine (*P. sibirica*) forests with firs (*Abies*) and larch stands with *Picea* and *P. sibirica* Usually these are bilberry – green moss or cowberry – green moss forests. Also numerous birch forests and yernik thickets form at burn sites and forest cuttings. In the eastern and southern parts of the ridge the forests become*Rhytidium* and cowberry – *Rhytidium* forests. Further down the slopes, larch stands and derived birch and aspen forests dominate, commonly mixed with rhododendron – forb, rhododendron – forb – cowberry, sedge – forb, and grass – forb forest types mainly of similar ages and quality.

The eastern part of the Khentii mountains are represented by medium altitude mountain ridges, piedmonts, and broad river valleys. The ridges stretch in a north-eastern direction and become gradually lower to the east. Maximum heights are near 1700 to 1800 m. The left tributaries of Onon River carry their waters through forests and rich forb steppes, i.e. through an exposed forest-steppe belt. Here larch stands with forb-sedge cover are common. The forest – steppe belt develops dry *Lespedeza*, *Spiraea*-forb-*Rhytidium*, *Rhododendron*-forb-*Rhytidium* steppized pine stands, substituted by birch stands at burn sites. At the southern slopes of medium and low altitude mountains the most common plant communities are forb-*Leymus* – tyrsa (*Stipa capillata*) and rich forb-sedge-bluegrass (*Poa*) steppes. The sand terraces of the broad Onon valley are occupied by xerophyte-forb steppe pine stands. Along the main course and streams of the Onon floodplain wood-bush riparian forests grow. These forest communities are composed of poplars, willows, sometimes larches, and often they grow in combination with swamps. At stony sites of low mountains rock-loving petrophyte plants of steppes are common with a dominance of xerophytes.

The western Khentii has low piedmonts and low mountains separated to different degrees with altitudes less than 2000 m and the rather broad valleys of the Orkhon River tributaries, Khara-Gol and Ero-Gol, are occupied mainly by steppes with small pine-tree stands containing young bush and forb cover. The detrital-stony slopes of the low mountains usually are covered by rich forb – sedge – bluegrass communities; blending downslope into pea shrub – small bunchgrass – tyrsa steppes. The rocky spots usually are inhabited by xerophyte petrophytes. River valleys have willow thickets among forb – grass meadows and sometimes they are swampy.

The Dagurian – East Mongolian region includes the East – Mongolian and Middle Khalkha provinces that experience a weak influence of monsoon moisture transfer.

The East-Mongolian province includes the lower reaches of the Uldza River and the East-Mongolian Plain. The middle and low reaches of the Uldza River mainly is a weakly undulating plain with numerous broad dry valleys laying in the north-eastern direction as a rule. Hills and uplands with low knolls have the same orientation. At the deepest parts of valley-like depressions one can see small salt lakes and salinized soils. Here an ancient hydrographic net is clearly seen (Murzaev, 1952). The north-eastern plain is covered mainly by forb – grass tansy steppes, while, at the low reaches of the Uldza River, one finds forb – fescue – tall grass meadows and forest – meadow – swamp complexes. Unstructured leached saline soil complexes are very typical. These vegetation complexes form small belts around little salt lakes and saline soil depressions (Vostokova, 1983).

The East-Mongolian Plain lies at heights between 800 – 1000 m, while the lowest part is at the southern shore of lake Buir-Nur (583 m). The highest summit is the isolated low knoll mass, Matad (1246 m). The plain has numerous valley-like depressions with saline soils and low knoll masses. The formers are characterized by combinations of halophilic-gypsophilic plant communities; the latter by stony low knolls and by scattered groups of petrophytes. The plain as a whole is covered by steppes with small sod grasses and *Aneurolepidium-Stipa* associations.

The Middle Khalkha province includes the valley of the Middle Kerulen tributaries, the plains of Dolongyn Gobi, and a lava plateau of Dariganga. The plains of Middle Kerulen are characterized by combinations of low knoll uplands and vast plains with long depressions. All the territory is covered by smallbunch – Stipa steppes with petrophilic plants on the rocky surfaces of knolls. At the deepest depressions halophilic communities develop often with a peripheral belt of *Hordeum brevisubulatum* and *Achnatherum splendens*.

The most characteristic features of Dolongyn Gobi are saline soil depressions laying from the south-west to the north-east that represent the north-eastern end of the East Mongolian depression. Dolongyn Gobi predominantly is a plain with small hills, ridges and steep slopes at the periphery of small depressions with clay-like saline soils or small salt lakes. The vegetation cover consists mainly of desertified sod-grass and dwarf semi-shrub – turf grass steppes represented by *Cleistogenes squarrosa* – tyrsa (*Stipa capillata*) -feathergrass and pasture sagebrush (*Artemisia frigida*) – *Cleistogenes squarrosa* – feathergrass communities. These small depressions are occupied by reomuria – feathergrass – and *Kalidium* communities.

It is necessary to note the volcanic plateau Dariganga, which reaches 1150 to 1300 m with some volcano cones near 1400 to 1760 m. The plateau is mainly occupied by small-sod grass steppes such as tyrsa (*Stipa capillata*), wormwood – *Cleistogenes squarrosa* – feathergrass and other communities often combined with scenic forbs and pea shrubs. On the stony slopes of volcanoes or at basalt outcrops the vegetation cover is scarce and is mostly petrophytes of dry or desert steppes. At the southern margin of the plateau along the river valley the sandy mounds of Moltsog-Els occur. On the mounded sands of Moltsog-Els there are depressions with small elm stands accompanied by tamarix and pea shrubs. The sand mounds are surrounded by areas of thin sandy cover with wormwood – low sod grass typical of desert steppes.

The Khyangan region is represented only by a small site at the far east of Mongolia, so it includes only one province with two districts: the Khalkhin-Gol valley and the Khyangan piedmonts.

The plain Khalkhin-Gol valley with most altitudes about 1000 m gradually declines to the flat Buir-Nur depression. Here one finds vast forb – sod-grass steppes of the Dagurian-Mongolian type. The dominant communities are grass – rich forb – tansy and forb – tansy – grasses. The following plants are most common: *Stipa krylovii*, *Poa attenuata* subsp. *botryoides*, *Koeleria cristata*, *Cleistogenes squarrosa*, *Artemisia frigida*, and some pea shrubs (*Caragana* spp.).

The heavily eroded Khyangan piedmont gradually increases its height up to 1600 m; it is occupied by forb-grass steppes with numerous Manchurian floristic elements (Gunin & Vostokova, 1995a). At the slopes one also can find small stands of pine (*Pinus sylvestris*) and birch (*Betula* spp.).

The Central Mongolian region comprises three provinces, which structurally belong to the transition zone between Altai-Sayan and Transbaikal structures.

The Khangai province lies at the west and includes all of the Khangai uplands with the Khankhukhiin ridge. The main geographic unit here is the Khangai ridge with altitudes above 3500 m. The ridge serves as the water divide between the Arctic ocean basin and the Central Asian basin.

The entire northern part of Khangai contains numerous outskirts laying in a north-north-east direction divided by a dense network of Selenga tributaries. The area reaches from the south-west to the north-east for almost 700 km. This complicated system of ridges, flat uplands, and separate low mountain masses is a large arched uplift (Korina, 1982). The heights of most ridges range between 2000 – 3500 m with the largest being mount Otgon-Tenger (3905 m) located at the west upland edge. Here an alternation between alpine relief and plateau-like summits is characteristic; and sometimes a well conserved glacial relief is seen clearly. The large northern slope is relatively more gentle with an extensive piedmont, a dense river network, and numerous lake depressions. The large southern slope goes steeply down to foothills at 2100 –

2300 m. This structure asymmetry seems to be the result not only of tectonics, but also due to different climatic conditions. The northern slope is relatively open to moist winds and has a more humid climate. The southern slope is in a wind shadow and, being separated from moisture transfer flows, has a dry desert climate. The structural asymmetry and climatic differences of Khangai (Beresneva, 1983) are reflected in the vegetation distribution and in the formation of two altitudinal belts. On southern slopes the vertical vegetation belts only contain high mountain and steppe belts. To the north of the waterdividing ridge, the altitudinal belt structure on side ridges is rather peculiar. On southern exposure slopes two belts, high mountain and steppe, are formed; while on northern slopes five to six defined belts can be discerned. Above 2700 m a high mountain belt exists, represented by tundras and balds. Within the range 2500 to 2700 m a narrow belt is formed by sub-bald open woodlands. Below, at 1750 to 2500 m, fragments of a forest belt occur formed mainly by larch stands and with a belt of mountain meadow steppes below (Bannikova & Khudyakov, 1976; Karamysheva & Banzragch, 1977). To the west is the highest mount with altitudes higher than 3100 m and with glaciers and perennial snow. The high mountain tundras usually are combined with *Kobresia* alpine meadows, sedge meadows, and yernik-willow thickets often with some *Potentilla fruticosa.* Some sub-bald open woodlands are formed by larch (*Larix* spp.) and Siberian stone pine. The forest belt in the range of 1800 to 2300 m is composed of larch stands, yernik-mossy and fescue-mosses sometimes with *Kobresia* spp. and *Potentilla fruticosa*. Below the forb – *Rhytidium* and cowberry-*Rhytidium*, larch stands are common. At the north-east, a forest-steppe belt on medium-altitude mountains is easily seen that contains combinations of larch stands with forb cover or forb – small sod grass or forb-fescue steppes (Bannikova, 1980, 1983). Broad river valleys are covered by thickets of *Salix* spp. and *Potentilla fruticosa* with occasional larch. The floodplains often are water-logged and covered by sedge and forb-sedge communities.

To the west the large slope of the Khangai ridge has a south-western orientation. Here the alteration of altitudinal belts is different. Thus, sites above 3700 m are covered by perennial snow; the belt of 3500 to 3700 m is occupied by yernik-dryad in combination with *Kobresia* alpine meadows and sedge communities. Below is belt of high mountain cryophyte meadows formed mainly by *Kobresia* and sedges. Occasionally they are combined with high mountain meadow steppes. The forest belt is only scattered larch stands. The belt of high mountain meadows and steppes change into forb-*Koeleria* steppes. Small sod grass – tyrsa (*Stipa capillata*) steppes are characteristic of the low mountains.

On the large southern slope of the Khangai ridge with its sub-latitudinal orientation, as already noted, the altitudinal belt structures are simple. The highest parts (above 3300 m) are covered by high mountain cryptophytes (Karamysheva & Banzragch, 1977). Below is a fragmentary belt of dwarf shrub and moss-lichen tundras that blends into a forb – fescue steppes. The lowest belt is complied of dry petrophyte – forb – tyrsa (*Stipa capillata*) steppes. The deeply cut river valleys have rather broad bottoms covered by detrital – pebble materials that only fill with water from rain showers in the high mountains. Willow forests are present in the wide parts of valleys just outside the mountains.

At the north-west part of Khangai the Khankhukhiin ridge occupies a prominent place. The ridge is broad, flat, and gently sloped. On the upper part are gentle slopes with broad valleys and traces of ancient glaciation. Relief forms on the northern slope are smoothed; but relief features on the southern slope are more sharp and notched. To the south-east the Khankhukhiin ridge descends gradually through a set of low mountains connected with Khangai. The highest mountain reaches 2928 m, but average heights do not exceed 2600 m. Three or four altitudinal belts are on the ridge

(Karamysheva & Banzragch, 1976). From 2150 to 2600 m bald and sub-bald belts occur. The first belt is comprised mainly of dryad tundras in combination with moist *Kobresia* alpine meadows or sedge meadows. Scattered stony places without continuous vegetation cover are common. The sub-bald complex of Siberian stone pine – larch open woodlands with yernik and dryad tundras occupies the narrow belt of 2150 to 2300 m. *Kobresia*-sedge communities do occur in stony places. Northern exposure slopes in the central part of the ridge have sub-bald open woodlands that blend with larch stands and Siberian stone pine. Their grass-moss cover is composed of reed grass, cowberry, and green mosses. Spruce trees are found only in cold moist ravines. Below 2150 m, forests exist only in river valleys and in weakly eroded floodplains. Mostly the forests are larch stands with forbs, forb-sedge, and reed grass communities. On southern exposure slopes a rather strict boundary occurs between steppe and high mountain communities. The rich forb – grass steppes mainly with petrophilic forbs occupy middle height mountains. The steppe vegetation is characterized by the presence of East-Kazakhstan floristic elements (Karamysheva & Banzragch, 1976). The south-eastern low mountains are covered mainly by dry petrophyte – forb – grass steppes; the flood plains are wormwood – grass – tyrsa (*Stipa capillata*) dry steppes.

The Orkhon-Tuul province in Central Mongolia is comprised of the middle and lower reaches of their rivers. It forms the outskirts of the Burget-Uul ridge and chains of middle and low altitude mountains in general laying in a north-east direction. Structurally these uplands are transition structures from the Transbaikal region, which includes three districts.

The northern part is characterized by heavily cut middle height mountains with average altitudes 1300 to 1500 m with the highest mountains at 2000 m. On summit surfaces one finds larch stands with forbs. At many sites larch is replaced by birch. On the slopes forb – small sod grass steppes dominate often along with petrophytes. The Orkhon valley cuts through mountains and at certain places the canyon-like shape snakes through weak tectonic areas. Downstream the valley broadens, and terraces appear on slopes. The floodplain is covered by meadows with occasional bushes and isolated elm trees.

The eastern part is characterized by a broad distribution of floodplains. The Orkhon valley is broad, after the Tuul mouth the river channel forks frequently and sand-pebble islands appear. The floodplain is covered by meadows that are sometimes swampy. On flat parts dry small sod grass steppes are typical often in combination with low knolls occupied by petrophyte steppes, and sometimes with birch and pine stands on northern slopes. These steppes to a great extent are cultivated or exist as idle lands for one or more years.

The low knoll uplands under the influence of the Orkhon and Tuul Rivers mostly lay latitudinally; but some lay in a north-eastern direction. These uplands are from 1000 to 1500 m high with the plain areas between them complicated by valley-like depressions. Here dry sod-grass steppes dominate and pea shrubs plus petrophilic forbs occur at stony sites. From place to place the areas between knolls are cultivated, but the fields are small and often are inhabited by weeds of steppe origin.

The Khalkha province of Central Mongolia combines rather typical Mongolia landscapes: low relief knolls, low mountains, vast flat areas with valley-like depressions, sometimes with intermittent small salt lakes or saline soils. This province includes two districts.

Mandal Gobi situated in the northern part of the province, is characterized by the prevalence of hilly and knob-and-basin denuded plains at 1000 to 1500 m in combination with separate low rocky knolls with altitudes near 1700 m with a few higher summits. This low knoll terrain has a variety of weathered granite forms that make the area peculiar. On the knobby plain small sod grass steppe prevails along with forbs and pea shrubs. Low knoll stony habitats are covered by petrophyte steppe varieties plus forbs and shrubs. Here one can find *Spiraea* (*Spiraea* spp.), sometimes Mongolian almond (*Amygdalus mongolica*), elm (*Ulmus* spp.) etc. At the saline soil valley-like depressions and around small lakes, small belt complexes form dominanted by deris (*Achnatherum splendens*) and forbs (Vostokova & Kazantseva, 1995).

The southern part is different in that low knoll masses are not wide apart in the knob and basin plain. Their dimensions are smaller and heights are quite low. Here there are also plenty of valley-like depressions, but instead of small lakes, saline chalk soil practically without vegetation cover are common. A belt around depressions is formed by *Reaumuria – Kalidium* complexes, often mixed with *Limonium*. This area can be considered as a transition to Gobi deserts. Predominating are desertified *Cleistogenes squarrosa* – tyrsa (*Stipa capillata*) – feathergrass and pasture sagebrush (*Artemisia frigida*) – *Cleistogenes squarrosa* – feathergrass steppes. On stony sites of the low small knoll terrain, petrophyte groups are common along with small sod grasses and forb communities.

The Central Asian region comprises almost one third of Mongolia and contains six provinces. According to its landscape-ecological conditions this territory is closely connected with other regions of the Central Asia basin located in Central and Western parts of China. Closed drainage basins are characteristic to all provinces of this region.

The Gobi-Altai mountain province includes the Gobi Altai and the transitional southern branch of Mongolian Altai both with latitudinal orientations. This mountain province consists of separate sets of ridges divided by broad intermountain depressions. The ridges can be connected by saddles or divided by depressions. Average heights are above 3000 m and the highest mountains achieve 3765 and 3769 m. Their altitudes decrease to the east. The characteristic feature of all Gobi Altai ridges is the presence of vast steeply inclined basal plain belts (Figure 1.7). The belts are heavily cut by numerous dry river channels. Permanent water only flows in valleys at heights above 2500 m. These mountain slopes are rocky and cut by numerous ravines, troughs, and erosion furrows. The slopes near belts often are abrupt. These mountains, due to their heights, have typical altitude belt patterns more complete to the west and simpler to the east (Volkova, 1994).

The altitudinal belt spectrum is almost completely comprised of grass-shrub communities. Above 3000 m the belt is composed of *Kobresia* alpine meadows in combination with cryophyte high mountain steppes. Below they change to cryophyte forb – sod grass steppes. Forb – fescue – fairway crested grass (*Agropyron cristatum*) steppes dominate here. Below 2000 m the belt is fairway crested grass sod grass steppes with some petrophilic forbs and an occasional juniper (Petukhov & Etminus, 1990). Down the slope these are replaced by desert petrophyte – grass – pasture sagebrush (*Artemisia frigida*) – feathergrass steppes with a few pea shrubs. At the steep detrital bases baglur (*Anabasis brevifolia*) – feathergrass and baglur – feathergrass – onion communities dominant.

Their intermountain depressions such as Alagnur, Burunnur and Tsagangol have much in common in their landscape structure and vegetation. All of them have a lake or saline soil at their lowest site. Around such depressions are found sedge – deris (*Achnatherum splendens*) communities, some reed thickets with saline soil forbs, as well as *Reaumuria* and *Kalidium* communities. The sloping banks of depressions are occupied by steppe baglur – feathergrass – desert onion. The dry river beds are distinguished due to the presence of sparse groups of bushes such as *Haloxylon*, *Nitraria*, and sometimes poplar.

The Gobi Tyan-Shan province unites the scattered mountain ridges of Gobi Tyan-Shan and the intermountain depressions which separate them. The low mountains of Tyan-Shan are closed at the east by a middle altitude mountain ridge, Gurvansaikhan that also can be related to the Gobi Altai as well acting like a closing structure for these mountain chains.

The dominant heights in the Gobi Tyan-Shan are 1500 to 2500 m. Separate ridges are connected by vast basal plains. Immediately at foothills one finds steeply sloped belts heavily and deeply cut by numerous dry rivers. Great areas are occupied by stony – detrital inclined alluvial plains partially covered by weakly fixed sands.

Vertical belt patterns are not obvious in these mountains. The only full spectrum of altitudinal belts is in the Gurvansaikhan Mountains. Here on summit surfaces there is a belt of mountain meadow steppes. The belt below is bunchgrass – fairway crested grass (*Agropyron cristatum*) steppes with a few junipers and petrophyte forbs. The lowest belt is desert steppes plus pasture sagebrush (*Artemisia frigida*) and pea shrubs. Their foothills and vast detrital inclined plains are occupied by desert steppes dominated by petrophyte forbs – feathergrass, and occasionally with baglur-feathergrass – onion communities. Small natural oasis are formed at ground water outcrops on mountain slopes (Lavrenko & Yunatov, 1960; Gunin et al., 1984). Here one finds poplar (at the west) and elm (at the east) stands alongside deris (*Achnatherum splendens*) plus some reed thickets and aquatic halophyte plant complexes.

The largest area in the Central Asian region is formed by the vast inner-mountain plains. A Great Lakes depression is located here between Mongolian Altai and the Khangai uplands. It has a complicated geography and is subdivided into a number of semi – isolated depressions by inner medium height ridges and low mountains. This province is more than 200 km wide and lays from the north-west to the south-east for more than 750 km. Generally it is characterized by desert landscapes, because it is closed from all sides by high mountain systems which intercept moist air flows. As a result, the atmospheric precipitation rate here does not exceed 200 mm/year, and is not more than 25 mm/year at the south end. During the last decade there were years without a raindrop. There is general climatic trend for the atmospheric precipitation rate to decrease as shown in meteorological station observations at Gurvan-Tes (Figure 1.8).

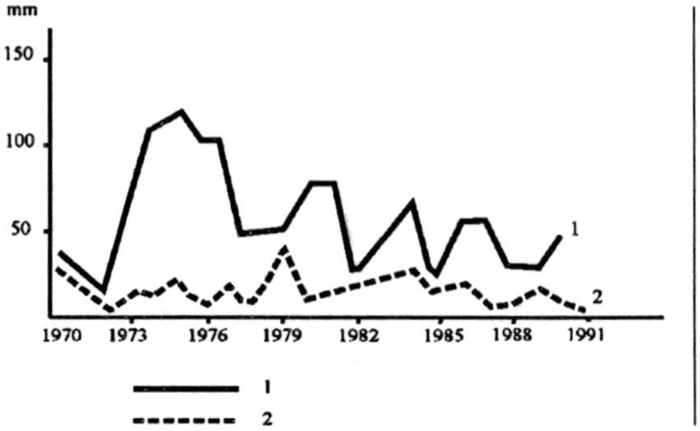

Figure 1.8. Changes in precipitation for 1970 – 1991. According to observed data at Gurvan-Tes station.

Precipitation amounts: 1 – for the rate more than 3 mm/day; 2 – less than 3 mm/day.

The outmost northern part is occupied by the Ubsu-Nur depression that is separated from the remaining province by the Khankhukhiin ridge and the outskirts of Mongolian Altai (Bugrovsky, 1987, 1996). This is a low altitude depression with its lake shoreline at 743 m. The Ubsu-Nur Lake is the greatest salt lake in Mongolia. It is not drained and water mineral content reaches 18 g/l. The lake is fed by several small rivers. The largest, River Tesiin-Gol, flows from the high plains between Sengilen ridges and the Khangai uplands. The lake is surrounded by reed swamps and saline soil deserts with halophilic forbs and abundant deris (*Achnatherum splendens*). Separated poplar stands are present along with willows (*Salix* spp.) and (*Hippophaë rhamnoides*) and (*Ribes* spp.). Near to the lake, poplar (*Populus* spp.) stands are disappearing and being replaced by saline soil meadows and halophyte reed thickets. To the west and north-west from the lake one mainly finds detrital inclined piedmont plains with a dominance of dry petrophyte – forbs – feathergrass steppes

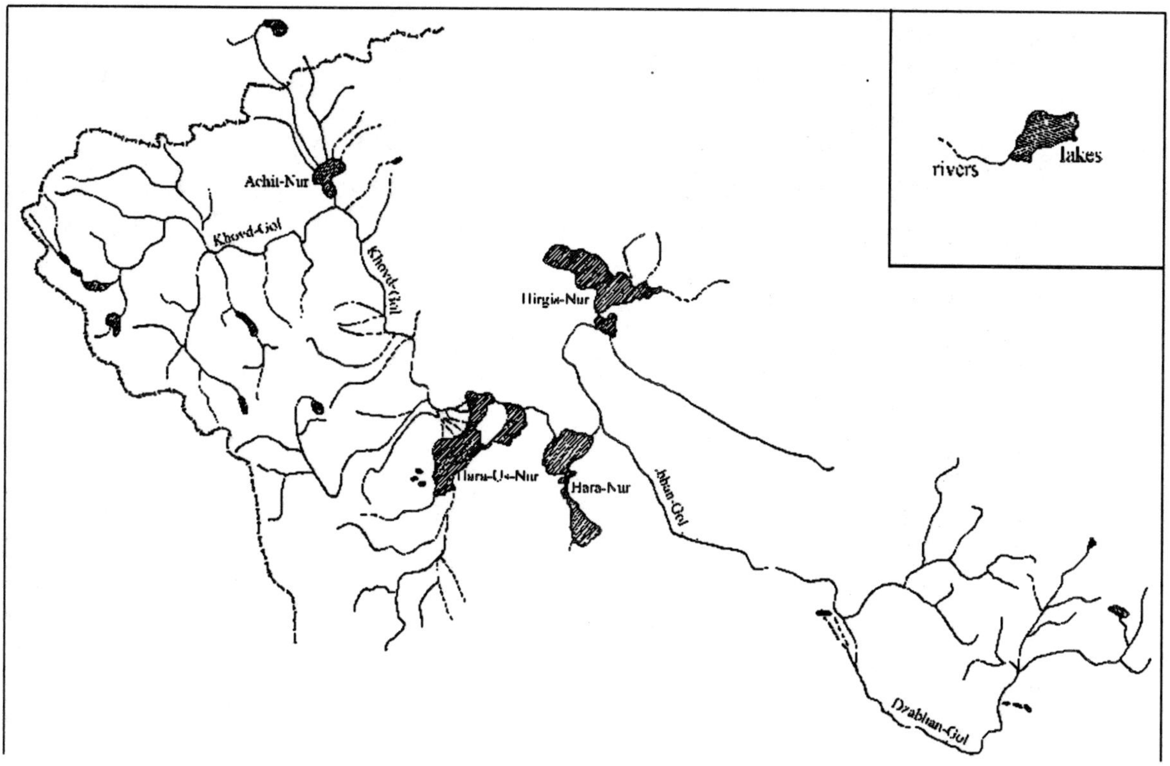

Figure 1.9. River network of the Hirgis-Nur – Hara-Nur – Hara-Us-Nur basin.

with *Spirea* and bushes being characteristic. Pasture sagebrush (*Artemisia frigida*), tyrsa (*Stipa capillata*), and fescue – tyrsa communities also are common. In low places one finds desert steppes and steppe deserts where tare – grass – wormwood communities are dominant. To the east of the lake sand sediments that are easily redeposited are common. They form knobby hills of different sizes. The largest one is Borig-Del-Els. Besides sand loving forbs and *Leymus chinensis*, sparse groups of teresken (*Krascheninnikovia ceratoides*), pea shrub (*Caragana* spp.), and *Spiraea* spp. are present (Buyan-Orshikh, 1977; Vostokova & Khrutsky, 1996).

The depression of Lake Achit-Nur is situated at the north-west of the province. It is almost surrounded by outskirts of Mongolian Altai and cut by the Khovd river valley. Mostly the depression is covered by desert semi-dwarf bunchgrass steppes (baglur – feathergrass, teresken – baglur – feathergrass communities). The lake itself has fresh water. It is surrounded by meadows and swampy vegetation with some willows and poplars. The latter are most common to the north from the lake and in river valleys.

A system of three large lakes is found in the central part of the Great Lakes depression: Hirgis-Nur, Hara-Us-Nur, and Hara-Nur. They obtain water from the Khovd, Zavkhan, and Khungii-Gol rivers (Figure 1.9). The lakes are surrounded by plains with altitudes from 1050 to 1200 m. Only a few low knoll hills are higher. A minimal altitudes is found at Hirgis-Nur lake (1034 m). So, all of this part of the Great Lakes depression is higher than the northern part with Ubsu-Nur lake. The central part has lakes with flat and weakly undulating relief along with desert steppes and deserts

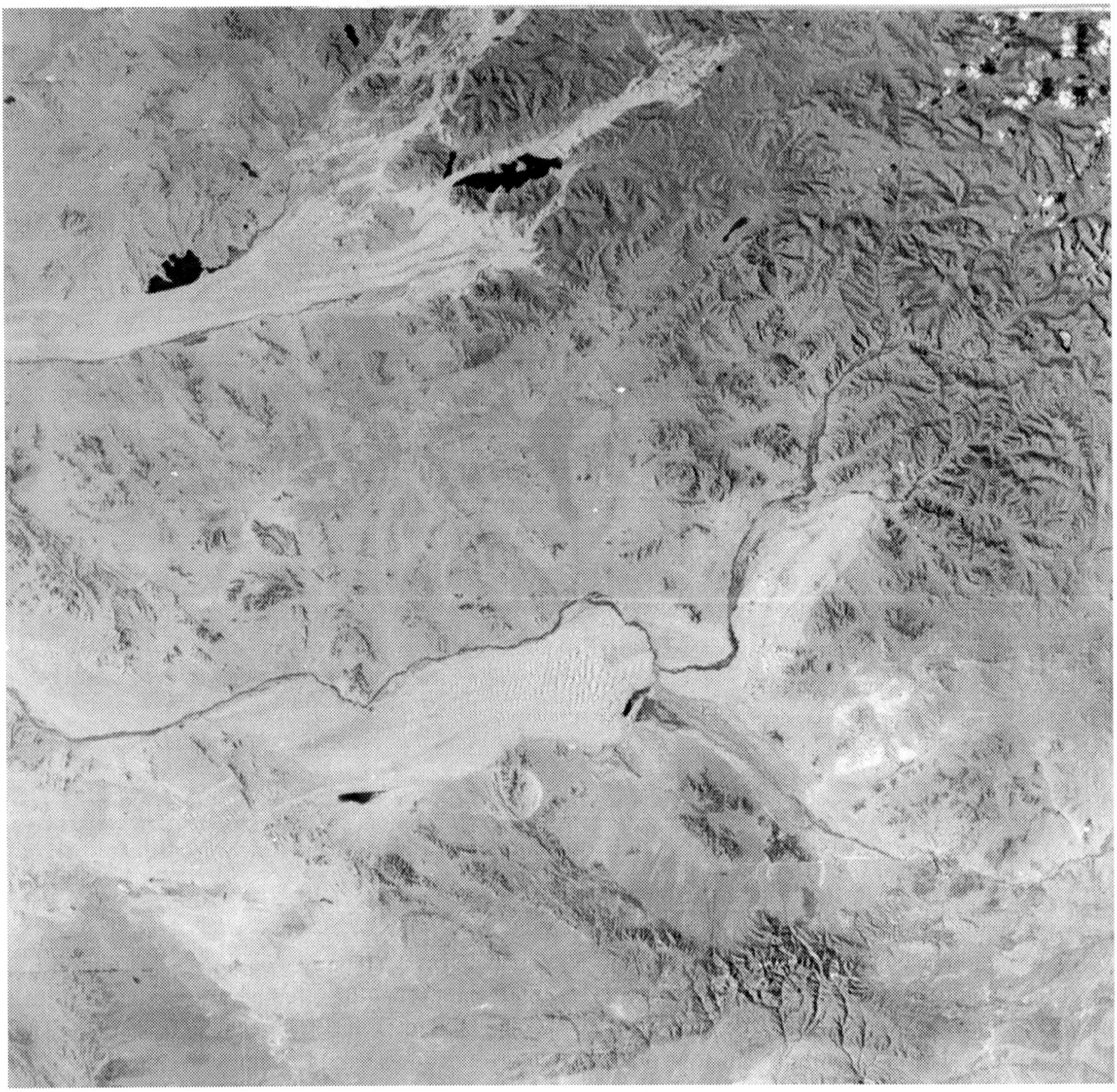

Figure 1.10. Mongol-Els sands in space image (scale 1:1 000 000).

dominated with semi-dwarf shrub – feathergrass, feathergrass-*Reaumuria*-baglur communities, commonly along with pea shrubs. Within the Khovd river delta, and at the southern part of Hara-Us-Nur Lake, large areas are occupied by reed swamps (*Phragmites australis*). Amidst swamps salinized plots with saltbush communities are frequent. East from the lakes the number of low knoll hills and residual mountains increases. Mostly they are occupied by desert steppes with *Cleistogenes squarrosa* – feathergrass communities and stone loving forbs. On stony slopes of knolls petrophyte varieties of dry small-sod – grass tyrsa steppes are common. Along the Zavkhan and Khunguii-Gol rivers a broad strip of large knob sandy hills, Mongol-Els and Borkhar-Els, stretches. They are weakly fixed by vegetation, and at some places, large areas are covered by moving sand sediments (Figure 1.10).

South from the lakes there are separate mountain ridges Zhargalant-Khaiin-Khan-Uul, Khassagt-Khaiirkhan, Khan-Taishiryn-Nuruu, and others, stretched in a south-eastern direction. The usual heights of these ridges are within 2500 to 3000 m. The highest, 3797 m, is in the Zhargalant-Khaiin-Khan-Uul Mountains. These isolated peaks have relatively similar types of altitude vegetation belts and geological structures. The highest mountain ridge Zhargalant-Khaiin-Khan has an almost complete spectrum of altitudinal belts, excluding a forest one. The summit surfaces are occupied by cryophyte *Kobresia* – sedge alpine meadows and high mountain cryophyte cushion-like forb steppes. The low mountain belt at the inclined base plain is occupied by dry bunchgrass steppes. A belt of cushion-like vegetation is absent in mount Khassagt-Khaiirkhan and Taishiryn-Uul. Their low summit surfaces and slopes are covered by dry bunchgrass steppes which are replaced by desert steppes and deserts at the inclined base plains.

The southern part of the Great Lakes depression has the Shargain Gobi depression, surrounded by steep mountains due to tectonic disjunctions on the slopes facing the plains (Murzaeva, 1982). In the center lies an intermittent salt lake. The small river Shargain-Gol feeds this lake as well as number of streams formed by ground water discharges at tectonic disjunctions. Almost all of the depression is occupied by baglur and feathergrass – baglur deserts. But, near the lake, *Haloxylon* woodlands plus *Tamarix* plants, perennial saltbushes and halophytes are common. The lake drying up stages are well identified by low bars covered with *Tamarix* thickets. Also halophylic communities occupy rather large areas along small streams of water.

To the south from Khangai the Lake Valley province is strongly distinguished geographically. The Lake Valley is subdivided into two distinct parts: the northern part is uplifted and the southern is gently inclined towards its southern boundary formed by ridges of Gobi Altai.

The northern uplifted part of Lake Valley is a hilly terrain sometimes with small rocky uplands. Desert steppes (baglur – wormwood – feathergrass and pea shrub – feathergrass) dominate on the plains. The plains are dissected rather deeply by valleys of small rivers originating in the Khangai Mountains and entering lakes Bon-Tsagan-Nur, Orog-Nur, and Tatsyn-Tsagan-Nur. The longest river, Baidrag-Gol, feeds Lake Bon-Tsagan-Nur.

The southern part, really the depression-lake, is limited at the south by a tectonic escarpment of the Gobi Altai Mountains. Almost at the escarpment there are several lakes: the nearly fresh water Bon-Tsagan-Nur, the intermittent Orog-Nur and Tatsyn-Tsagan-Nur. Alongside gently inclined plains with feathergrass-onion steppe deserts there are sandy hills covered by wormwood – dwarf bush communities. Reed and halophyte thickets form near the lakes. Between lake Orog-Nur and Tatsyn-Tsagan-Nur one finds semi-fixed sands with single *Haloxylon* trees. Judging by such landscape features it is possible to conclude that lakes formerly occupied large areas here.

To the east from Lake Valley is the Middle Gobi province that stretches in a latituninal direction for more than 200 km. It is characterized by undulated-ravine plains, drainage depressions, numerous low knoll hills and low mountain ridges. Also there are hills of non-fixed or weakly fixed sands. The province occupies an intermediate position between typical Mongolian steppes and true deserts. The flat topography of the territory with heights between 1000 to 1500 m (1913 m maximum) ensures a latitudinal zoning of vegetation cover. Thus, the northern parts are covered by desert steppes and steppe deserts dominated by feathergrass – baglur, feathergrass – pea shrub communities. In the southern part true deserts are more characteristic, covered by baglur (*Anabasis*) and semi-dwarf shrubs. At the west large areas are covered by sand hills (Figure 1.11). Everywhere one can find closed depressions, often valley shaped, with saline soil deserts.

Figure 1.11. Barkhans of Khangoryn massif (photo by A.V.Prishchepa).

The largest saline soil areas are located in the Ongiin-Gol dry delta.

Almost all of south Mongolia is occupied by the Gobi province, from the western boundary to the eastern. This province is subdivided into several separate intermountain depressions by ridges from the Mongolian and Gobi Altai, the Gobi Tyan-Shan Mountains, and their outskirts. These territories are arid or extra-arid with rather similar soil-vegetation complexes. Although each has specific features due to its position and the influences of neighboring territories.

The Dzungarian Gobi is situated at the outermost west. It experiences the influences of Dzungarian landscapes because it is practically closed from the east by outskirts of the Mongolian Altai. That is why many Dzungarian species are found here, thereby greatly increasing the diversity its desert communities (Volkova & Rachkovskaya, 1980; Rachkovskaya, 1993). For example, *Artemisia terrae-albae*, *Anabasis aphylla* and other Turano-Dzungarian species are common (Gubanov, 1996). But at the east true Gobi deserts with *Haloxylon* and baglur communities dominate. Alongside stony-detrital rocky surfaces where vegetation is scarce, only near riverbeds, are sites with *Reaumuria*, *Achnatherum splendens*, reed – chingil (*Halimodendron*) and *Tamarix* thickets, particularly at sites where ground waters outcrop or is close to the surface. At the upper piedmont plains there are steppe deserts with a prevalence of *Anabasis brevifolia*, *Stipa gobica* communities. The lowest part is the Barun-Khurai depression with a vegetation cover composed mainly of perennial saltbushes communities, and occasionally *Haloxylon*. This depression is the base for erosion for many streams originating in the Mongolian Altai Mountains.

To the west from Dzungarian Gobi, between the Gobi Altai and Gobi Tyan-Shan, a vast area is occupied by Transaltai Gobi. It stretches from west to east for more than 700 km and represents true and extremely arid deserts (Sokolov & Shagdarsuren, 1983; Sokolov & Gunin, 1986; Gunin & Vostokova, 1995a). True deserts occupy small spots at the outermost south-west (Rachkovskaya, 1977, 1993). Characteristic features of the Transaltai Gobi are broadly spread, structurally denuded hilly-ravines, with residual mountain plains sometimes up to 2000 m high, and stony low knoll hills. On the plains and low knoll hills *Haloxylon*, baglur, simpegma (*Sympegma*) and other desert communities spread. In the river bottoms amidst stony surfaces only occasional *Haloxylon ammodendron*, *Ephedra* spp., and *Iljinia regelii* are found. Near a few water springs, knobs of *Nitraria* spp. and *Haloxylon*; and sometimes poplar thickets occur accompanied by meadow-saline soil vegetation and *Tamarix* spp.

To the south-east a vast plain is occupied by Alashan Gobi that forms the northern margin of China's Alashan desert. The Alashan Gobi is characterized by rather monotonous stony-detrital deserts crossed by numerous shallow dry sandy river channels. On the plain, baglur communities are typical. The dry channel bottoms are characterized by groups of *Zygophyllum* spp., *Ephedra* spp., *Amygdalus mongolica* thus distinguishing these dry channels from the desert background. A small contribution that increases the diversity of desert landscapes is made by the sand hills of Borzongiin Gobi. Here on gently sloped knobby sands, high *Haloxylon* grows. Along water springs of tectonic origin, as everywhere in the Central Asia region, one finds broad strips of knobby *Nitraria* spp. stands accompanied by meadow and halophylic *Achnatherum splendens* stands, chains of knobs with *Tamarix* and *Haloxylon*, and sometimes poplar forests.

The Gashun Gobi is situated at the south-west of the region. It stretches from south-west toward the north-east to the vast tectonic East Mongolian depression, with elevations below 1000 m. The surrounding flat and hilly-ravine plains rise gradually in a south-east direction and reach altitudes of 1400 m and more. The depression is characterized by combinations of saline soils and sand deserts, separated by stony-detrital mountain residuals dominated by baglur communities. At the south-east the true deserts switch to grass-bush steppe communities of feathergrass – pea shrub, boialych (*Salsola arbuscula*), feathergrass – baglur or ayania at upper levels. A majority of the small sand dunes are held by *Haloxylon* forests. The lowest closed depressions are surrounded by *Kalidium* and perennial saltbush – *Haloxylon* communities, because the center of a low depression has no vegetation as a rule.

1.4 LANDSCAPE AND ECOLOGICAL FACTORS OF VEGETATION DYNAMICS

The contemporary dynamics of Mongolian vegetation is controlled both by natural and anthropogenic factors.

Of the natural factors, a main role is played by the precipitation rate because both long-term and annual precipitation oscillations are important determinants of vegetation cover behavior. The dynamic changes in vegetation cover due to long-term oscillations of climatic conditions are studied by paleobotanical methods, considered in Chapter 2. The climatic changes of historic periods are studied from the viewpoint of contemporary vegetation dynamics. Recent decades are characterized by a steady trend toward a decrease in total annual precipitation. For instance, the data of observations obtained for Transaltai Gobi (meteorological station Gurvan-Tes) show a steady decrease both in total precipitation and in the number of days with effective rainfall (Table 1.1). In addition the number of days with dust storms has increased.

A decrease in precipitation leads not only to progressive desiccation of soil horizons, but also to changes in surface and ground waters. Sequentially all of these factors cause changes in vegetation cover and, as a rule lead to desertification. These changes are very slow, so they cannot be defined accurately as a result of a field route, or singular observations, or according to short-time period data. For example, climax plant communities first experience the appearance of more drought-resistant species, which gradually expel the less adapted mesophilic species and communities. Thus, at the denuded plains of Gashun Gobi *Ephedra sinica* is penetrating into the existing *Cleistogenes squarrosa* – wormwood – dwarf bush vegetation (Gunin et al., 1993). The appearance and active multiplication of *Ephedra* seems to be connected with its well-developed root system that can reach additional water from ground water.

Obvious changes in vegetation cover have been observed at the shores of intermittent lakes without a satisfactory water inflow. Thus, in 1978–1979 Lake Tatsyn-Tsagan-Nur in the Lake Valley was surrounded by reed thickets alternating with spots of saline soil swamps. In the small hills a physiological association was created by deris (*Achnatherum splendens*) communities. By 1989 the lake dried up, and its bed was occupied by saltbush groups; at the shores, reed thickets were replaced by dispersed saltbush-deris plants. This decade was remarkable due to the absence of rain for five years. The same substitution of hydromorphic vegetation by haloxeromorphic communities is characteristic also of the Khalkha Plains. Small lakes existed here before that turned into dry saline soil depressions surrounded by a small belt of complex halophytes and haloxerophytes (Vostokova, 1980).

So, a decrease in the rate of precipitation became the primary reason for haloxerophytization processes not only in steppized deserts and dry steppes, but also in true steppes and caused a considerable northward penetration of species formerly inherent in the desert zone.

Climatic changes resulted not only in a natural substitution of hydromorphic vegetation by haloxeromorphic species, but these changes also stimulated some negative exogenic processes influencing the dynamics of plant communities. The most important are wind processes on the vast Mongolian plains and water erosion in the foothills and mountains. Desiccation of upper soil horizons produced additional materials for the wind transfer of dust and sand. As a result of such redistribution of loose materials, perennial plants can accumulate small sandy mounds and, in contrast, in some places blowing sandy particles leads to the exposure and death of plant root systems (Figure 1.12). Also strong wind flows can transport sandy material for a distance of many kilometers inside mountain areas often burying local vegetation (Figure 1.13). These phenomena are observed not only in southern regions (Transaltai Gobi, Gashun Gobi and others), but also in northern areas. An intensification of wind transportation of dust-sand material leads to disturbances of the normal cycle of vegetation development on sand hills too. Sometimes even pioneer plant habitats disappear under a thick sandy mantle. In some regions sand-dust sediments completely cover the steppe communities particularly in mountain valleys open to dominant north-west winds. Such phenomena are seen even in space images (Figure 1.14). Under «wind shadow» conditions such sand sediments are covered gradually and fixed by vegetation. First, single plant habitat species appear (for example *Leymus chinensis*), than more closed plant communities form a more xeromorphic type as a rule. These differ greatly from previously existing plant ecosystems due to drastic changes in soil-ground conditions and moisture.

Changes in precipitation rates and consequent changes in habitat water regimes leads to a soil redistribution of easily soluble salts. Chloride-sulphate salts, due to increasing insolation, are forced up to the surface forming

Figure 1.12. Roots of *Haloxylon ammodendron* exposed by wind erosion (photo by A.V.Prishchepa).

Figure 1.13. Wind erosion in Transaltai Gobi (photo by A.V.Prishchepa).

Figure 1.14. Wind transportation of sand deposits to mountains (photo by A.V Prishchepa).

saline soils in microdepressions. Such changes in salt balance in turn cause salt plant communities. So, in many areas steppe zone plant communities are changed to salt ecosystems.

Drought in mountain steppe and forest belts leads to an increase in the number of thunderstorms, and consequently to the occurrence of fires. (Although a majority of the fires in steppes and forests are due to man). Eitherway the after-fire development of vegetation leads to drastic changes in forest plant successions.

Natural vegetation successions also are influenced by processes of denudation and soil formation. For example, the development of plant cover at blown away sand sites and in dried beds of lake depressions or at outcrops of crystalline rocks and their eliuvians. The first stages of such successions are formed by non-dominant rock loving plants; the final by typical zone vegetation communities. The specific composition of vegetation groups is strongly influenced by the geochemical properties of the parent rocks (Volkova, 1976). As always in nature the processes in successions are very slow and substitution of a group member by others can only be distinguished by extrapolation of spatial ecological groups over time.

One natural factor strongly influencing vegetation dynamics is animal wildlife. The most obvious changes are the results of fossorial animal activity. This factor is very important and broad spread in Mongolia (Table 1.2 & 1.3). Such animal activity has resulted in the creation of absolutely new vegetation communities over considerable areas in steppes, deserts, and even in mountain steppe belts (Guricheva & Dmitriev, 1983; Dmitriev, 1985; Dmitriev et al., 1990).

Table 1.2. Intensity of influence of main species of fossorial animals on plant cover (Guricheva & Dmitriev, 1983).

Main features of influence	Species of fossorial animals			
	Lasiopodomys brandtii	*Ochotona daurica*	*Lasiopodomys mandarinus*	*Marmota sibirica*
Area of interference, m^2	3–25	20–200	5–2000	5–900
Depth of main system of burrows, cm	20–50	30–100	5–15	200–500
Height of above-surface construction, cm	2–20	5–30	50–100	10–150
Weight of digged-out material, kg	10–30	50–500	up to 6000	50–2000
Intensity of influence on vegetation:				
number of animals per 1 km^2	200	40	5–30	1–10
number of holes per 1 km^2	30–60	2–8	1–5	1–3
maximal period of hole existence, years	10	10	1–2	3
Degrees of influence on plant cover	Reversible disturbance		Disturbance of short duration	Irreversible distortion

Table 1.3. Volume of material dug-out from holes by the principal fossorial mammals of Khangai (Dmitriev et al., 1992).

Species	Mean volumes of single dig-out action		Quantity of yearly dug-out volumes	Mean dimensions of joined dug-out material	
	Area, dm^2	Height, cm	per 1 hole	Area, dm^2	Height, cm
Marmota sibirica Pall.	10	15	7	2000	100
Citellus undulatus Pall.	1	2	1.5	9	20
Myospalax aspalax Pall.	9	25	20	–	–
Lasiopodomys brandtii Radde	1	4	30	400	3
Microtus gregalis Pall.	1	2	15–40	–	–
Lasiopodomys mandarinus Miene-Edw.	2	8	60	–	–
Ellobius talpinus Pall.	4	10	30	–	–
Ochotona daurica Pall.	3	7	25	–	–

Obviously in nature we observe the influence of several factors simultaneously. But in general, changes connected with xerophitization and halophytization of vegetation cover are examples of negative consequences in plant cover changes due to the impact of climatic factors. Nowadays such changes are enhanced due to intensive anthropogenic influences. Anthropogenic factors influence the contemporary dynamics of vegetation cover either by direct disturbance or through changes in habitat ecological conditions. But the combined action of anthropogenic factors and climatic changes causes the most drastic consequences.

In Mongolia there is a strong movement to increase agricultural production and to improve living conditions. These results in a worsening ecological situation not only due to intensification of agriculture, but due as well to an extreme development of technologies for nature management in this country with its high degree of natural vulnerability. The greater part of Mongolia experiences extreme ecological pressure, specifically an extraordinary uncontrolled exploitation of resources (without a proper attitude toward production or careful management) undermines the basis of ecological equilibrium and practically sentences the population to a low level of life (Gunin, 1990a, 1991). When combined, these result in an increasing pressure on biological cycles, on trophic relations, and especially on soils and vegetation cover.

The main anthropogenic factors which cause changes in structural-functional organization of plant ecosystems controlling the dynamics of vegetation can be defined as direct and indirect (Table 1.4). The first includes the direct anthropogenic impact on vegetation cover from its use as a natural resource. Indirect anthropogenic factors which change plant ecosystems are the use of natural resources (land, water, minerals, etc.), or activity such as electric energy production, heavy machinery construction, etc.

Table 1.4. Human-associated factors disturbing natural equilibriums.

Type of land use	Disturbing factors	Some specific features	Intensity
Industries and settlements	Dangerous emissions to atmosphere. Pollution of surface waters and soils by municipal and industrial sewages.	Local disturbance, sometimes with trends of regional transfer.	From weak to very strong.
	Construction works. Pollution of waters and soils by municipal sewages and garbage. Destruction of spontaneous vegetation. Destruction of upper soil horizons.	Local, within the settlements and at their margins; the area of influence depends on dimensions of a settlement and number of industrial objects.	Strong and very strong.
Agriculture:			
Transhumant grazing	Grazing, stock concentrations, along migration routes, seasonal pasture rotation.	Locally near yurtas, watering points, linear and strip, of regional character.	Strong and very strong from weak to strong.
Hay cutting	Phytomass removal leading to substitution of dominants in plant communities.	Local, still restricted.	From weak to moderate.
Rainfed agriculture	Ploughing leading to substitution of natural soil and plant cover by agroecosystems.	Regional and local (only at the plains of steppe zone).	Strong and very strong.
Irrigated agriculture	Ploughing and changes in water balance leading to substitution of natural ecosystems by unstable agroecosystems.	Local with weakly expressed contact interaction.	Very strong

Table 1.4. Continued.

Type of land use	Disturbing factors	Some specific features	Intensity
Forestry	Forest fires leading to appearance of plant successions at burns.	Local and regional, with impact increasing when repeated.	From moderate to very strong.
	Industrial and domestic forest cutting, including *Haloxylon* break down in deserts.		
Transport	Vehicle movement along ground roads and outside leading to disturbances in soil and vegetation cover. Soil and vegetation pollution along roads by heavy metals and garbage.	Linear – regional, rapidly increasing in intensity near large settlements.	Strong and very strong
Geological survey and research activity	Reconnaissance mining.	Local, local – regional	Strong and very strong
	Vehicle movement outside roads,	Linear – regional	From moderate to strong
	excavations and soil test pits,	Local	
	collection of plant and mineral specimens.		
Special objects	Training military sites, aerodromes, etc.	Local	Strong and very strong

All kinds of extensive nature management are direct anthropogenic influences on the dynamics of plant cover. Stock-rising based on grazing management plans is of primary importance.

Traditionally in Mongolia grazing practices plays a central role in the economic activity of the population. Indeed the extensive agricultural grazing influences natural vegetation everywhere. In fact the contemporary vegetation cover of steppes and steppe deserts in Mongolia was formed under the influence of grazing. This type of climax vegetation was formed during the last 4 millennia (Dinesman & Bold, 1992). And nowadays grazing often leads to pasture overload in certain places and underload of others (Bold, 1989). To a certain degree this situation is connected with the creation of permanent settlements with schools and with developing a cultural infrastructure. Thus the herds now are concentrated near settlements and the traditional migration and seasonal pasture rotations are out of use. In addition, a contribution was made by the creation of artificial administrative boundaries between adjacent farms. But, it should be noted that, due to privatization processes, one can see changes in these trends that seems to be influencing the dynamics of vegetation even now. Overgrazing is easily observed near settlements along with undergrazing at far ranges. Sometimes one can observe absolutely undisturbed vegetation near aimak boundaries. Overgrazing leads: to a decreased survival of forage plants; to the appearance of species with less forage value; to changes in the dominant plant ecosystems; and, finally to the complete destruction of original vegetation and its substitution either by ruderal plants or the appearance of bare land. During such long-term changes one can observe in some years a decrease in plant biomass with temporary overgrazing. Fortunately under favorable moisture conditions the initial state of vegetation cover can be restored rapidly with the cessation of grazing.

The influence of grazing on grass cover has a more complicated character when combined with natural shortages of water. Under such conditions even a light grazing load can cause serious changes in plant cover because decreases in precipitation also influences drinking water availability on grazing lands. For example, during continuous droughts: the seasonal rotation of grazing lands is disturbed; stock concentrates during summer periods at winter pastures which leads a complete vegetation devastation of several tens of square meters near drinking water sources; and to general decreases in pasture productivity, thus causing winter forage shortages. Uncontrolled grazing leads to changes in vegetation cover and also to soil degradation, hence to the appearance of negative erosion processes, wind and soil erosion, and to the occurrence of micro-mounds, slopes, terraces, etc.

Grazing in Mongolia is accompanied by herd migrations along long routes to sites of seasonal consumption. Along such routes the animal to pasture load increases greatly. This leads to pasture degradation along permanent herd routes. Here the aboveground parts of plants are destroyed, but in addition sod and root systems of plants suffer from trampling. Such activity enables accelerated erosion, micro terracing of slopes and the enlargement of devegetated areas around water sources and near temporary herd stay sites.

Great damage can be caused by grazing in forests, in forest clearings and in burnt places. Grazing here leads to destruction of tree undergrowth, worsening forest growth conditions and the substitution of forest plant ecosystems by bush-grass ones. The most dangerous sites are on the forest-steppe contact zones at the boundary of boreal vegetation.

Man made changes in forests are relatively restricted both according to species and to the land area involved. But in Mongolia the problem is important due to the high economic and ecological significance of forests, especially taking into account the small area (~4%) occupied by forests (Dorzhsuren, 1979).

Industrial forest harvests in Mongolia are on rather small areas, but felling for local consumption is prevalent (Figure 1.15). With forest fires, which are common here, forest

Table 1.5. Ecological changes with forb larch stands in clearings and burns of Eastern Khentii (Krasnoshchekov & Gombosuren, 1988a).

Ecological Parameters	Mean annual data measured at sample sites								
	Non-disturbed larch stand (*Larix dahurica* + *Rosa acicularis, Spiraea media, Potentilla fruticosa* + *Festuca ovina, Vicia venosa, Aegopodium alpestre*)			Clearing, the cover if formed by: *Poa sibirica, Geranium eriostemon, Thalictrum minus, Valeriana alternifolia*			Burn (12 years), renewal of *Betula* spp., ground cover: *Chamaenerion angustifolium, Artemisia tanacetifolia, Carex lanceolata*		
Grass cover, %	60–85			95			50–75		
Illumination of the soil surface (% from illumination of an open site)	19			32			63		
Air temperature, ^{0}C, 2 m above land surface	13.1	19.9	2.8	14.1	22.6	0.9	14.1	22.4	1.5
Temperature at land surface, ^{0}C	14.0	22.9	3.6	19.1	32.8	1.5	18.6	30.9	1.6
Relative air humidity, %	77.6	80.7	73.5	72.6	75.7	69.7	–	–	–

Table 1.6. Liquid surface runoff during snow thawing (Krasnoshchekov & Gombosuren, 1988b).

Observation site	Year	Runoff period (days)	Steepness, degrees	Maximal snow store, mm	Runoff, mm	Runoff coefficient
Larch stand with forbs (test area 1)	1983	24.III–28.III	5	11.2	0.020	0.0017
		24.III–28.III	20	12.6	0.088	0.0069
	1985	9.IV–11.IV	5	24.0	1.1	0.045
		9.IV–11.IV	20	38.0	3.0	0.079
Burn (test area 1–g)	1984	27.III– 2.IV	15	38.0	4.1	0.107
	1985	4.IV–10.IV	5	24.0	2.8	0.117
		4.IV–10.IV	15	51.3	7.7	0.150
Selective tree cutting (test area 4)						
Non-damaged stand	1983	24.III–30.III	15	25.8	1.4	0.054
forest clearing	1985	9.IV–11.IV	15	28.8	8.9	0.309
main trailing rout	1983	24.III–30.III	15	23.8	2.5	0.105
	1985	4.IV–8.IV	15	32.4	23.5	0.725

Table 1.7. Surface runoff and soil erosion processes from larch stands with forbs during warm periods (Krasnoshchekov & Gombosuren, 1988b).

Year	Period of observations	Slope steepness, degrees	Precipitati on rate (mm)	Runoff (mm)	Runoff coefficient	Solid runoff module	Erosion coefficient
Test area 1							
1982	20.VI–20.IX	5	273	0.720	0.0026	0.02	32×10^{-5}
1983	20.VI–20.IX		337	0.956	0.0028	0.07	79×10^{-5}
1984	20.VI–10.IX		256	0.819	0.0032	0.04	61×10^{-5}
1985	1.VII–20.VIII		135	0.567	0.0042	0.02	40×10^{-5}
Mean			250	0.765	0.0031	0.04	53×10^{-5}
1982	20.VI–20.IX	20	273	1.074	0.0039	0.15	38×10^{-5}
1983	20.VI–20.IX		337	1.582	0.0046	0.62	108×10^{-5}
1984	20.VI–10.IX		256	1.152	0.0045	0.37	88×10^{-5}
1985	1.VII–20.VIII		135	0.729	0.0054	0.22	82×10^{-5}
Mean			250	1.134	0.0045	0.34	79×10^{-5}
Test area 5							
1983	20.VI–20.IX	5	303	0.902	0.0030	0.93	118×10^{-4}
1984	20.VI–10.IX		256	1.997	0.0078	0.98	56×10^{-4}
1985	1.VII–20.VIII		135	0.864	0.0064	0.72	95×10^{-4}
Mean			231	1.254	0.0054	0.88	90×10^{-4}

fells cause large changes in forest vegetation, often stimulating changes in its dynamics in the direction of forming less productive plant ecosystems. (Table 1.5, 1.6 & 1.7; Figure 1.16). Original coniferous species are replaced by small- leaved birches or aspens or bush thickets, and in some cases the forested area decreases. At forest clearings and burnt forests, meadow and steppe plant ecosystem replacements can occur. In medium height mountains such sites are usually occupied by thickets of yernik, *Potentilla fruticosa*, etc.

Figure 1.15. Forest fells in mountains of Khangai (above) and Khentii (below) (photos by A.V.Prishchepa).

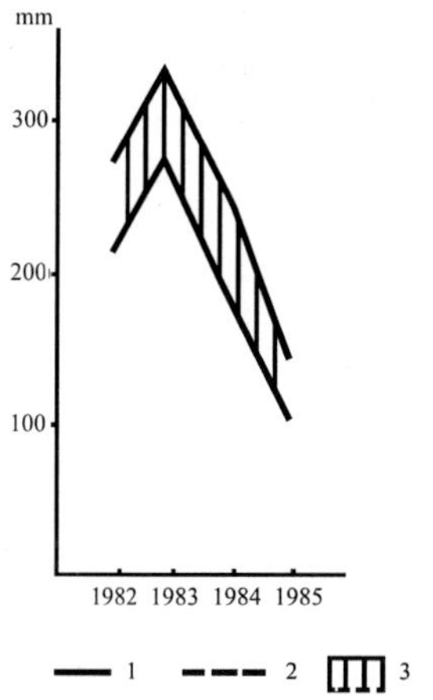

Figure 1.16. Precipitation retained by forest canopy (according to Krasnoshchekov & Gombosuren, 1988b).

Precipitation amounts: 1 – at forest fell, 2 – under tree canopy, 3 – retained by forest canopy.

The renewal of arboreal plant communities can require decades or even hundreds of years. Steppe communities appearing on previously forested areas almost completely prevents the renewal of forest species.

Disturbances of boreal vegetation by cutting or fire increases the danger of accelerated erosion on mountain slopes, increases surface runoff, and decreases the infiltration of moisture for recharging ground waters. The application of heavy vehicles and tractors for timber harvesting and transportation also increases the danger of accelerated erosion. Deep ruts after each passing of such a machine are triggers for the occurrence of erosion furrows and gulleys (Figure 1.17). A strong negative influence on forest vegetation particularly on regeneration is exerted by the grazing of domestic animals, especially goats, in eating young forest tree shoots. This damage commonly is observed near settlements, in low mountains, and along herd migration routes.

All types of economic activity connected with the use of land, water, and mineral resources, including construction and exploitation at municipal, industrial and transport sites, leads not only to a complete destruction of the original site vegetation, but also to large changes in the surrounding plant communities. The scale of such violations on vegetation cover can exceed, by many times, the scale of the economic activity itself (Table 1.8).

Figure 1.17. Slope erosion along timber transport routes (photo by A.V.Prishchepa).

Table 1.8. Radiuses of influence on vegetation by different administrative – economic centers (Gunin & Vostokova, 1995a).

Administrative - economic center	Radius of the area with disturbed plant cover (km) and degree of disturbance		
	Very strong (up to complete destruction)	Strong (up to appearance of weed-grass communities)	Medium (up to weed appearance)
Aimak (oblast)	1.0	10	15
Somon (region)	0.5	5	10
Brigade (small industrial settlement)	0.3	3	5
Single buildings, yurtas and watering points	0.1–0.2	1	3

Today a stable population growth and its increasing demands, has resulted in concentrating the population in towns and large settlements (now up to 60% of the total population). Until recently there were only a few cities in Mongolia with populations near 100,000 including Ulan-Bator, Erdenet, and Choibalsan. But their impact on the adjacent territory exceeds by 2–3 times when compared with a village settlement. Vegetation near cities is subject to different types of negative impact, i.e., direct elimination of vegetation cover at places of chaotic storage of industrial and domestic wastes, concentrations of automobile ruts without paved roads, pastoral cover degradation due to non-regulated grazing of domestic animals, and large environmental losses from pollution by oil products plus municipal and industrial sewages.

The distribution of settlements and industrial sites over Mongolia is very uneven. The greatest population density, up to 6 persons per sq. km, is in the central aimak around the capital – Ulan-Bator. The north-western and northern parts, at elevations not higher than 1500–1800 m, have a population density near 3 persons per sq.km. A permanent population in mountains and deserts either is absent or its density does not exceed 1 person per sq.km.

A drastic influence is being exerted by mining including vegetation destruction at mining sites, the creation of absolutely new relief forms such as quarries, embankments, and so on, with each possessing specific ecological features. During the processes of extracting mineral deposits toxic materials often are brought to the surface, which must be meliorated for plants to grow. The most dangerous is gold mining in river valleys when dredges are used. As a result all of the valley is absolutely modified: not only are plant

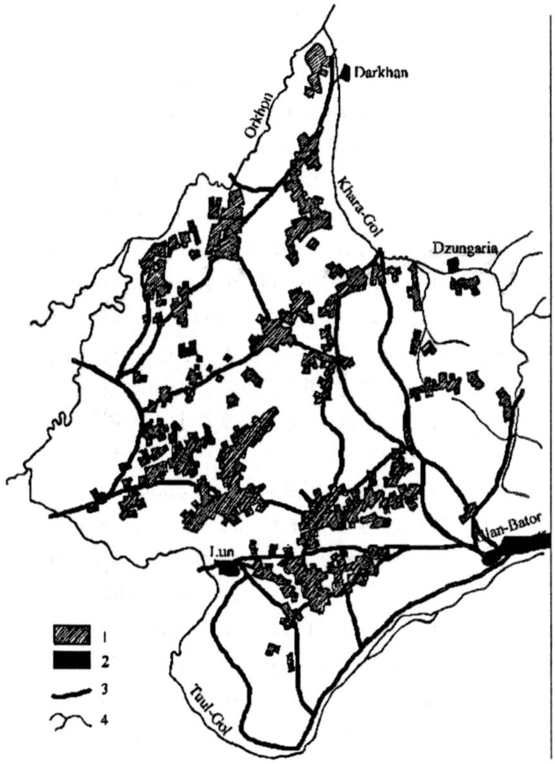

Figure 1.18. Allocation of arable lands in the main agricultural zone (north-west from Ulan-Bator).

1 – arable lands, 2 – settlements, 3 – roads, 4 – rivers.

Table 1.9. Changes in properties of dark steppe chestnut soils due to ploughing (Vostokova et al., 1992).

Changes in soil properties	Signs of degrading
Humus losses (as compared with virgin lands)	Appearance of spot pattern at the soil surface.
Increase in carbonate content due to ploughing up of carbonate horizon to the surface	Appearance of signs of blowing out and formation of eolian accumulation (sand eolian forms).
Increase in soil alkalinity due to ploughing up of salinized horizons.	Sheet and linear erosion forms (from shallow rill erosion furrows to well-formed sinkholes).
Loss of nutrients (phosphorus, potassium) Soil structure destruction Changes in water-physical properties.	Appearance of water accumulation grounds.

communities destroyed, but hydrological regimes and river channel processes are drastically changed. Then all such sites are inhabited by ruderal plant species. The renewal of more valuable plant communities requires determined reclamation measures. Often such sites have completely new landscapes with cultivated vegetation.

Grazing influences vegetation cover almost everywhere. In contrast, agriculture only causes local changes in vegetation. Rainfed agriculture disturbes vegetation cover only in steppes (Table 1.9). Irrigated agriculture occupies lesser areas in steppe and desert zones. Plowed sites are used for three to four years, then abandoned and new virgin steppe lands are rotated into agriculture. So, the steppe vegetation areas disturbed by plowing are considerably larger than the actual production areas. The vegetation cover on these idle lands demonstrates all vegetation successional stages. The greatest steppe areas used for rainfed agriculture are concentrated to the north and to the west from Ulan-Bator (Figure 1.18). In general, the influence of rainfed agriculture on vegetation still is rather insignificant. But at some local sites changes in plant cover can be considerable. For example, in desert conditions, irrigation by ground waters supports new hydromorphic plant communities at the periphery of irrigated fields (Table 1.10).

Clearly any kind of nature management intensification or settlements, increases transportation networks. In the absence of a paved cover on many roads the width of disturbed strips is very large. The greatest damage is experienced by vegetation on plain and mountain steppes. Here the width of vegetation damaged strips can reach 10–15 km (Figure 1.19). Such extensive damage occurs at practically all towns. The greater damage to

Table 1.10. Peculiarities of hydromorphic soils in oases of Transaltai Gobi (Pankova, 1992).

Parameters	Indices
Critical depth of groundwater table	200–250 cm
Upper limit of groundwater mineral content	15–25 g/l
Character of soil accumulation	Surface, with formation of salt crust
Salt content in salt crusts	40–60%
Hydrogenic accumulation of salts due to evaporation	About 10 g/ha (for ground waters at the depth 1.8 m and mineral content 5 g/l)
Salt income with atmospheric precipitation in extreme arid deserts	20–50 kg/ha per year

Figure 1.19. Roads in steppe of Central Khangai (above) and South Khangai (below) (photo by E.N.Matyushkin).

vegetation caused by unpaved roads is observed along the roads from Ulan-Bator: to the east towns of Baganur, Underkhan, Choibalsan); to the west Erdenet; to the south-west Arvaikhere, Altai, Ulangom; and to the south Mandal Gobi, Dalanzadgad. Only now is there an intense effort to pave these roads and as a result disturbed vegetation in these areas is decreasing and vegetation renewal is occurring on former dirt roads.

The general scope of human-associated disturbances on contemporary vegetation cover is given in Figure 1.20, which is derived from the multi-sheet map «Ecosystems of

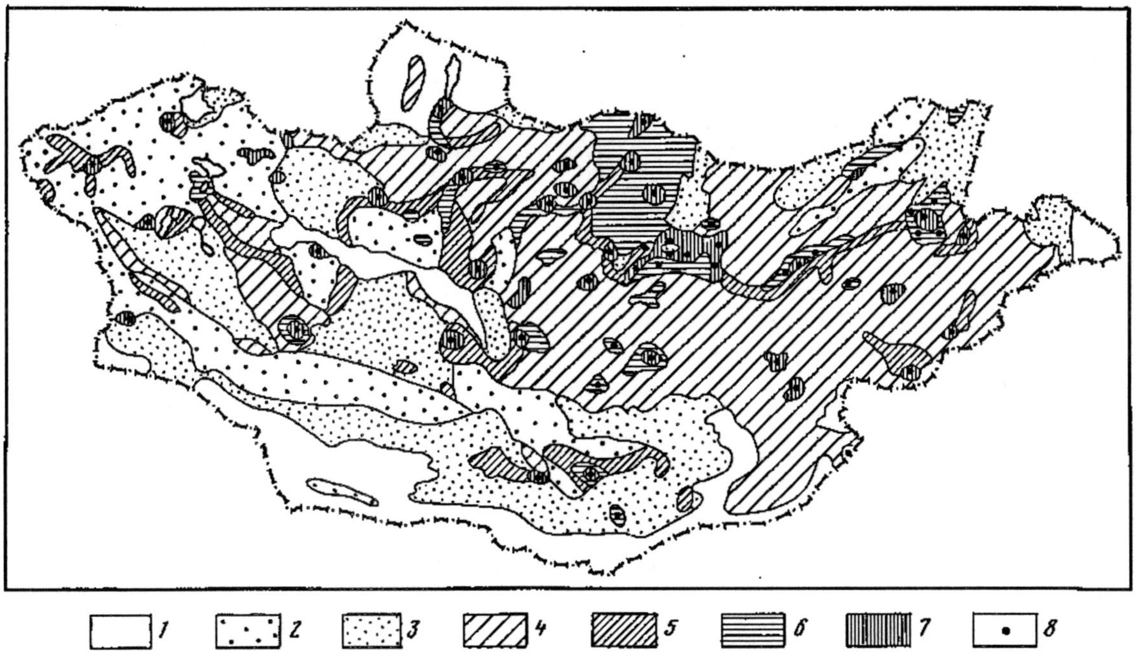

Figure 1.20. Degrees of anthropogenic vegetation cover modification.

1–7. Degrees of modification of vegetation cover: 1 – weak or absent; 2 – from weak to medium; 3 – medium, 4 – from medium to high; 5 – high; 6 – from high to very high; 7 – very high; 8 – industrial complexes and settlements.

Mongolia» (1995) compiled in 1 : 1000,000 scale. The scheme shows territories where vegetation dynamics were influenced first by anthropogenic factors. Their effects increase when they act in combination with changes in natural factors, for example, a decreased precipitation.

1.5 CONCLUSION

The central position of Mongolia in the Asian continent pre-determines most of its natural features. The semi-isolation of the territory from moist-carrying air flows results in a severe continental climate, that in turn, determines the distribution and structure of its vegetation cover.

The contemporary dynamics of vegetation is stimulated both by natural, mainly climatic fluctuations, and by several anthropogenic factors. The impact of human-associated factors increases in the presence of unfavorable climatic changes (mainly decreases in precipitation).

In brief, the dynamics of Mongolia's vegetation are controlled by – changes in climatic parameters (a decrease and long-term oscillations in precipitation), an intensification of natural destructive processes that are subject to the influence of economic activity, improper nature management practices, and numerous direct human-associated disturbances and changes in vegetation cover. These destructive processes are vital issues for the Mongolian society.

CHAPTER 2

LATE QUATERNARY VEGETATION HISTORY OF MONGOLIA

2.1 INTRODUCTION

Studies on past changes in vegetation are a central key to understanding modern plant distribution, to reconstructing past climates, and to testing biome and climate models. This chapter is an initial attempt to reconstruct and to map spatial changes in the vegetation across Mongolia since ca 15,000 14C yr B.P. Our work is a logical continuation of extensive studies on the late Quaternary vegetation and environments of Mongolia done by Soviet and Mongolian scientists during recent decades (Giterman et al., 1968; Lavrenko & Rachkovskaya, 1976; Dinesman et al., 1989; Logatchov, 1989; Sevastyanov et al., 1994; etc.). Since the main publications only are published in Russian and poorly accessible for the English speaking scientific community, we first briefly will summarize previous studies, which reconstructed regional vegetation changes in Mongolia. These vegetation reconstructions are broadly correct, but some major discrepancies do appear. This summary demonstrates that Mongolian pollen records, rather than other sources of paleobotanical data, are the main information source for vegetation reconstructions and that these are based on relationships between modern pollen spectra and actual vegetation. Moreover, we will describe some potential problems that different types of sediments can cause in the interpretation of fossil pollen spectra.

A substantial number of radiocarbon-dated continuous pollen records from Mongolian lakes were obtained by research teams of the Joint Russian (Soviet)-Mongolian Biological Expedition from 1972 to 1980. Most of these records still are unpublished. In this chapter, we try to fill this gap and to describe the most interesting pollen diagrams from different geographical regions in Mongolia.

Pollen data from 25 radiocarbon-dated sediment cores taken from 20 lake basins across Mongolia are used to reconstruct Mongolian vegetation patterns and to draw a series of paleovegetation maps for the late glacial and Holocene time slices with the most pronounced changes. For that task we applied a recently developed method of «biomization», i.e. an objective quantitative biome reconstruction from pollen samples, as described by Prentice et al. (1996). Modern pollen data from 102 surface spectra, which reflect the present-day vegetation of Mongolia, provide a strong test of the biomization method.

A set of 24 radiocarbon-dated tree and shrub macrofossils from 4 sites in southern Mongolia was compiled from published sources and mapped together with modern taxa distributions to show changes in the distribution of the selected plants during the second part of Holocene. These data provide additional information on Mongolia's vegetation dynamics when compared to pollen-based reconstructions.

Then results are compared with data from surrounding Asian regions. Qualitative estimations of climate are given in each reconstruction of Mongolian vegetation. However, no previous attempt exists to explain the causes and mechanisms of suggested climate changes nor do clear explanations exist about why one or another climate parameter is responsible for certain vegetation changes. We try to interpret observed changes in vegetation with the idea that the distribution of different functional types of plants is influenced by climate parameters which provide physiological limits for plant growth (Prentice et al., 1992). These reconstructed changes in the vegetation across Mongolia then are discussed in terms of possible changes in the regional climate, caused by broad-scale orbitally induced insolation changes.

2.2 AN OVERVIEW OF PREVIOUS STUDIES

Studies on the vegetation history of Mongolia

Steppe and forest steppe dominate modern Mongolian vegetation (Lavrenko, 1979) and occupy, respectively, about 25% and 53% of the projected surface area of the country (Hilbig, 1995). Other vegetation types, e.g. northern mountain Siberian forests and southern deserts, are less important and only cover ca 4% and 15% of the land. Pollen records from western and northern Mongolia suggest that intermountain depressions were covered by steppe and that mountain slopes were already occupied by forest-steppe communities by 1,000,000 to 700,000 yr B.P. (Devyatkin & Shilova, 1970; Shilova, 1973).

Taxa which are widely distributed in Mongolia today, e.g., *Pinus*, *Betula*, *Artemisia* and Chenopodiaceae dominate pollen spectra during Pleistocene (Malaeva, 1989a), suggesting that the general composition of Mongolian vegetation then was similar to the present one. However, the re-appearance of broad-leaved taxa (e.g. *Juglans*, *Acer*, *Carpinus*, *Tilia*, *Quercus*, *Corylus*) in pollen spectra during the most favourable intervals of Pleistocene (Devyatkin & Shilova, 1970; Malaeva, 1989a,b), indicate some environmental and climatic variability and a decrease in the number of arboreal taxa in Mongolia towards present days. Also Malaeva & Murzaeva (1987) used undated pollen records from lower river terraces to suggest that *Corylus*, *Tilia*, *Alnus* and *Ulmus* grew in some river valleys in the northern part of the country at least in middle Holocene (ca 4500 to 4000 yr B.P.). A shrub-like *Ulmus pumila* and *Alnus fruticosa* grow in Mongolia until the present (Grubov, 1982; Gubanov, 1996). However more data are needed to prove that other temperate deciduous taxa grew in the country during Holocene.

Pollen-based qualitative reconstructions of late Quaternary vegetation in different parts of Mongolia are presented in a limited number of Russian publications poorly accessible to western readers (Golubeva, 1976, 1978; Vipper et al., 1976, 1989; Savina et al., 1981; Dinesman et al., 1989; Malaeva, 1989b; Sevastyanov et al., 1993). These studies are based on different methods applied to various types of sediments (Table 2.1). Most of these published records are undated or poorly dated, creating a problem for correlations between regions and between sites within the same region even if they are situated in different localities (e.g. slopes of northern and southern exposure) or altitudes. However, a chronology based on a limited number of radiocarbon dates (Vipper et al., 1976) has been used as a basis for long-distant correlation by Golubeva (1976, 1978) and Savina et al. (1981).

In most previous studies of Late Quaternary vegetation changes in Mongolia, late glacial time was described as a phase when treeless vegetation types (dry steppe and desert) extended north of their modern position, suggesting a drier and colder (because of crioturbations in the sediments) climate than that today (Vipper et al., 1989; Golubeva, 1976; 1978; Sevastyanov et al., 1993). However, Malaeva (1989), in re-interpreting pollen data from published sources (Golubeva, 1976; 1978), suggested that, during the late glacial, forestation of the plains was even greater than today, caused by cool summers and lower evaporation. The early Holocene (10,000 to 8000 yr B.P.) is characterised by a

Table 2.1. Records used for reconstructions of Late Quaternary vegetation changes in Mongolia.

Record type	Site and sediment type	Study area	Record length (yr B.P.)	Chronology based on	Continuous or discontinuous	Reference
Pollen	lacustrine sediment	Khangai-Khentii, Altai, Great Lakes Basin	~ 11,500	Radiocarbon dates	continuous	Vipper et al., 1976; 1989
		Uvs-Nur Basin	~ 11,000	radiocarbon dates	discontinuous	Sevastyanov et al., 1993
	river terrace	plains of northern and eastern Mongolia	>15,000	correlation with regional stratigraphical schemes	discontinuous	Golubeva, 1976, 1978; Malaeva & Murzaeva, 1987; Malaeva, 1989b
	peat sequence	Gobi Altai	~ 3500	radiocarbon dates	discontinuous	Dinesman et al., 1989
	soil profile	forest-covered regions of northern Mongolia	~ 9000	correlation with other types of records	continuous	Savina et al., 1981
	cave coprolites	Gobi Altai	~ 8000	radiocarbon dates	discontinuous	Dinesman et al., 1989
Plant macrofossil	buried plant macrofossills	Gobi Altai, Gobi desert	~ 4500	radiocarbon dates	discontinuous	Dinesman et al., 1989

drier than modern vegetation and climate (Vipper et al., 1976; 1989). Savina et al. (1981) reconstructed that a *Kobresia-Carex* tundra was more extended in the Altai and Khangai mountains at 9000 to 8000 yr B.P. and suggested that the climate was colder than present. The middle Holocene (8000 to 4000 yr B.P.) was characterised as an interval when forestation of the country gradually increased to the maximum. If so conditions then were wetter and milder than present. Dating of the maximum distribution of tree species in Mongolia varies in different publications: 7500 to 5500 yr B.P. (Vipper et al., 1989), 6150 to 4400 yr B.P. (Dinesman et al., 1989), 4500 to 2500 yr B.P. (Vipper et al., 1976; Savina et al., 1981; Sevastyanov et al., 1993). However, some authors (Golubeva, 1976; 1978; Malaeva, 1989b) reported that steppe was more important than today in the eastern part of Mongolia in the middle Holocene, when the climate may have been warmer and drier than today. They attributed a spreading of arboreal taxa in Mongolia to the interval between 4000 and 2000 yr B.P. After 2000 yr B.P., the vegetation became similar to the vegetation of today in all reconstructions. Beside these broad patterns of vegetation and climate change, millennium-to-century-scale changes are reported in some publications (Dinesman et al., 1989; Vipper et al., 1989).

Data in Table 2.1 demonstrate that pollen studies are the major source of records on the vegetation history of Mongolia. However, pollen records from lake sediments provide the best opportunity for a reconstruction of continuous vegetation changes in the country where semi-arid and arid climate conditions were unfavourable for the accumulation and preservation of peat and other organic material. All vegetation reconstructions from pollen in Mongolia (see Table 2.1) were based on an objective relation between modern pollen spectra and vegetation. However, our review of those publications shows that authors often do not provide well-defined criteria for these reconstructions. Moreover, authors interpret fossil pollen spectra in different ways from one another and refer to different studies for modern pollen spectra. The latter makes it difficult to compare the results of different authors. Here we decided to synthesize the most important results coming from studies of surface pollen spectra in Mongolia as a background for our subsequent reconstructions.

Modern pollen spectra: problems of interpretation

Several attempts have been undertaken (Malgina, 1971; Golubeva, 1976; Vipper et al., 1976; Savina & Burenina, 1981; Shilova, 1984; Malaeva, 1989a; Dinesman et al., 1989; Chernova & Dirksen, 1995) to analyse the relationship of modern-pollen spectra to present-day vegetation in Mongolia with the idea of showing the importance of these data for vegetation and climate reconstructions from fossil pollen spectra. The main conclusion reported by these publications was that the spatial patterns in modern pollen data reflect the current vegetation. However possible problems that may occur during the interpretation include: poor preservation of some taxa pollen, long distant pollen transport, and re-deposition of pollen from older sediments.

Pollen preservation

The poor preservation of *Larix* pollen, the main tree taxa in Mongolia, creates a problem for interpreting data. Analyzing recent pollen spectra from river deposits (Malaeva, 1989) found that *Larix* pollen was easily destroyed by water as well as during the chemical preparation of sediment for pollen analysis. On the other hand, *Larix* pollen dominated in samples collected from surface soil in larch forests and varied from 10 to >90% of the total arboreal pollen depending on its composition in the vegetation (Savina & Burenina, 1981). *Larix* has high pollen productivity similar to other boreal conifers, but dispersal from the tree is usually less than 200–350 m (Savina & Burenina, 1981). This fact may explain the low concentration of *Larix* in fossil pollen spectra. Indeed Vipper et al. (1976) and Malaeva (1989) suggested that low but constant presence of *Larix* in fossil pollen spectra should be interpreted as the evidence of its growing close to the site. *Populus* is the only other arboreal taxon whose pollen has not been found in surface pollen spectra from Mongolia where it forms floodplain forests (Malgina, 1971; Savina & Burenina, 1981).

Nonarboreal Gentianaceae, Scrophulariaceae, Campanulaceae, Crassulaceae, *Myosotis* and *Patrinia* species are important in the vegetation of high-elevation sites now, but were not found in recent pollen spectra from alluvial and soil sediments from north-western Mongolia. These taxa may have low pollen production or poor preservation (Chernova & Dirksen, 1995).

The conditions for pollen preservation also are connected with the type of sediment. Bottom deposits in fresh-water lakes have much better pollen preservation compared with other types of sediments in Mongolia. Thus Savina et al. (1981) established that only humic layers in soil profiles contain enough pollen for analysis and layers with no pollen were systematically recorded in pollen diagrams from river terraces (Golubeva, 1976; Malaeva, 1989a).

Long distant transport and re-deposition of pollen

Gravis & Lisun (1974) suggested that a single «pollen cloud» of *Betula* and *Pinus* pollen come from the Transbaical region over Mongolia. However, that suggestion was not accepted by other paleontologists (Malgina, 1971; Savina & Burenina, 1981; Malaeva, 1989a). Pollen transport distances, as mentioned by Sladkov (1967), are different among taxa: from several hundred metres (*Larix*), 250 to 300 km (*Betula*, *Alnus*), up to 500 km (*Picea*), and even to 1000 km and more (*Pinus*). However, surface pollen spectra from forest usually contain more than 50% arboreal pollen (Malaeva, 1989a), and this amount progressively decreases to less than 10% in the spectra from dry steppe and desert (Malgina, 1971). Pollen of *Pinus sibirica* was not found in the surface soil samples from north-eastern Mongolia, and *Pinus sylvestris* pollen was rare in the pollen spectra from western Mongolia, in the regions where these

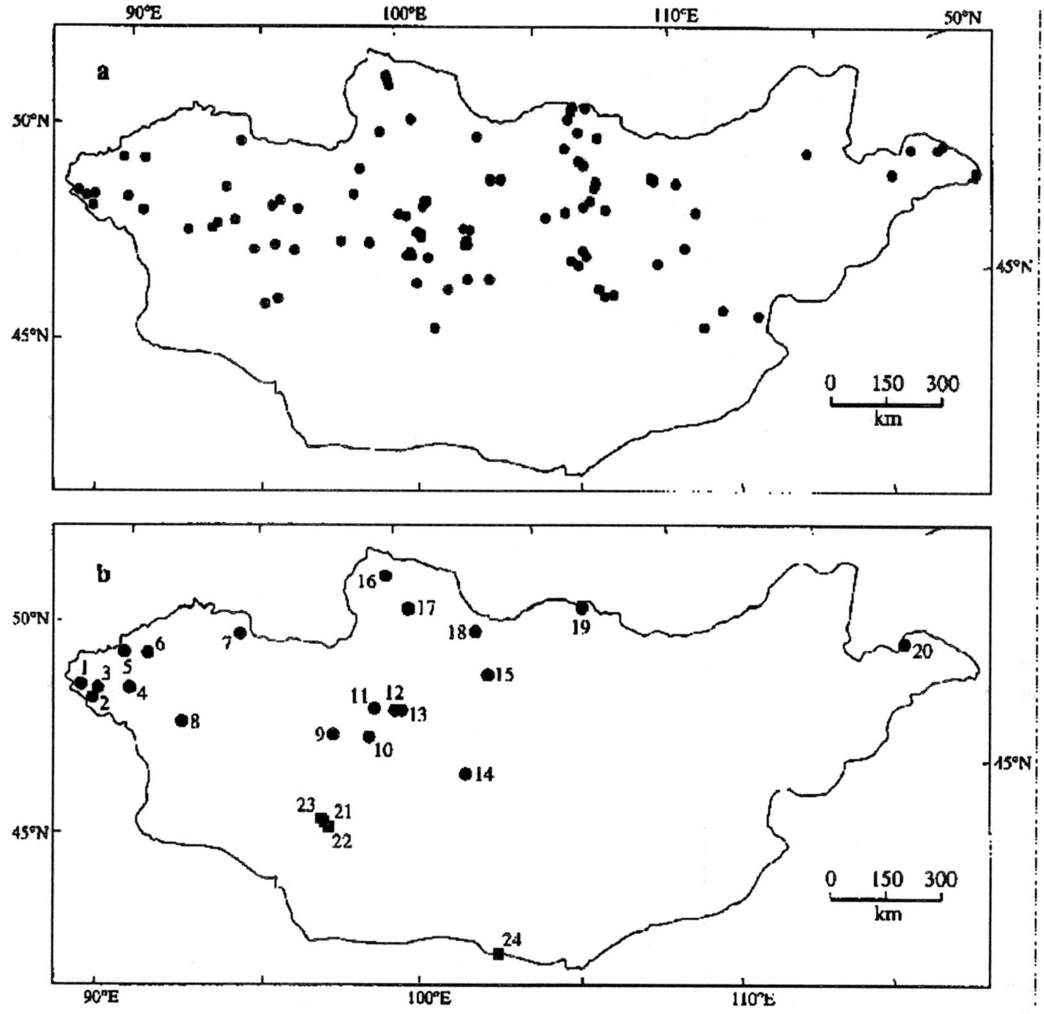

Figure 2.1. Distribution of (a) modern surface pollen spectra and (b) sites with fossil pollen (closed circle) and plant macrofossil (closed square) data from Mongolia.

species do not grow today (Savina & Burenina, 1981). However, pollen of *Pinus* often was present in pollen spectra from the alpine meadows and tundra belts in the mountains, suggesting its transport by upslope airflow (Malaeva, 1989a; Chernova & Dirksen, 1995). Pollen from older sediments in Mongolia may have been redeposited in alluvial deposits (Malaeva, 1989a). Shilova, (1984) mentioned the presence of *Tsuga* and other exotic coniferous pollen reworked from the Pliocene deposits for the Darkhat Depression. Redeposition was not mentioned as a factor in lake bottom sediments in Mongolia.

2.3 DATA USED IN THIS STUDY

Modern data

We have assembled a set of 102 pollen surface samples from Mongolia (Appendix 2) with 23 samples from Malgina (1971) and 79 samples analysed by E.Meteltseva and V.Sokolovskaya (Sokolovskaya, 1996, personal communication) (Figure 2.1a). All samples were collected from a soil surface in the northern and central parts of the country

Table 2.2. Summary of sites with pollen and macrofossil data used in this study.

Site NN	Site name	Lat. (°N)	Lon. (°E)	Elev. (m)	No. ^{14}C dates used	Source of evidence	Reference
1	Hoton-Nur	48,67	88,30	2083	6	pollen	unpublished
2	Dayan-Nur	48,37	88,83	2232	2	pollen	unpublished
3	Danyagiin-Hara-Nur	48,62	88,95	2493	1	pollen	unpublished
4	Tolbo-Nur	48,65	90,08	2079	1	pollen	unpublished
5	Dund-Nur 2	49,50	89,79	2097	2	pollen	Vipper et al., 1976
	Dund-Nur 8				2	pollen	unpublished
6	Achit-Nur 2	49,50	90,60	1435	1	pollen	Vipper et al., 1976
	Achit-Nur 6				3	pollen	unpublished
	Achit-Nur 8				4	pollen	unpublished
7	Hoit-Gol	50,05	94,03	925	3	pollen	Sevastyanov et al., 1993
8	Hara-Us-Nur 2	47,92	92,00	1156	1	pollen	unpublished
9	Tsagan-Nur	47,65	97,27	2236	3	pollen	unpublished
10	Huh-Nur	47,53	98,52	2649	1	pollen	unpublished
11	Daba-Nur 3	48,20	98,79	2465	2	pollen	unpublished
	Daba-Nur 8				6	pollen	unpublished
12	Hudo-Nur 3	48,13	99,53	2061	2	pollen	unpublished
	Hudo-Nur 8				4	pollen	unpublished
13	Terkhiin-Tsagan-Nur 8	48,15	99,70	2060	8	pollen	unpublished
14	Shiret-Nur	46,53	101,82	2500	3	pollen	unpublished
15	Urmiin-Tsagan-Nur	48,84	102,93	1450	2	pollen	unpublished
16	Dood-Nur 4	51,33	99,38	1538	2	pollen	Dorofeyuk & Tarasov, 1998
17	Hubsugul	50,53	100,17	1645	2	pollen	Dorofeyuk & Tarasov, 1998
18	Yamant-Nur	49,90	102,60	1000	1	pollen	unpublished
19	Gun-Nur	50,25	106,60	600	7	pollen	Dorofeyuk & Tarasov, 1998
20	Buir-Nur	47,75	117,70	583	3	pollen	unpublished
21	Bayan-Sair	45,57	96,91	2600	14	macrof.	Dinesman et al., 1989
22	Tsakhir-Khalgyn-Nuruu	45,43	97,08	2800	7	macrof.	Dinesman et al., 1989
23	Uert-Am	45,63	96,83	2600	1	macrof.	Dinesman et al., 1989
24	Sudzhiin-Khuduk	42,17	102,45	1180	2	macrof.	Dinesman et al., 1989

Note: Site numbers are the same for the figures 2.1b, 2.13, 2.14.

between elevations of 550 to 2500 m. There are no data from the southern desert region or from the upper mountain belt occupied by tundra.

Fossil data

Pollen and macrofossil records

We compiled a set of pollen diagrams from 19 fresh-water lakes in Mongolia (Figure 2.1b). Samples from core sediments were processed for pollen analysis following standard procedures (Grichuk & Zaklinskaya, 1948; Berglund & Ralska-Jasiewiczowa, 1986). Pollen sums generally exceed 200 to 1200 identified grains. Seven samples from the basal part of the Hoton-Nur core contained fewer than 20 grains per sample. They, however, were not used to interpret vegetation changes. A published pollen diagram of the Hoit-Gol lacustrine sequence (Sevastyanov et al., 1993) has been used here to reconstruct a vegetation history in the Great Lakes Basin of north-western Mongolia. The primary pollen counts (kindly provided by D. Sevastyanov) let us analyse the Hoit-Gol pollen records in the same way as the other records we used. Table 2.2 presents details for each core.

Dating

All cores are radiocarbon dated, but the quality of dating control is unequal for different cores (Table 2.3). Some of them have only one or two radiocarbon dates. All radiocarbon dates are conventional dates obtained from bulk sediment except for three recently obtained AMS dates from Hoton-Nur Lake (Kershaw, 1997 personal communication). The latter came from the organic fraction of clay sediment samples collected for diatom and pollen analyses. All three dates appear too old and are not used. The lack of calcareous rocks around the basin makes it possible that the clay was contaminated by microscopic charcoal particles that were deposited in the mountain glacials in the catchment area and came into the lake with melting water during the late glacial time. Such has occurred in the southern French Alps (Nakagawa, 1997, personal communication).

Table 2.3. List of available radiocarbon dates from sites used in this study

Site NN	Site name	Sample depth (cm)	^{14}C years B.P.	Laboratory ref.	Material	Reference
1	Hoton-Nur	70–95	2950±80	TA-1471	gyttja	Tarasov et al., 1996
		147–170	3900±140	TA-1440	gyttja	Tarasov et al., 1996
		195–220	5360±80	TA-1472	gyttja	Tarasov et al., 1996
		245–270	5975±150	TA-1439	gyttja	Tarasov et al., 1996
		295–320	7910±120	TA-1473	gyttja	Tarasov et al., 1996
		350–375	9070±150	TA-1419	gyttja	Tarasov et al., 1996
		390–400	*14,250±200	ANSTO OZB556U1OA1A	clay	Kershaw, 1997, pers. com.
		410–420	*15,550±630	ANSTO OZB556U1OA1A	clay	Kershaw, 1997, pers. com.
		480–490	*20,900±1160	ANSTO OZB556U1OA1A	clay	Kershaw, 1997, pers. com.
2	Dayan-Nur	380–385	3340±70	TA-963	clay	unpublished
		435–440	3970±80	TA-961	clay	unpublished
3	Danyagiin-Hara-Nur	115–135	2450±100	TA-1032	gyttja	unpublished
4	Tolbo-Nur	117–123	680±80	Vib-5	gyttja	unpublished
5	Dund-Nur 2	155–160	4240±100	Vib-17	gyttja	Vipper et al., 1976
		184–190	5140±100	Vib-18	gyttja	Vipper et al., 1976
	Dund-Nur 8	177–190	4830±70	TA-1190	gyttja	unpublished
		195–200	5220±60	TA-1189	gyttja	unpublished
6	Achit-Nur 2	140–162	5270±80	Vib-14	gyttja	Vipper et al., 1976
	Achit-Nur 6	52–65	1000±120	TA-1030	gyttja	unpublished
		215–227	3270±90	TA-1029	gyttja	unpublished
		302–320	9470±100	TA-960	clay	unpublished

Table 2.3. Continued.

Site NN	Site name	Sample depth (cm)	^{14}C years B.P.	Laboratory ref.	Material	Reference
	Achit-Nur 8	152–165	6540±100	TA-1832	gyttja	Tarasov et al., 1996
		175–190	9397±80	TA-1192	clay	Tarasov et al., 1996
		205–220	10,150±100	TA-1191	clay	Tarasov et al., 1996
		235–245	11,500±100	TA-1183	clay	Tarasov et al., 1996
7	Hoit-Gol	20–30	2930±90	LU-2662	soil	Sevastyanov et al., 1993
		230–240	6510±170	TA-2286	silty sand	Sevastyanov et al., 1993
		280–290	7440±90	TA-2284	silty sand	Sevastyanov et al., 1993
8	Hara-Us-Nur 2	32–45	660±90	Vib-13	gyttja	Tarasov et al., 1996
	Hara-Us-Nur 6	90–115	1550±100	TA-1080	gyttja	Tarasov et al., 1996
		115–125	1960±200	TA-1067	gyttja	Tarasov et al., 1996
9	Tsagan-Nur	75–90	1050±100	TA-1064	gyttja	unpublished
		315–340	2660±80	TA-1063	gyttja	unpublished
		340–370	2850±90	TA-1033	gyttja	unpublished
10	Huh-Nur	80–85	3430±90	Vib-105	gyttja	unpublished
11	Daba-Nur 3	190–195	5590±120	TA-968	gyttja	unpublished
		228–233	7750±80	TA-1027	gyttja	unpublished
	Daba-Nur 8	75–100	3100±120	TA-1355	gyttja	Tarasov et al., 1996
		250–275	5600±80	TA-1356	gyttja	Tarasov et al., 1996
		325–350	7680±100	TA-1357	gyttja	Tarasov et al., 1996
		375–400	9400±100	TA-1358	gyttja	Tarasov et al., 1996
		460–475	10,580±100	TA-1188	gyttja	Tarasov et al., 1996
		490–505	11,180±120	TA-1028	peat	Tarasov et al., 1996
12	Hudo-Nur 3	440–445	7760±100	TA-969	gyttja	unpublished
		640–645	9360±100	TA-970	gyttja	unpublished
	Hudo-Nur 8	380–400	5450±80	TA-1346	gyttja	unpublished
		610–630	7740±80	TA-1345	gyttja	unpublished
		760–780	9230±110	TA-1344	gyttja	unpublished
		850–875	9800±100	TA-1247	gyttja	unpublished
13	Terkhiin-Tsagan-Nur 8	175–200	2740±60	TA-1538	gyttja	Tarasov et al., 1996
		275–300	3840±50	TA-1539	gyttja	Tarasov et al., 1996
		300–325	4150±80	TA-1351	gyttja	Tarasov et al., 1996
		325–350	4230±50	TA-1340	gyttja	Tarasov et al., 1996
		400–425	5050±80	TA-1352	gyttja	Tarasov et al., 1996
		450–475	5950±120	TA-1353	gyttja	Tarasov et al., 1996
		525–550	6690±60	TA-1246	gyttja	Tarasov et al., 1996
		550–575	6890±100	TA-1248	gyttja	Tarasov et al., 1996

Table 2.3. Continued.

Site NN	Site name	Sample depth (cm)	^{14}C years B.P.	Laboratory ref.	Material	Reference
14	Shiret-Nur	50–75	1320±100	TA-1474	gyttja	unpublished
		125–150	3150±120	TA-1775	gyttja	unpublished
		200–220	8360±100	TA-1420	gyttja	unpublished
15	Urmiin-Tsagan-Nur	240–250	5200±80	TA-1484	gyttja	unpublished
		260–280	7050±150	TA-1485A	gyttja	unpublished
16	Dood-Nur 4	120–125	8150±100	Vib-112	gyttja	Tarasov et al., 1996
		235–240	11470±100	Vib-113	clay	Tarasov et al., 1996
17	Hubsugul	100–112	3910±60	TA-670	peat	Tarasov et al., 1996
		190–202	5800±100	TA-671	peat	Tarasov et al., 1996
18	Yamant-Nur	425–450	14300±200	TA-1483A	clay	unpublished
19	Gun-Nur	50–75	1395±80	TA-1441	gyttja	Tarasov et al., 1996
		150–175	2870±70	TA-1481A	gyttja	Tarasov et al., 1996
		200–225	3300±80	TA-1482A	gyttja	Tarasov et al., 1996
		270–300	7970±100	TA-1417A	gyttja	Tarasov et al., 1996
		325–350	8150±100	TA-1350A	gyttja	Tarasov et al., 1996
		467–470	8760±150	TA-1348A	lime	Tarasov et al., 1996
		480–483	9550±150	TA-1349A	clay	Tarasov et al., 1996
20	Buir-Nur	80–100	1320±80	TA-1831	gyttja	Tarasov et al., 1996
		100–115	1650±70	TA-1830	gyttja	Tarasov et al., 1996
		235–260	3290±80	TA-1193	gyttja	Tarasov et al., 1996
21	Bayan-Sair		2797±125	IEMEZH-225	*Larix*, wood	Dinesman et al., 1989
			2923±148	IEMEZH-224	*Larix*, wood	Dinesman et al., 1989
			3245±163	IEMEZH-222	*Larix*, wood	Dinesman et al., 1989
			3310±104	IEMEZH-218	*Larix*, wood	Dinesman et al., 1989
			3420±210	IEMEZH-233	*Larix*, wood	Dinesman et al., 1989
			3520±96	IEMEZH-221	*Larix*, wood	Dinesman et al., 1989
			3530±123	IEMEZH-245	*Abies*, wood	Dinesman et al., 1989
			3570±170	IEMEZH-242	*Abies*, wood	Dinesman et al., 1989
			3815±243	IEMEZH-234	*Picea*, wood	Dinesman et al., 1989
			3855±168	IEMEZH-223	*Larix*, wood	Dinesman et al., 1989
			3930±182	IEMEZH-244	*Abies*, wood	Dinesman et al., 1989
			4113±183	IEMEZH-243	*Abies*, wood	Dinesman et al., 1989
			4170±183	IEMEZH-229	*Picea*, wood	Dinesman et al., 1989
			4358±171	IEMEZH-227	*Picea*, wood	Dinesman et al., 1989

Table 2.3. Continued.

Site NN	Site name	Sample depth (cm)	^{14}C years B.P.	Laboratory ref.	Material	Reference
22	Tsakhir-Khalgyn-Nuruu		3428±181	IEMEZH-303	*Larix,* stump	Dinesman et al., 1989
			3463±113	IEMEZH-314	*Larix,* stump	Dinesman et al., 1989
			3622±164	IEMEZH-305	*Larix,* wood	Dinesman et al., 1989
			3663±106	IEMEZH-311	*Larix,* wood	Dinesman et al., 1989
			3724±107	IEMEZH-312	*Larix,* stump	Dinesman et al., 1989
			3834±117	IEMEZH-301	*Larix,* stump	Dinesman et al., 1989
			4165±219	IEMEZH-300	*Larix,* stump	Dinesman et al., 1989
23	Uert-Am		2171±92	IEMEZH-302	*Larix,* stump	Dinesman et al., 1989
24	Sudzhiin-Khuduk		541±84	IEMEZH-	*Caragana,* brunch	Dinesman et al., 1989
			548±78	IEMEZH-	*Haloxylon,* wood	Dinesman et al., 1989

Note: Dates marked by a star appear too old.

We used linear interpolation between uncalibrated radiocarbon dates to estimate the age of the observed pollen events and reconstructed vegetation changes, and to construct a regional pollen stratigraphy of the Mongolian late Quaternary in the same way as in other syntheses of pollen (Peterson, 1993; Tarasov et al., in press) and lake-level (Tarasov et al., 1994, 1996) data from northern Eurasia. For the mapping, we always selected the pollen sample closest to the mapped interval (within a ±500-year range) in the profile rather than interpolating between pollen spectra. To analyze changes in the distribution of the main tree and shrub taxa, we compiled available plant macrofossil data with radiocarbondates (Dinesman et al., 1989) from three locations in Mongolian Altai and from one location in the Gobi desert (Tables 2.2 & 2.3).

2.4 REGIONAL POLLEN RECORDS FROM INDIVIDUAL SITES

Grubov (1955, 1982) divided Mongolia into 16 plant geographical regions, belonging broadly to four traditionally known landscape units: Mongolian Altai and Khangai-Khentii mountain where most forests and open woodlands (forest steppe) are concentrated; plains of the Gobi desert; and steppe of Eastern Mongolia (Murzaev, 1952). We chose previously unpublished pollen diagrams with the most pronounced changes to illustrate late Quaternary vegetation history across Mongolia.

Hoton-Nur Lake

Site description

Hoton-Nur Lake (48° 40' N, 88° 18' E, 2083 m) is the westernmost site with a pollen record from the Mongolian Altai (Figure 2.1b: site 1). It partly occupies an intermountain depression dammed by moraine sediment attributed to the Late Pleistocene glaciation (Khilko & Kurushin, 1982). The history of the last glaciation in Mongolia still is not well defined. Geomorphological and sedimentary studies suggest that mountain glaciers extended to lower elevations than they do today, but with no large extension in the region (Malaeva, 1989b; Devyatkin, 1993).

The lake has a water area of 50.1 km2 and is 21.5 km long and 4.0 km wide (Tarasov et al., 1994). A maximum water depth is 58 m and an average depth is ca 26.6 m. A catchment area is ca 2670 km^2. The lake has inflow from the Karatyr and Ak-Su rivers and from several small streams. The Khovd River, which is the biggest river of the Mongolian Altai, flows out of the lake.

Climatic data from the region are limited. The area experiences cold winters with mean January temperatures between –20 and –25°C. Mean July temperature is about 15°C and mean annual precipitation is ca 300 mm (Zhambazhamts & Bat, 1985). Most of the precipitation comes during the warm season.

Today steppe and meadow-steppe vegetation communities dominates the area close to the site (Lavrenko, 1983). Herbaceous communities represented by sedges and grasses are widely distributed (e.g. *Kobresia myosuroides*, *Carex rupestris*, *Festuca lenensis*, *Poa attenuata* etc.) as well as *Artemisia* species. A mountain tundra dominated by sedges and shrub *Betula rotundifolia* is characteristic of high elevations (>3000 m) in the lake catchment area. An open *Picea obovata-Larix sibirica* forest with *Lonicera altaica* in the understory grows at the mountain slopes south-west of the lake and in the river valleys. Single trees of*Pinus sibirica* exist in the upper part of the Ak-Su River north-west of the lake. Only small patches and single trees of *Larix sibirica* grow at the north-eastern edge of the lake basin. *Juniperus sibirica* and *J. pseudosabina* grow in the lake coastal zone.

Pollen record

A 920 cm core was taken in the north-eastern part of the lake ca 200 m from the lake shore at a water depth of 4.8 m. Dorofeyuk (1988) published the results of diatom and lithology studies of the core sediments and Tarasov et al. (1994) used them to reconstruct lake level changes. The chronology is based on 6 radiocarbon dates (Table 2.3). The pollen diagram (Figure 2.2) is visually divided into

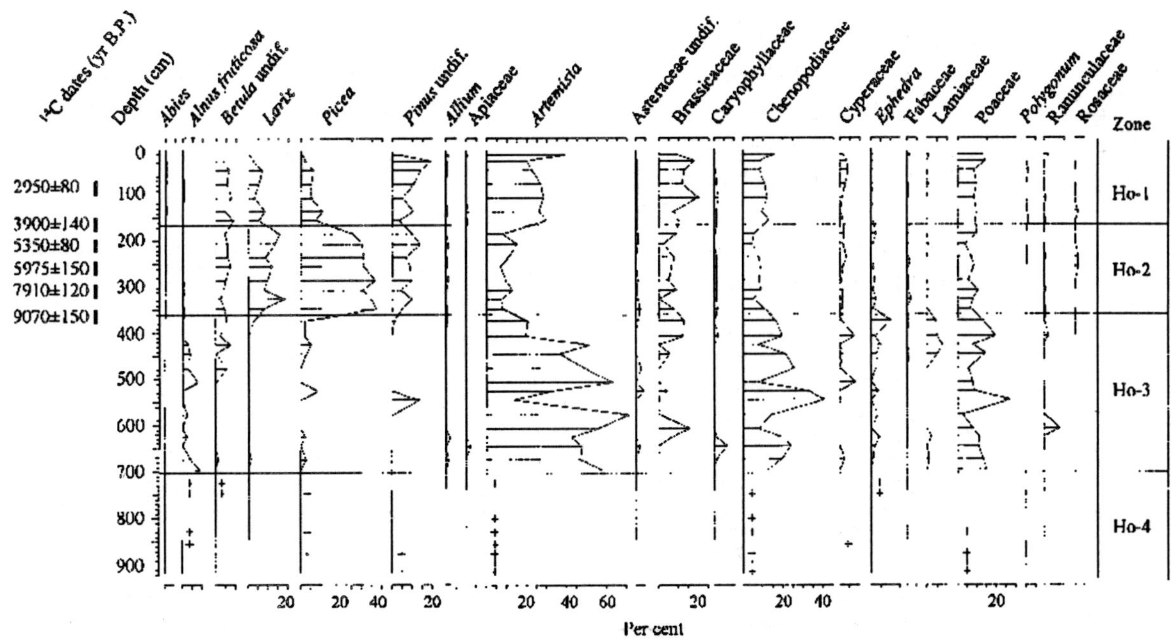

Figure 2.2. Pollen diagram from Hoton-Nur Lake (48° 40' N, 88° 18' E, 2083 m a.s.l.).

In figures 2.2–2.7 pollen of individual taxa are expressed as percentage of terrestrial pollen sum, excluding aquatics.

4 major local pollen zones (LPZ) to simplify the description of pollen record.

LPZ Ho-4 (700–920 cm) contains fewer than 20 grains per sample mostly of *Artemisia*, Chenopodiaceae and Poaceae. The small number of pollen grains makes it impossible to interpret the pollen spectra. The absence of diatoms or other algae species and the inorganic nature of the clay sediment (Dorofeyuk, 1988) suggest deposition during the cold phase of the last glacial maximum ca 16,000 to 21,000 yr B.P., when mountain glaciers were situated close to the site. End moraine forms can be found in the relief of the basin (Khilko & Kurushin, 1982).

In LPZ Ho-3 (360–700 cm) the total pollen sum reaches 50 to 100 grains in each sample, and dominant pollen percentages were calculated as *Artemisia* 50–80%, Chenopodiaceae 10–20%, Poaceae 5–15% and *Ephedra* 1–5%. *Betula* and shrub-like *Alnus fruticosa* pollen never exceed of 5%. Nonarboreal (NAP) taxa are distributed in Mongolia today in a very wide range but mostly in steppe and desert (Grubov, 1982). Chenopodiaceae and *Artemisia* also are known as pioneer taxa with a high pollen production, dominating pollen spectra from northern Eurasia from late glacial time (Grichuk, 1984). *Ephedra* pollen provides evidence of drier conditions (Grubov, 1982). It is also abundant in the glacial and late glacial spectra from northern Eurasia (Grichuk, 1984). *Betula rotundifolia* shrubs are typical plants of the transitional zone between an alpine meadow-steppe and mountain tundra or as an understory in an open larch forest (Lavrenko, 1979). An extrapolation from the lower-most radiocarbon date suggests the age of LPZ Ho-3 is ca 9000 to 16,000 yr B.P.

LPZ Ho-2 (170–360 cm; 4250 to 9000 yr B.P.) is characterised by a sharp increase in the arboreal pollen (AP) up to 70%. *Picea* 30–40%, *Larix* 10–20% and *Pinus*, more likely *Pinus sibirica*, 5–15% pollen become very important in these pollen spectra. Poaceae pollen 5–10% still dominate the NAP, but *Artemisia* pollen are 3 times and Chenopodiaceae pollen are 2 times less abundant than in LPZ Ho-3. Brassicaceae, Caryophyllaceae, Cyperaceae, *Ephedra* and Ranunculaceae as well as *Betula* pollen occur consistently but in comparatively low percentages. Traces (<1%) of *Abies* pollen are recorded in the middle part of this zone.

LPZ Ho-1 (0–170 cm; 0 to 4250 yr B.P.) is characterised by a decrease in the total AP to less then 35%. *Picea* and *Larix* pollen percentages decline more or less gradually to <2% at the core top. The amount of *Betula* (5–7%) pollen is similar to that in LPZ Ho-2, although *Pinus* pollen exhibits higher percentages (20%) at 20 cm. *Artemisia* pollen became abundant again and reached 40% in the uppermost sample. Except for the appearance of Apiaceae and a slight increase in Brassicaceae, up to 20%, the variety and percentages of other herb and tree taxa are the same as those in LPZ Ho-2.

Vegetation history

The late Pleistocene records from Hoton-Nur Lake suggest that prior to 9000 yr B.P. the region was characterised by cold steppes or tundra-steppes dominated by *Artemisia*, Chenopodiaceae and Poaceae species, and shrubs of *Betula* and *Alnus*.

Forests developed in the middle Holocene between 4250 and 9000 yr B.P. At that time Siberian-like forests dominated, e.g., *Picea*, *Pinus sibirica*, and *Larix* which occupied a much larger area close to Hoton-Nur Lake than they do today.

The latest part of the records demonstrates that vegetation around the lake became similar to modern steppes with small forest patches dominated by *Larix* after 4250 yr B.P. The reduction of areas occupied by woods was associated with a pronounced decrease in the amount of *Picea* trees that currently grow only in their most favourable microhabitats close to the lake.

Climatic interpretation

Since human impact in the Hoton-Nur region is unimportant up until now, we interpret the reconstructed pattern of vegetation in terms of possible climate changes. According to the known present-day bioclimatic limits of plants growing in the intercontinental area (Woodward, 1987; Prentice et al., 1992), the transition from cool or cold grass/shrub communities to taiga dominated by boreal evergreen and summergreen trees would require an increase in the available moisture (which is ratio of actual over potential evapotranspiration) from <65% to >75% and a mean winter temperature ranging from –35 to –19°C. Because meteorological stations nearest to the study site (Ulangom and Khovd) systematically register mean temperature now for the coldest month of –34.0 and –25.3°C respectively (Shver & Zhadambaa, 1971), we assume that winter temperatures below –35°C also could be a limiting factor for boreal conifers during the late glacial-early Holocene when winter insolation was ca 11% lower than its present value at the latitude of 50° N (Berger & Loutre, 1991). A sharp reduction of boreal forest in the Hoton-Nur region and the retreat of *Picea* trees into local moist habitats soon after 4500 yr B.P. would suggest, according to Prentice et al. (1992), conditions drier than in the previous phase, but still wetter than at 9000 to 16,000 yr B.P.

Achit-Nur Lake

Site description

Achit-Nur lake (49°30'N, 90°36'E, 1435 m) is situated in a large depression of north-western Mongolia (Figure 2.1b: site 6) as the northernmost of several interconnected basins dividing Altai from the Khangai-Khentii mountain area. The fresh-water lake occupies an area of about 297 km^2, with an average depth of ca 2 m and a maximum depth of 5 m (Dorofeyuk, 1988; Tarasov et al., 1994). The Khuv-Usny-Gol River and the Mogen-Buren River flow into the northern part of the lake and the Usun-Kholoi River flows out southwards into the Khovd River.

The climate of the Achit-Nur depression is continental, it is relatively dry, and it is strongly influenced by its position close to the center of the Asian High. Ulangom meteorological station, which is nearest to the site, records a mean January temperature of –34°C, mean July temperature of 19.7°C, and mean annual precipitation ca 140 mm (Shver & Zhadambaa, 1971). Precipitation increases to over 300 mm per year at elevations above 3000 m.

Dry steppe or desert steppe vegetation with *Haloxylon ammodendron*, *Anabasis brevifolia* and *Stipa gobica* (Lavrenko, 1979) dominate in the Achit-Nur region reflecting the climate conditions. Open patches of *Larix sibirica* with shrub-like *Betula rotundifolia* and *Salix glauca* grow at 1900 to 2500 m ca 65 km north-westwards and north-eastwards of the site (Lavrenko, 1983). Small stands of *Pinus sibirica* and *Picea obovata* are situated ca 150 km west-south-west of the lake in the river valley. *Larix*, 1700–2500 m, *Pinus sibirica* and *Betula rotundifolia*, 2200 to 2500 m, also grow ca 165 to 200 km eastwards of the lake.

Pollen record

A 275 cm core was taken in a water depth of 2.75 m, in the western part of the lake. Dorofeyuk (1988) published sedimentary and diatom records from the core. These records then were used to reconstruct changes in the relative water depth since the late glacial (Tarasov et al., 1994). The chronology is based on 4 radiocarbon dates (Table 2.3).

Three local pollen zones can be distinguished in the pollen diagram (Figure 2.3).

LPZ An-3 (240–260 cm; 11,500 to 12,500 yr B.P.) is characterised by low amount 12–17% of AP represented by *Pinus*, *Picea*, *Larix* and *Betula*. *Artemisia* 40–58%, Chenopodiaceae 18–38% and *Ephedra* 3–10% pollen

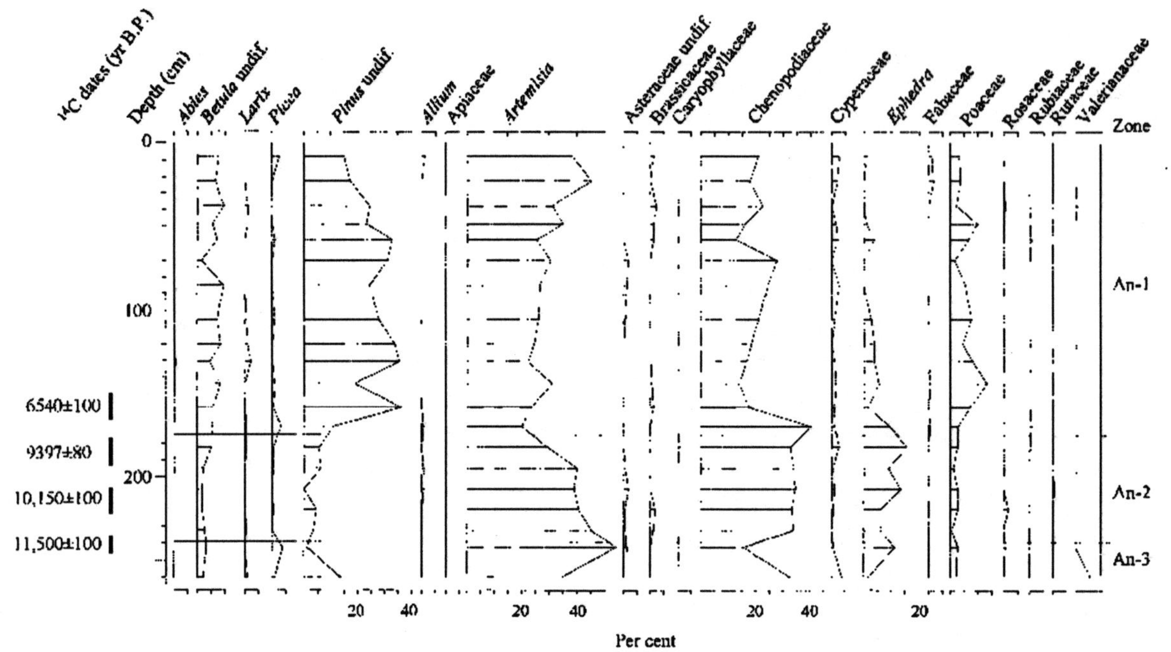

Figure 2.3. Pollen diagram from Achit-Nur Lake (49°30'N, 90°36'E, 1435 m a.s.l.).

dominate the pollen spectra and Poaceae pollen is a minor (3%) component.

In LPZ An-2 (175–240 cm; 9000 to 11,500 yr B.P.) pollen of arboreal taxa decrease to less than 5%, and pollen of *Pinus*, *Picea* and *Larix* completely disappear at 210 cm. *Ephedra*, Chenopodiaceae and *Artemisia* pollen still dominate the spectra, but *Artemisia* gradually decreases, 55 to 25%, in the top of this zone.

The pollen assemblages of LPZ An-1 (0–175cm; 0 to 7200 yr B.P.) differ from those in zones An-2 and An-3. Tarasov et al. (1994) suggested a sedimentary hiatus between 7200 and 9000 yr B.P. that is indicated by a change in the lithology (gravely sand interlayer at 165–175 cm) and in the sedimentation rate. An abrupt change in the pollen spectra is consistent with that interpretation. *Pinus*pollen rise to 38% at 158 cm becoming the most abundant taxon and then go down to 17% in the topmost sample. *Artemisia* pollen change in the opposite way. *Betula* pollen becomes more abundant 5–10%, and traces of *Picea* and *Larix* (systematically) and *Abies* (at 135 cm) are recorded in this zone. Chenopodiaceae 15–25% pollen is still important, but *Ephedra* pollen decreases to almost zero and Poaceae pollen reaches 5–15%.

Vegetation history

The changes in pollen composition shown in the pollen diagram indicate a broad pattern of vegetation change close to the study site. The vegetation history consists of: 1) dry steppe or desert steppe similar to the modern one in the basin ca 11,500 to 12,500 yr B.P. Small stands of boreal and deciduous conifers likely grew at the mountain slopes north of the lake in the humid conditions as they do today; 2) desert vegetation around the lake and the absence of conifers in the Achit-Nur basin between 11,500 and 9000 yr B.P.; 3) a spreading of patchy forests dominated by *Larix* and *Pinus sibirica* in the Achit-Nur region after 7200 yr B.P. The forest-covered area was higher than today in the middle Holocene and became gradually similar to modern forests in the late Holocene; but forests never dominated in the region.

Climatic interpretation

The existence of dry steppe similar to modern ca 11,500 to 12,500 yr B.P. implies moisture conditions similar to today. The presence of boreal evergreen conifers in forest patches on mountain slopes bordering the Achit-Nur depression suggest winter temperatures were not colder than –35°C according to Prentice et al. (1992). Climatic data from northern Mongolia systematically register winter temperatures in the intermountain depressions collecting cold and dry air masses that are a few degrees lower than the surrounding slopes (Zhambazhamts & Bat, 1985). The establishment of desert vegetation around the site at 9000 to 11,500 yr B.P. suggests that the moisture index was less than 28%. Also the drier conditions themselves are enough to explain the disappearance of conifers. There is no limitation from winter temperatures at that time. The re-establishment of steppe vegetation in the depression and the increase in forest-covered area to levels higher than today at 6500 yr B.P. suggest conditions wetter than present in the Achit-Nur region during the middle Holocene.

Dood-Nur Lake

Site description

The lake (51°20'N, 99°23'E, 1538 m) is the northernmost site with a late Quaternary pollen record from Mongolia (Figure 2.1b: site 16). It is located in a large depression within the Khangai-Khentii mountain area. The lake covers 64 km^2 with an average depth of ca 6 m and a maximum depth of ca 15 m (Tarasov et al., 1996). There is no clear evidence of mountain glaciation during Late Quaternary (Sevastyanov et al., 1994). However, the melting of mid-Pleistocene glaciation formed a lake over twice as large as today. Old lake terraces can be found at 1700 m today (Uflyand et al., 1969). The modern lake is subdivided into a three interconnected subbasins, named Targan, Dund and Kharmai. The lake is fed by many rivers and springs, coming from the catchment area of 1100 km^2. It has an outflow via the Shishkhid-Gol River.

The climate of the basin is characterised by a mean January temperature below –25°C, mean July temperatures ca 15–20°C and mean annual precipitation of 250–400 mm, increasing with elevation (Zhambazhamts & Bat, 1985).

The vegetation in the basin is forest-steppe. Patchy open *Larix sibirica* forests cover the upper level terraces and slopes of northern exposure. Forests on mountain slopes with more precipitation contain *Picea obovata* and *Pinus sibirica* (Lavrenko, 1983). An isolated stand of *Abies sibirica* grows close to the lake (Lavrenko, 1983).

Pollen record

A 270 cm core was taken at the south-eastern part of the lake from the Harmai subbasin ca 200 m from the shore in water 3.3 m deep. It provides a sedimentary record since ca 12,500 yr B.P. These data were used to reconstruct the relative changes in water depth (Tarasov et al., 1994). Two radiocarbon dates (Table 2.3) are a basis for chronology. The pollen diagram (Figure 2.4) is subdivided into 3 local pollen zones.

LPZ Dn-3 (260–270 cm; 12,100 to 12,500 yr B.P.) is dominated by tree taxa, e.g., *Pinus* 40%, *Picea* 30%, *Betula* 10% and *Larix* 3% pollen. *Artemisia* 10% pollen is the most important nonarboreal taxon.

LPZ Dn-2 (205–260cm; 10,500–12,100 yr B.P.) is characterised by a substantial decrease in *Pinus* 10% and *Picea* 7–10% pollen, but *Betula* and *Larix* pollen percentages are unchanged. *Artemisia* pollen reached up to 45% in this zone, and other herbaceous taxa, e.g., Chenopodiaceae, *Ephedra*, Asteraceae, Poaceae, and Ranunculaceae became more abundant than in Dn-3.

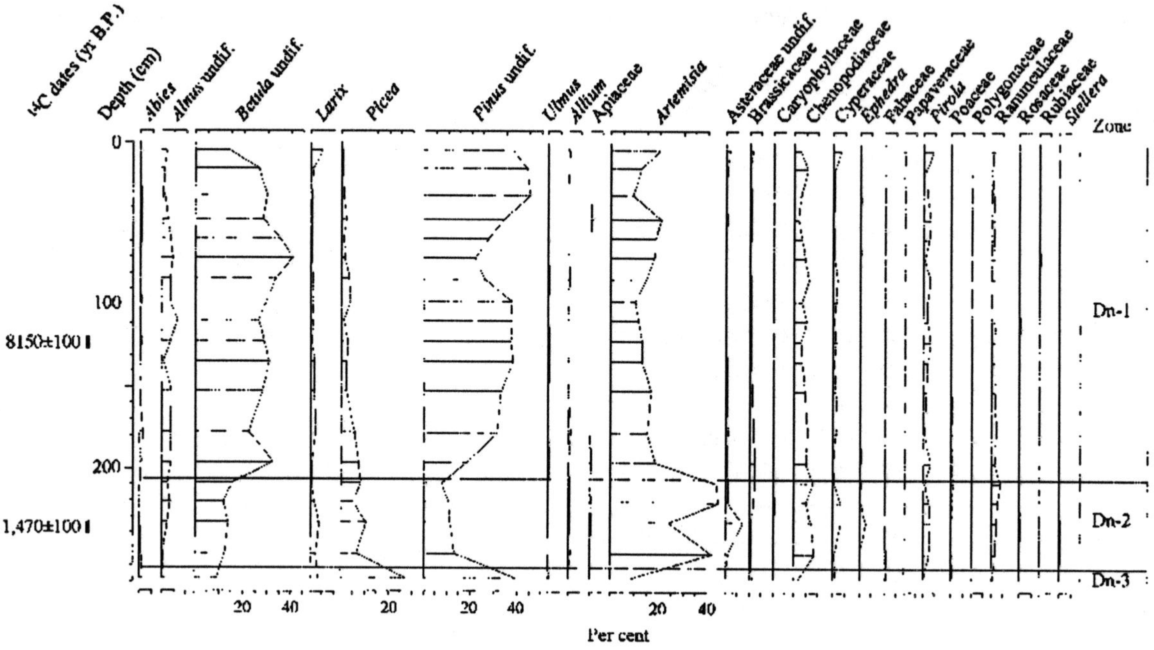

Figure 2.4. Pollen diagram from Dood-Nur Lake (51°20'N, 99°23'E, 1538 m a.s.l.).

LPZ Dn-1 (0–205 cm; 0–10,500 yr B.P.) is characterised by a co-dominance of *Betula* 18–42% and *Pinus* 25–48% pollen. *Larix* pollen was present continuously in a small 1–5% amount, and *Picea* pollen decreased gradually to <0,5%. Traces of *Abies* and *Ulmus* are recorded in the lower part of this zone. Among NAP, *Artemisia* pollen decreased to 10–20%.

Vegetation history

The Dood-Nur pollen records suggest that forest vegetation was an important vegetation component around the study site at ca 12,500 yr B.P. and again after ca 10,500 yr B.P. It was largely replaced by a steppe vegetation between ca 12,000 and 10,500 yr B.P. However conifer trees were able to survive in local refuges even during that time interval. The importance of *Picea* in the patchy forests gradually decreased through this time.

Climatic interpretation

The most pronounced change in the vegetation is a steppe phase dated to 10,500–12,100 yr B.P. This episode suggests a decrease in available moisture and conditions drier than at ca 12,500 yr B.P. and after ca 10,500 yr B.P. However, the arid period was not strong enough to kill boreal evergreen conifers in the Dood-Nur region.

Daba-Nur Lake

Site description

Daba-Nur Lake (48°12'N, 98°47'40"E, 2465 m) lies in the Dzagastain-Daba pass through the Tarbagatai Ridge of the Khangai-Khentii Mountain area (Figure 2.1b: site 11). The basin origin is connected with a natural damming of the small river valley by moraine deposits attributed to the middle Pleistocene. It has an area of about 3 km^2, an average depth of

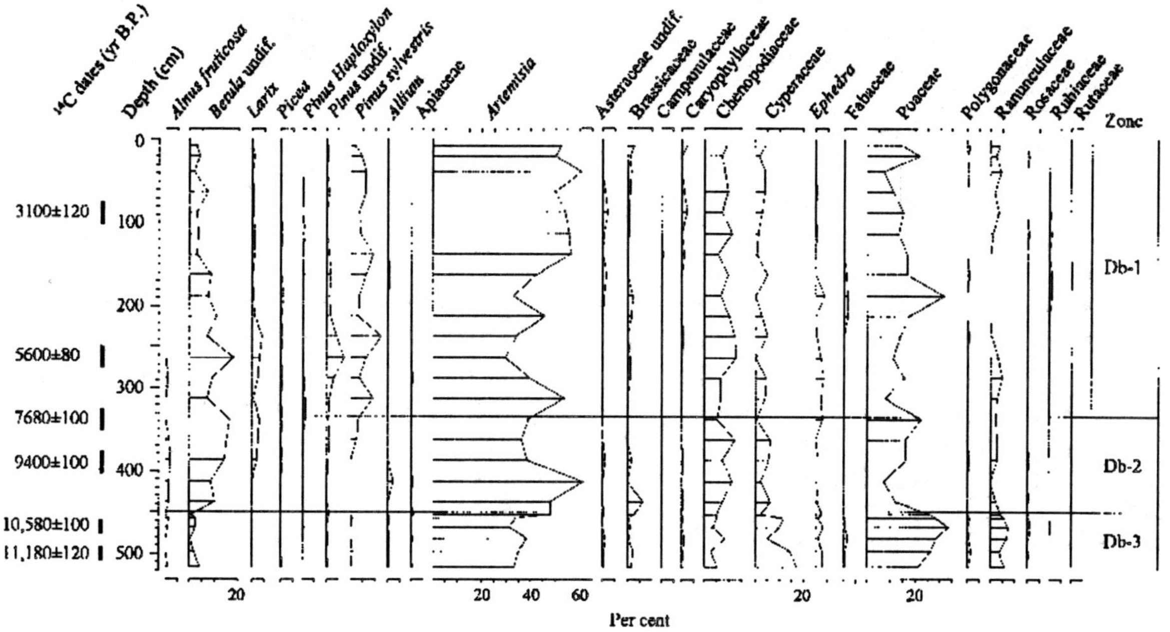

Figure 2.5. Pollen diagram from Daba-Nur Lake (48°12'N, 98°47'40"E, 2465 m a.s.l.).

ca 2 m and a maximum depth of 4.5 m (Dorofeyuk, 1988; Tarasov et al., 1994). The shores are flat and swampy. The lake is fed by small streams and by plus direct precipitation and has no outflow.

Data from the nearest meteorological station Uliastai give a mean July temperature of 13.8°C, a mean January temperature of –25.3°C, and mean annual precipitation of 217 mm at the lower elevations and of ca 400 mm at an elevation of 3000 m (Shver & Zhadambaa, 1971).

Flat and swampy shores of the lake are covered by *Carex*, *Kobresia* and shrub-like *Betula rotundifolia*, *Betula humilis* and *Salix* species. Single *Larix sibirica* and *Pinus sibirica* trees grow on the western slope of the basin just 50 to 100 m above lake level. Mountain tundra communities appear on the eastern slope of the basin going up to 3128 m. In the region, dry steppes form on the foothills and are replaced by light forests with*Larix* and shrub birch at 1750 to 2500 m (Lavrenko, 1979). *Pinus sibirica* appears in the upper part of this forest belt. Mountain forests dominated by *Pinus sibirica* with a mixture of *Larix* and *Picea* grow ca 225 km northwards. The closest habitat with *Pinus sylvestris* is ca 300 km to the northeast in Khentii at 700 to 1100 m (Lavrenko, 1983).

Pollen record

Continuous sedimentary and diatom records from a 525 cm core were taken in a water depth of 1.75 m, in the central part of the lake (Dorofeyuk, 1988). They were used in the reconstruction of lake-level changes (Tarasov et al., 1994). The chronology is based on 6 radiocarbon dates (Table 2.3). The pollen diagram contains three pollen zones (Figure 2.5).

LPZ Db-3 (450–515 cm; 10,300–11,500 yr B.P.) is characterised by only a small amount of AP 1–5% with *Betula* (more likely shrub-like forms) the main taxon. Traces of *Alnus fruticosa* were recorded. In the NAP assemblage *Artemisia* 31–47% and Poaceae

23–33% are dominants and Chenopodiaceae pollen is less abundant 3–9%. The highest values for Cyperaceae pollen 10–20% occur in this zone.

In LPZ Db-2 (335–450 cm; 7600 to 10,300 yr B.P.), AP, represented mainly by *Betula*, increased in abundance up to 10–23%. Small quantities 1–3% of *Alnus fruticosa* pollen exist throughout the zone. A few grains of *Larix*, *Pinus* and *Picea* pollen appear in the upper part of this zone above 390 cm. *Artemisia* 37–64% and Chenopodiaceae 5–12% pollen became more abundant, and Poaceae pollen decreased to 8 % at 410 cm and then gradually increased to 22%.

The most pronounced change occurred in LPZ Db-1 (0–335 cm; 0 to 7600 yr B.P.). AP dominated by *Betula*, *Pinus* and *Larix* increased up to 34% at 260 cm and then gradually decreased to ca 5% at the topmost part of the diagram. *Alnus fruticosa* pollen was absent above 270 cm. However, in this zone, as in the whole diagram, *Artemisia* 30–64%, Poaceae 8–33% and Chenopodiaceae 6–13% dominate pollen spectra.

Vegetation history

The pollen record from Daba-Nur Lake shows that open types of vegetation dominated the catchment area from 11,500 yr B.P. to the present. The area was probably treeless between 11,500 and 9500 yr B.P. and *Larix* grew close to the site only after 9500 yr B.P. Herbaceous Cyperaceae-Poaceae tundra dominated the area close to the site at the late glacial. An increase in abundance of *Betula* and *Alnus fruticosa* pollen suggests that it was replaced by a dwarf-shrub tundra from 10,300 to 7600 yr B.P. Vegetation became similar to present day steppe after 6000 yr B.P. However forest patches probably had wider distribution in the region during the mid-Holocene than today.

Climatic interpretation

Growth of tundra around sites where steppe and open woodland dominate today indicate a climate colder than the present during the glacial and the early Holocene. The physiological limits listed in Prentice et al. (1992) for plants in semi-arid regions show that moisture requirements are the same for tundra and cool grass/shrub, 28–65% of demand, but tundra grows when the climate is too cold to support cool steppe grass and shrubs. The temperatures became similar to present after ca 7500 yr B.P. but available moisture was greater than present after 9500. This moisture increase at first allowed growth of boreal summer green (*Larix*) plus eurythermic conifer (*Pinus sylvestris*) trees; and then boreal conifers (*Pinus sibirica*).

Yamant-Nur Lake

Site description

Yamant-Nur (49°54'N, 102°36'E, 1000 m) is an overflowing lake, situated in a former-river valley belonging to the Selenga River system, at the boundary between the mountains of central Mongolia and the eastern Mongolian plains (Figure 2.1b: site 18). It has a basin oriented north and south with a maximum length of 2.8 km and a maximum width of 1.3 km. An average water depth is about 1 m and a maximum depth is of 1.6 m. According to the nearest meteorological station at Muren, the climate is characterised by a mean January temperature of –24.4°C, a mean July temperature of 16.8°C and a mean annual precipitation of 228 mm (Shver & Zhadambaa, 1971). Meadow-Poaceae steppe dominates mid-elevations in the catchment area. Open *Larix sibirica* woodland appears in the upper part of the mountain belt.

Pollen record

A 460 cm core was taken at ca 150 m from the south-western coast in water 1 m deep and its sediments were analysed for pollen. Only one date from the late glacial provides a basis for the chronology. The pollen diagram from this core (Figure 2.6) is visually divided into 3 pollen zones.

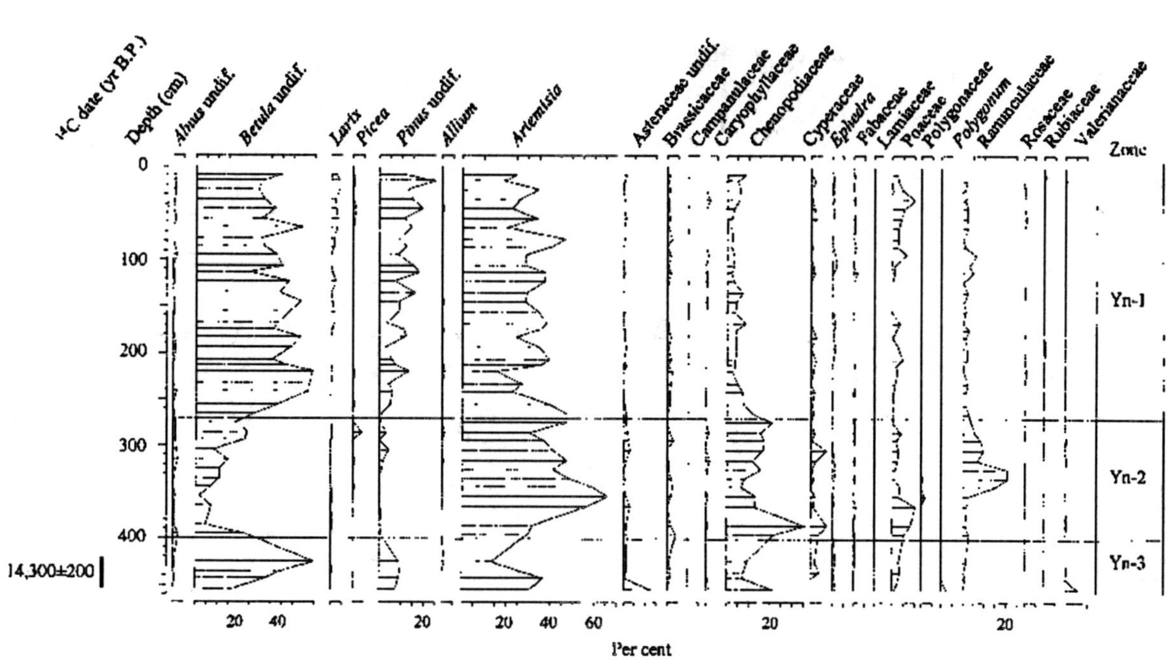

Figure 2.6. Pollen diagram from Yamant-Nur Lake (49°54'N, 102°36'E, 1000 m a.s.l.).

LPZ Yn-3 (400–460 cm; 13,000–15,000 yr B.P.) is characterised by a dominance of *Betula* 20–60%, *Artemisia* 10–40%, Chenopodiaceae 10–25%, *Pinus* 5–10% and Poaceae 5% pollen.

LPZ Yn-2 (270–400 cm; 8800–13,000 yr B.P.) is dominated by nonarboreal taxa, e.g., *Artemisia* 30–70% and Chenopodiaceae 10–40%, and arboreal pollen decreased to less 5% at 355 cm. Minor taxa include Poaceae and Cyperaceae at systematically less than 8% and Ranunculaceae pollen with a peak up to 20% in the middle of this zone. LPZ Yn-1 (0–270 cm; 0–8800 yr B.P.) has nearly the same composition as Yn-3, but with slightly greater values of *Larix* and *Pinus* in the topmost part of this zone, up to 5% and 30% respectively.

Vegetation history

The most pronounced change in vegetation is the phase with almost no arboreal vegetation dated to the end of the late glacial and the early Holocene (8800–13,000 yr B.P.). At that time the area was covered by cool steppe or even tundra-steppe when Cyperaceae, Poaceae and Ranunculaceae pollen were relatively abundant and Valerianaceae and Polygonaceae pollen was present. These taxa that represent colder conditions are in the present-day vegetation. Pollen of *Betula* are identified only at the genus level. However we suggest that they belong mostly to *Betula rotundifolia*, which grows currently here. If so open *Artemisia*-Chenopodiaceae-Poaceae vegetation with shrub *Betula* dominated the region from 15,000 to 13,000 yr B.P. Discontinuous traces of *Larix* pollen are recorded in the late glacial and early Holocene part of the records and suggest that this taxon existed in the region continuously but became a more important vegetation component in late Holocene.

Climatic interpretation

The absence of tree vegetation close to the study site between 13,000 and 8800 yr B.P. suggests conditions drier than the present, the pollen composition indicates the possibility

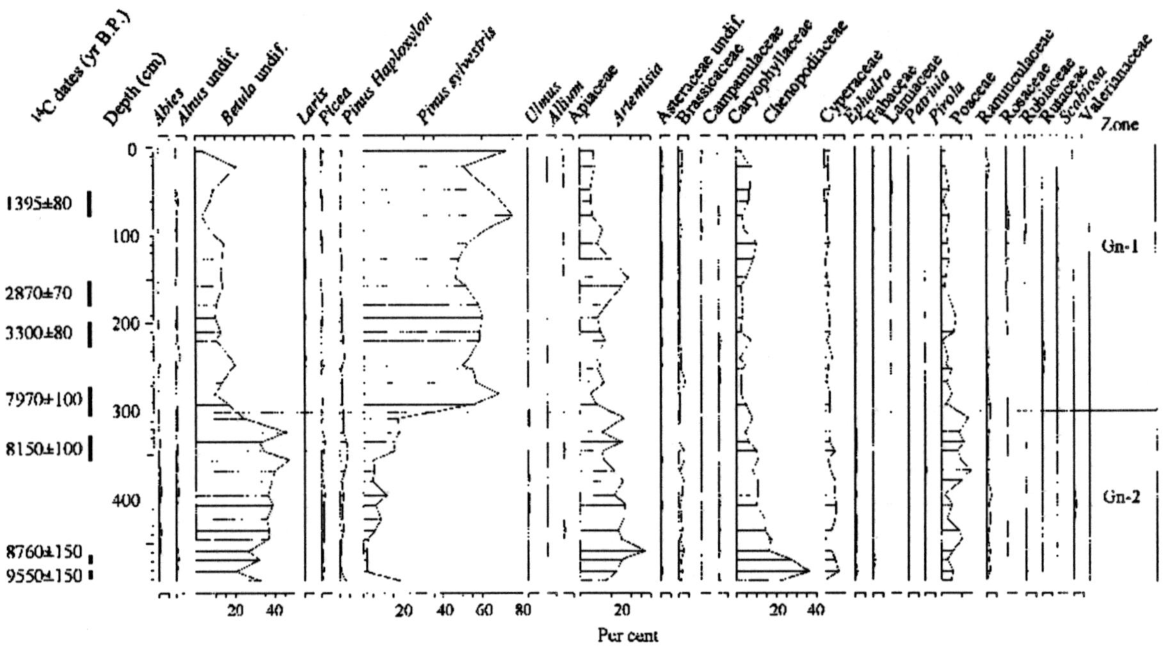

Figure 2.7. Pollen diagram from Gun-Nur Lake (50°15'N, 106°36'E, 600 m a.s.l.).

that this drier climate was associated with colder winters. The moisture conditions were similar to present from 15,000 to 13,000 and 0–8800 yr B.P.

Gun-Nur Lake

Site description

Lake Gun-Nur (50°15'N, 106°36'E, 600 m) is the easternmost Mongolia site with a continuous Holocene pollen record (Figure 2.1b: site 19). It occupies a topographic depression of a flat limno-fluvial plain drained by the Selenga River (Tarasov et al., 1996). The plain is bounded by mountain ridges going up to 1554 m. The lake has a water area of about 2.5 km^2, an average depth of 2 m and a maximum depth of ca 5 m. The lake is closed most of the year and has an outflow only during the summer season, when most of precipitation comes. Climate is characterised by a mean January temperature of –25 to –27°C, mean July temperature of 16°C and mean annual precipitation of 300–350 mm (Zhambazhamts & Bat, 1985).

The area is dominated now by steppe and forest associations. Sand dunes around the lake are occupied by light forest with *Pinus sylvestris* and a mixture of *Ulmus pumila* and *Salix*. The depressions are covered by steppe communities with *Koeleria glauca*, *Agropyron cristatum*, *Festuca lenensis*, *Artemisia frigida*, *Thermopsis lanceolata* and *Caragana microphylla* (Lavrenko, 1979). The river terraces are largely occupied by pine forests, and mountain slopes at 700–1200 m are covered by forest patches containing larch, pine and birch (Lavrenko, 1983).

Pollen record

A 493 cm core, taken in the south-eastern part of the lake, ca 100 m from the coast in water 4.5 m deep provides a sediment, pollen, and diatom record back to ca 10,000 yr B.P. (Dorofeyuk & Tarasov, 1998). The pollen diagram is divided into two zones (Figure 2.7).

In the basal sample of the LPZ Gn-2 (493–300 cm; ca 8000–10,000 yr B.P.), AP is moderately abundant ca 54% and represented by *Betula* 33% and *Pinus* 21% pollen. Chenopodiaceae 22% and *Artemisia* 13% dominate among the NAP taxa. At 490–450 cm, the amount of *Pinus* decreases to almost zero and only *Betula* 25–35% is present in the AP while *Artemisia* and Chenopodiaceae pollen reach a maximum. The abundance of AP up to 50–66%, including *Pinus* up to 19%, increased gradually in the upper part of this zone (300–450 cm; 8000–8700 yr B.P.). *Artemisia* and Chenopodiaceae pollen became less abundant, and Poaceae pollen increased in abundance up to 10–15%.

In LPZ Gn-1 (0–300 cm; 0–8000 yr B.P.), AP reached a maximum 63–83% and *Pinus* becomes a dominant species in the pollen assemblages 47–76%. The content of *Betula* pollen varied between 3 and 21% and *Larix* pollen was present continuously as a small content. *Artemisia* 5–25%, Chenopodiaceae 2–10% and Poaceae 1–7% dominate in the NAP.

Vegetation history

The Gun-Nur pollen record disclose a co-dominance of steppe and forest vegetation during the Holocene. However, steppe was the most important vegetation type in the region between 10,000 and 8000 yr B.P. and patchy forests became an important landscape feature after 8000 yr B.P. *Betula* was a key tree taxon in the early Holocene and then was replaced by *Pinus sylvestris*.

Climatic interpretation

A dominance of steppe over forest steppe in the Gun-Nur region between and 10,000 and 8000 yr B.P. suggests that the climate was drier then than during the other part of Holocene. *Pinus sylvestris* and *Betula* trees have the same moisture requirement, but *Betula* grows at lower winter temperatures (Woodward et al., 1987; Prentice et al., 1992). Therefore, taking in account that low winter temperature can limit pine growth, we accept that winters were at least 10°C colder in early Holocene than today.

2.5 HOLOCENE CHANGES IN THE DISTRIBUTION OF TREE AND SHRUB TAXA IN MONGOLIA VALIDATED BY PLANT MACROFOSSIL RECORDS

Results

Radiocarbon dated tree and shrub macrofossils from Mongolia are rather limited (Table 2.3). However, they were obtained from the southern regions where radiocarbon dated pollen records are absent. Thus they help increase our knowledge of Mongolia's vegetation history. Here we demonstrate the most pronounced distribution changes for several of the most important arboreal taxa during late Holocene.

Abies

Abies sibirica is the most sensitive tree of Mongolia to available moisture, winter temperatures, and soil richness. Today it only grows in a very few places in northern Mongolia (Figure 2.8) as the second level in

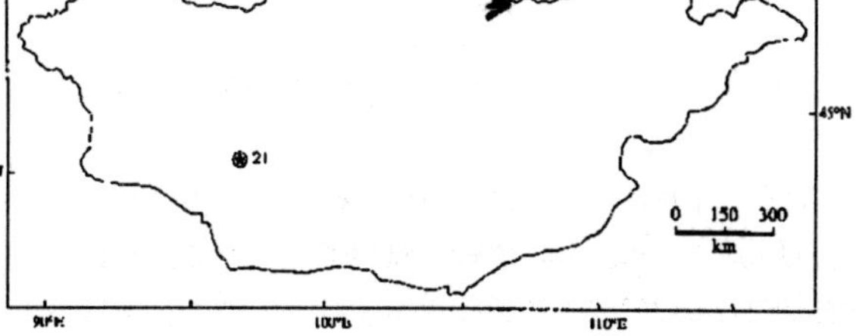

Figure 2.8. Modern distribution (single stand is marked by sign of tree) of *Abies sibirica* (after Lavrenko, 1983) and fossil site with radiocarbon-dated *Abies* macrofossils (after Dinesman et al., 1989). Sites are numbered as in Table 2.2.

boreal mountain forests dominated by *Pinus sibirica* and *Picea obovata*, and sometimes by *Larix sibirica* (Savin et al., 1978). The modern area occupied by *Abies* comprises only 1.9 ha or 0.02% of the forested area of Mongolia (Savin et al., 1978).

Abies macrofossils have been found at Bayan-Sair in the Gobi Altai Mountains (Dinesman et al., 1989). This site is situated ca 600 km south from the closest place in Mongolia where *Abies* grows today (Figure 2.8). The fossil site lies at the bottom part of a small depression currently occupied by a peat bog. These abundant tree macrofossils likely were redeposited from the surrounding mountain slopes and then buried by sedimentary and peat deposits. Dinesman et al. (1989) excavated the peat sequence and identified 160 pieces of wood. Among them 11% belong to *Abies*, 22% to *Picea*, 12% to *Pinus sibirica* and 55% to *Larix*, suggesting forest dominated by boreal and deciduous conifers existed close to the site in a region dominated today by steppe. Four radiocarbon-dated samples of *Abies* show a growth age time interval between ca 4113±183 and 3530±123 yr B.P. (Table 2.3).

Picea

Picea obovata also is a tree very sensitive to moisture conditions, but it is less sensitive than *Abies*. Today it grows in an area of 24 ha (0.25% of the forested area) mostly in river valleys north of 48°N latitude (Figure 2.9). In some of those places (e.g. the western slope of Khentii) it even dominates the forests. However, more often *Picea* trees grow as a mixture in forests dominated by *Pinus sibirica*, or it appears in the most favourable microhabitats in river valleys (Savin et al., 1978).

The presence of *Picea* macrofossils at the Bayan-Sair site, which is situated ca 450 km south from its closest modern stand (Figure 2.9), suggests that *Picea* grew in Mongolia over a much larger area than today. Three of the thirty six *Picea* macrofossils

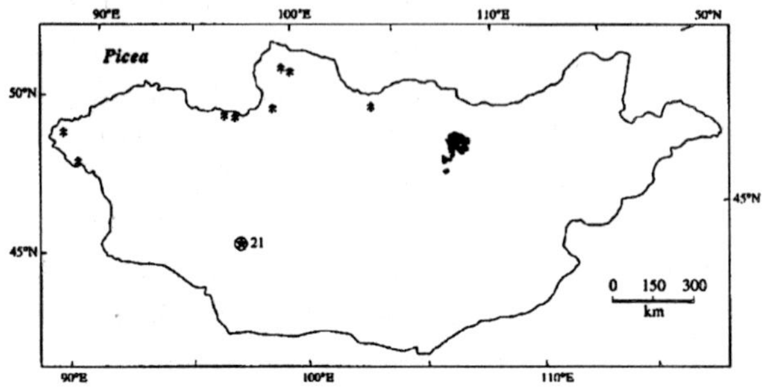

Figure 2.9. Modern distribution (single stands are marked by sign of tree) of *Picea obovata* (after Lavrenko, 1983) and fossil site with radiocarbon-dated *Picea* macrofossils (after Dinesman et al., 1989). Sites are numbered as in Table 2.2.

found at Bayan-Sair have been radiocarbon dated (Dinesman et al., 1989). The boundary dates are 4358±171 and 3815±243 yr B.P. (Table 2.3). The first date suggests that *Picea* appeared at the site at least 250 years before *Abies*. However, this is only a suggestion and more precise data are needed to pen point the growing time of each taxa.

Larix

Larix sibirica is the most important and widely distributed tree taxa in Mongolia. That can be explained by its broad tolerance in moisture, temperature and soil requirements. Pure *Larix* forests and open woodlands now occupy ca 82% of the forest-covered area of Mongolia and single trees appear in the most favourable habitats both in steppe and mountain tundra (Savin et al., 1978). *Larix* (Figure 2.10) penetrates much further south and east than any other conifer in Mongolia.

Larix macrofossils are the most abundant tree in the data set we used. Dinesman et al. (1989) describe 3 sites from Mongolian Altai where trunks and stumps of *Larix* have been found (Table 2.2). *Larix*, at 55%, dominates the macrofossil assemblage in the Bayan-Sair record. Seven dates from Bayan-Sair (Table 2.3), suggest it grew close to the site together with *Picea* and *Abies* (date 3855±168 yr B.P.); but it only disappeared about

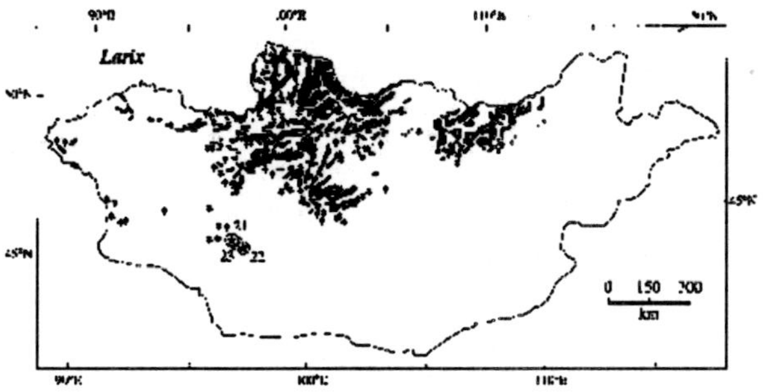

Figure 2.10. Modern distribution (single stands are marked by sign of tree) of *Larix sibirica* (after Lavrenko, 1983) and fossil sites with radiocarbon-dated *Larix* macrofossils (after Dinesman et al., 1989). Sites are numbered as in Table 2.2.

2797±125 yr B.P. Radiocarbon dated stumps of *Larix* excavated from slope deposits at two sites situated close to Bayan-Sair suggest that *Larix* patches were abundant in the Gobi Altai between 4165±219 and 2171±92 yr B.P. Now *Larix* does not grow at these sites of fossil records. However, individual trees of *Larix* and *Betula* exist in the region until today (Figure 2.10).

Haloxylon

Haloxylon ammodendron is a shrub or small tree of 0.7–2.0 m height with very broad ecological tolerance, which today dominates the vegetation of Mongolia's southern Gobi desert (Figure 2.11). It mostly occupies depression bottom parts with sandy, salty or clay soils, and open sands (Rachkovskaya, 1993). At higher elevations, with hardened stony soils, these deserts are dominated by *Haloxylon-Sympegma regelii* shrubby associations (Baitulin, 1993).

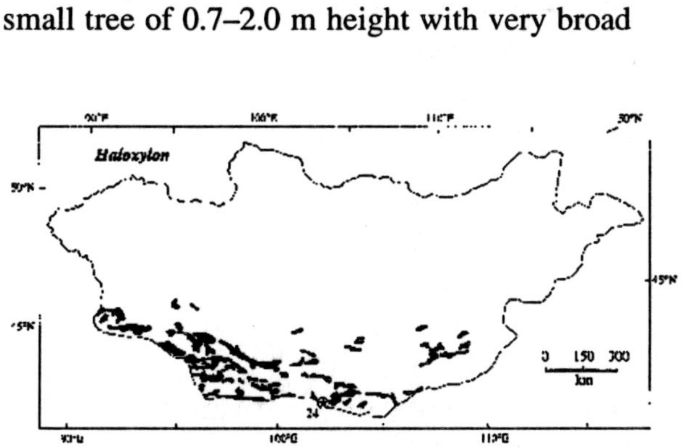

Figure 2.11. Modern distribution of *Haloxylon ammodendron* (after Lavrenko, 1983) and fossil site with radiocarbon-dated *Haloxylon* macrofossils (after Dinesman et al., 1989). Sites are numbered as in Table 2.2

An engineering construction from the middle ages known as «Dam of Chingiz-Khan» has been excavated (Dinesman et al., 1989) near Sudzhiin-Khuduk Well in southern Mongolia (Figure 2.11). Branches and trunks of *Haloxylon* and other desert and semi-desert taxa, e.g., *Caragana* and *Sympegma* commonly were used to consolidate the dam soil and sand material. Dating a *Haloxylon* trunk gave a radiocarbon age of 548±78 yr B.P. (Table 2.3), suggesting the composition of desert vegetation in southern Mongolia was similar ca 500 yr B.P.

Interpretation and discussion

Even though sparse and discontinuous, radiocarbon dated macrofossil records from southern Mongolia provide a nice pattern of vegetation changes at ca. 4500 yr B.P. that are consistent with pollen data from Mongolia and China. These pollen and macrofossils together provide clear evidence that mid-Holocene environments and climate of Mongolia substantially differ from the present. Thus the presence of *Abies* and *Picea* in the Gobi Altai between 4500 and 3500 yr B.P. suggests that climate was much wetter than today. The disappearance of boreal conifers from the forest at Bayan-Sair at about 3500 yr B.P. implies a decrease in available moisture. However, the presence of *Larix* macrofossils in 3 Gobi Altai sites, where they are not present today, suggests the climate was still wetter than now at least until 2000 yr B.P. The presence of abundant desert shrub macrofossils, e.g., *Haloxylon* and *Sympegma* in southern Mongolia where these taxa are dominate today suggests moisture conditions there were similar to today by at least 500 yr B.P.

A sparse network of meteorological stations in Mongolia and a large variability of temperatures and precipitation depending on the local geomorphology (Shver & Zhadambaa, 1971), make it difficult to estimate climate characteristics at both modern and fossil stands of *Picea* and *Abies* or to reconstruct palaeoclimate. However, at Bayan-Sair an increase in mean annual precipitation should be at least 100 mm over modern values to support *Picea* growth and at least 150 mm greater to support *Abies*.

Macrofossil results are in a good agreement with pollen records from regions situated north and south of the Gobi Altai sites. Thus at Hoton-Nur (Mongolian Altai) taiga forests, dominated by *Picea* during the mid Holocene, retreated dramatically after ca 4000 yr B.P. Pollen record from the Red May Bridge (elevation 2516 m) in Northwest China shows that pollen assemblages were completely dominated by *Picea* at 7320±200 yr B.P. and by *Ephedra* and *Artemisia* at 3950±150 yr B.P. There are no *Picea* forests in that area today and only single trees grow at lower elevations in the Urumqi River Valley (Winkler & Wang, 1993). Pollen spectra, from the peat deposits overlying the layer with woody macrofossils at Bayan-Sair (Dinesman et al., 1989), contain only occasional arboreal taxa pollen suggesting aforestation close to the site became similar to present after 2500 yr B.P.

2.6 SPATIAL RECONSTRUCTION AND MAPPING OF MONGOLIAN VEGETATION DURING THE LAST 15,000 YEARS USING A «BIOMIZATION» METHOD

Summary of the method

An objective quantitative method known as «biomization», was developed recently to reconstruct past vegetation (Prentice et al., 1996), that is based on the concept of assigning pollen taxa to one or more plant functional types which group taxa by their stature, phenology, leaf form, and bioclimatic tolerance. This method was successfully tested with modern surface pollen data and then applied to 6000 yr B.P. pollen data sets from Europe (Prentice et al., 1996), Africa (Jolly et al., in press), eastern North America (Summers et al., in press), China (Yu et al., in press) and the Former Soviet Union and Mongolia (Tarasov et al., in press). Each of these attempts demonstrate coherent regional patterns of biome distribution consistent with previous quantitative vegetation reconstructions where these were done. A recent compilation of pollen data from Mongolia provides an excellent opportunity to use the biomization procedure for an objective spatial and temporal reconstruction of Mongolian vegetation since the late Quaternary.

The detailed method for reconstructing biomes from pollen data was described in the Prentice et al. (1996) paper. The key steps can be summarised as: 1) each pollen taxon used in biomization is assigned to one or more plant functional types (PFT) according to the known biology and biogeography of the plant species included; 2) assignment to biomes of characteristic PFTs according to their bioclimatic range and actual distribution; 3) construction of a biome by taxon matrix illustrating which taxa may occur in each biome; 4) calculation of affinity scores for all pollen samples by the equation:

$$A_{ik} = \Sigma_j \delta_{ij} \sqrt{\{\max[0,(p_{jk} - \theta_j)]\}} \qquad (1)$$

where A_{ik} is the affinity of pollen sample k for biome i; summation over all taxa j, δ_{ij} is the entry in the biome × taxon matrix for biome i and taxon j; p_{jk} are pollen percentages, and θ_j is a threshold pollen percentage in order to reduce the noise due to occasional pollen grains (as a result of redeposition or long distance pollen transport). Here we use a universal threshold of 0.5% for all pollen taxa, following Prentice et al. (1996).

For each pollen sample, the biome with the highest score is assigned to that sample and this biome is then mapped at that site. If two biomes have exactly the same affinity score, the sample is assigned to that biome which is defined by the smaller number of taxa.

Implementation for Mongolia

The assignment of pollen taxa to the PFTs and PFTs to biomes was performed taking in account PFT definitions following Prentice et al. (1996) and Tarasov et al. (in press), and known biology of plants described in several studies of Mongolian flora (Yunatov, 1950; Grubov, 1955 1982; Gubanov, 1996). Table 2.4 lists all available pollen taxa (after exclusion of aquatics, exotic taxa, and taxa restricted to local microhabitats) in the modern surface sample sets from Mongolia and their assignment to a set of PFTs. Plotting the present-day distribution of pollen abundances for individual taxa helps to visualise relationships between actual vegetation and modern pollen spectra plus providing a quick test of the assignment procedure. Figure 2.12 illustrates such tests with several important pollen taxa from Mongolia. All maps show strong geographic patterns that clearly reflect zonal vegetation patterns. *Artemisia* and Chenopodiaceae have high abundances in steppe and desert zones. These taxa however were included in both steppe-forb and desert-forb PFTs because they are co-dominate in the steppe and desert environments (Yunatov, 1950; Lavrenko, 1979). *Ephedra* has a minor content in forest and steppe zones and high abundances only at a few sites in the transitional zone between steppe and desert and at one high elevation site in Mongolian Altai where it likely reflects soil disturbance. We kept *Ephedra* in desert-forb PFT similar to Prentice et al. (1996). Poaceae has high pollen abundances primarily in steppe and forest-steppe. We only placed Poaceae in tundra and steppe biomes where grasses dominate as key-taxon. *Larix* is strongly represented throughout taiga and cold deciduous forest including patchy deciduous forests within forest-steppes of central Mongolia. *Picea* is only represented at sites where forests dominated by boreal conifer (taiga) grow. These arboreal pollen taxa generally are restricted to areas of mother plant distributions. The truth of the last conclusion immediately appears from a comparison of Figure 2.12 with Figures 2.9 and 2.10.

Table 2.4. Plant functional types and pollen taxa from Mongolia assigned to them.

Codes	PFTs	Pollen taxa
	Trees and shrubs	
bec	Boreal evergreen conifer	*Picea*
bs	Boreal summergreen	*Betula, Larix*
cbc	Cool-boreal conifer	*Pinus* subgen. *Haploxylon*
bec/ctc	Boreal evergreen/cool-temperate conifer	*Abies*
ec	Eurythermic conifer	*Pinus* subgen. *Diploxylon* (e.g. *P. sylvestris*)
bts	Boreal-temperate summergreen shrub	*Lonicera*
bs/ts	Boreal/temperate summergreen	*Alnus*
bs/ts/aa	Boreal/temperate summergreen/arctic-alpine shrub	*Salix*

Table 2.4. Continued.

Codes	PFTs	Pollen taxa
ts	Temperate summergreen	*Quercus* (deciduous)
ts_1	Cool-temperate summergreen	*Corylus*, *Ulmus*
	Others	
sf	Steppe forb	*Allium*, Apiaceae, Asteraceae (Asteroideae), Asteraceae (Cichorioideae), Brassicaceae, Campanulaceae, *Cannabis*, Caryophyllaceae, *Centaurea*, *Chamaenerion*, Convolvulaceae, Dipsacaceae, Euphorbiaceae, Fabaceae, *Galium*, Geraniaceae, *Goniolimon*, *Hippophaë*, Lamiaceae, Liliaceae, *Linaria*, Papaveraceae, *Patrinia*, *Plantago*, Plumbaginaceae, Polygalaceae, *Potentilla*, Ranunculaceae, Rosaceae, Rubiaceae, Rutaceae, *Scabiosa*, *Stellera*
sf/df	Steppe/desert forb	*Artemisia*, Boraginaceae, Chenopodiaceae, *Kochia*
df	Desert forb	*Ephedra*, *Salsola*, Zygophyllaceae
aa	Arctic-alpine dwarf shrub	*Alnus fruticosa*-type, *Betula nana*-type, Gentianaceae, *Pedicularis*, Saxifragaceae
sf/aa	Steppe/arctic-alpine forb	Scrophulariaceae, Valerianaceae
sf/df/aa	Steppe/desert/arctic-alpine forb	Polygonaceae
g	Grass	Poaceae
s	Sedge	Cyperaceae
h	Heath	Ericales, Pirolaceae

Several modifications of the taxa × PFT table given by Prentice et al. (1996) were made by Tarasov et al. (in press) in an attempt i) to increase the number of taxa available for identification in forest biomes and ii) for the treatment of nonarboreal PFTs in northern Eurasia. Of these those, which have sense for Mongolia can be summarised as follow:

1. Prentice et al. (1996) assigned *Pinus* subgen. *Haploxylon* to the boreal evergreen conifer PFT in Europe and used it as a main characteristic of the taiga biome. However, in northern Asia this taxon only is represented by two species, each with a different ecology. *Pinus pumila*, a shrub-like form of *Pinus* subgen. *Haploxylon*, can survive under snow cover in the cold continental climate of eastern Siberia with mean temperature of the coldest month below –35°C (Davitaya, 1960); boreal conifer trees can not survive this cold. In Mongolia this taxon now grows above the tree line in the Khentii Mountains (Grubov, 1982; Gubanov, 1996). Pollen of *Pinus pumila* can not be distinguished from *Pinus sibirica*, a tree form of *Pinus* subgen. *Haploxylon*, which is widely distributed in the taiga. However we assigned *Pinus* subgen. *Haploxylon* to a cool-boreal conifer PFT.

2. Russian studies on pollen morphology (Kupriyanova, 1965; Kupriyanova & Aleshina, 1972) demonstrated that the pollen of dwarf birch (*Betula nana*, sensu lato) and shrub alder (*Alnus fruticosa*) can be distinguished from the corresponding tree forms, e.g., *Betula* sect. *Albae*, *Alnus glutinosa*, and *Alnus incana*. Where these distinctions were made, we assigned these taxa to appropriate PFTs, e.g.,

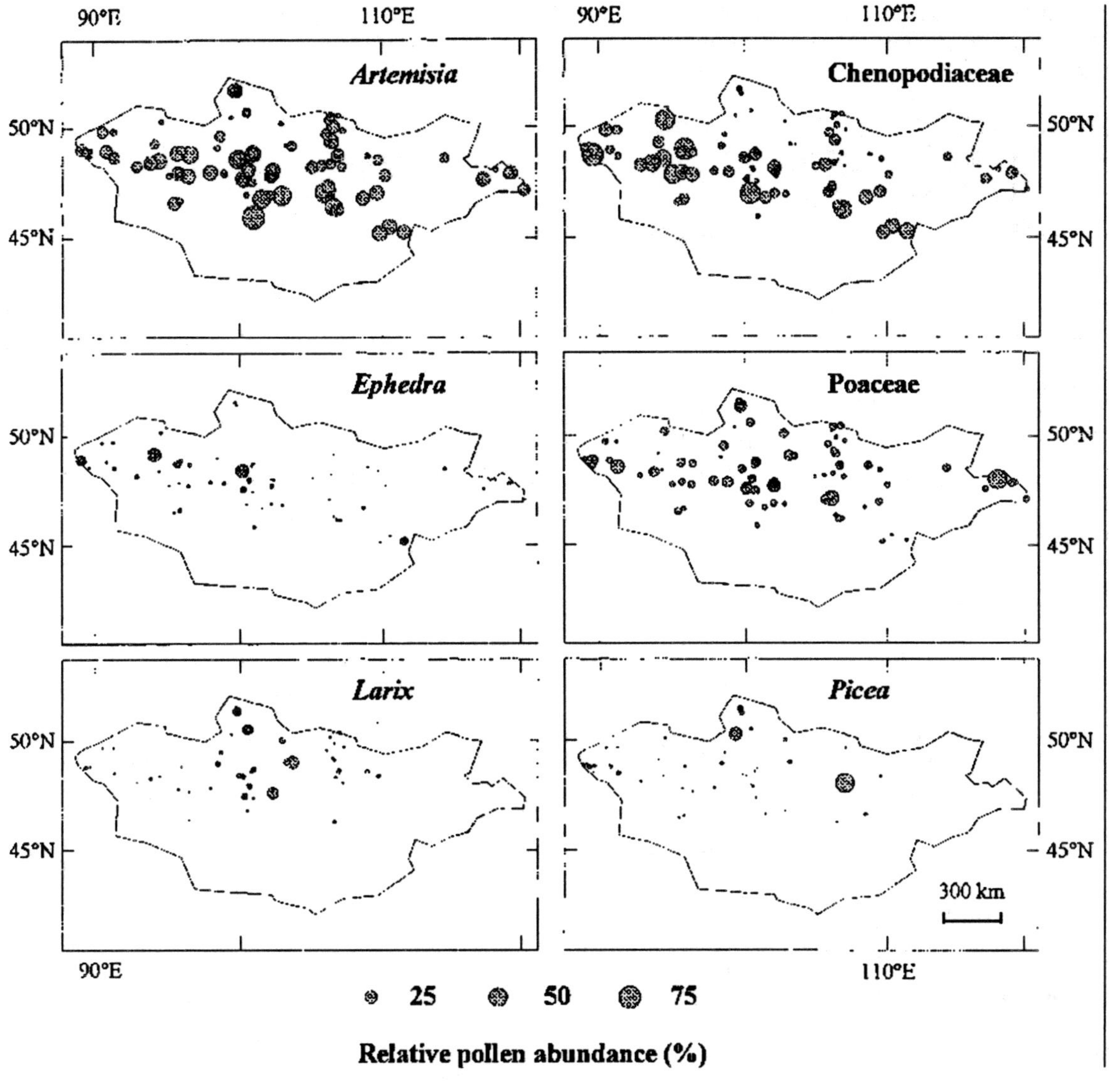

Figure 2.12. Distributions of pollen percentages for the main taxa from Mongolia based on modern surface pollen data.

arctic-alpine dwarf shrub and boreal or temperate summergreen trees.

3. The number of taxa assigned to steppe forbs and desert forbs was drastically increased compared to Prentice et al. (1996) in order to improve the distinction between tree and treeless biomes and the distinction among the herbaceous biomes. We used the same kind of empirical decisions as in Prentice et al. (1996) to assign nonarboreal taxa to appropriate PFTs. Most of them can appear in each biome, but certain taxa have a useful diagnostic value. For example, Brassicaceae, Caryophyllaceae, Rubiaceae have higher percentages in steppe as does *Ephedra* in desert or Cyperaceae in tundra.

Mongolian biomes then were defined as very simple combinations of the newly adopted PFTs (Table 2.5). Data from Tables 2.4 and 2.5 provide a basis for

Table 2.5. Mongolian biomes and their characteristic plant functional types (PFTs). PFTs in parentheses are restricted to part of their biome. Abbreviations for PFTs as in Table 2.4.

Tundra	(aa), g, s, (h)
Cold deciduous forest	bs, (cbc), ec, (h)
Taiga	bs, bec, (bts), cbc, ec, h
Cool conifer forest	bs, bec, cbc, ctc, ec, (bts), (ts_l), (h)
Desert	df
Steppe	sf, g

constructing a biome × taxon matrix used to calculate affinity scores.

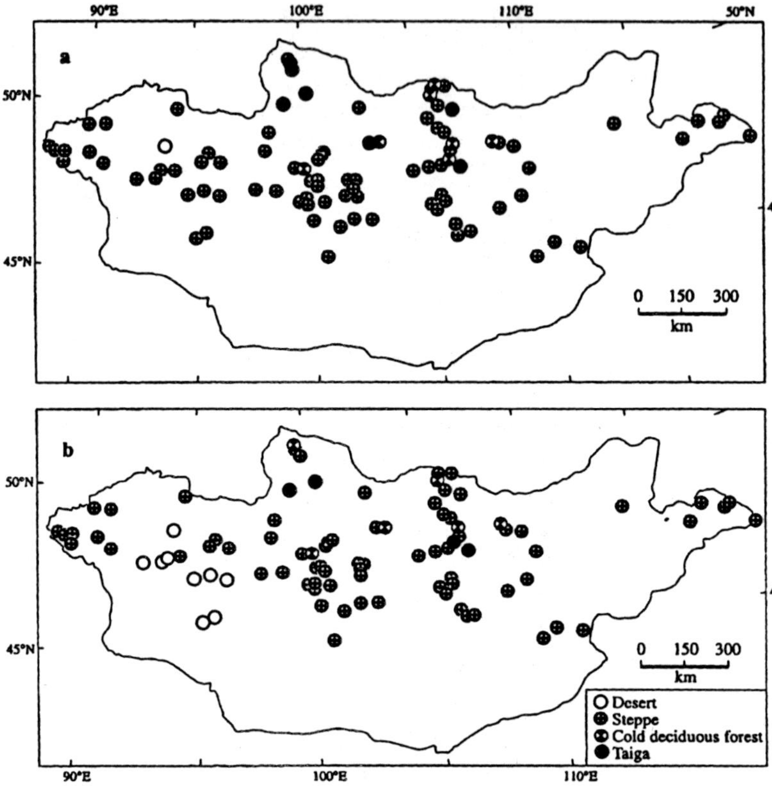

Figure 2.13. Pollen-derived biomes at present (a) and modern biome distribution (b) derived from vegetation maps of Mongolia (Lavrenko, 1979, 1983).

Validation of the method: present-day pollen-derived biome reconstruction

Modern surface pollen assemblages were used to reconstruct present-day biome distributions in Mongolia (Figure 2.13a). Maps of modern biomes, recorded in actual vegetation maps of Mongolia (Lavrenko, 1979, 1983), also were produced for all sites from available modern pollen data (Figure 2.13b). For that purpose the names of vegetation types in vegetation maps of Mongolia (Lavrenko, 1979, 1983), were assigned to comparable biomes (Table 2.6).

Table 2.6. Simplified vegetation types used by Lavrenko (1979; 1983), Grubov (1982) and Gubanov (1996) and their allocation to the biomes used by Prentice et al. (1992).

Biome name	Vegetation type
Tundra	Hugh mountain tundra and meadows: moss-lichen, dwarf shrub (*Betula rotundifolia, B. humilus, Salix glauca, S. rectijulis, Dryas*), heath (*Empetrum sibiricum*), sedge (*Kobresia myosuroides, K sibirica, Carex* sp.) and criopetrophitic (*Draba, Saxifraga, Bistorta vivipara* etc.) associations
Cold deciduous forest	Thin high mountain forests: larch (*Larix sibirica*) forest with *Betula* and *Salix* shrubby weeds; *Pinus pumila* shrubby weeds
	Taiga-like forests: larch or pine-birch (*P. sylvestris, B. platyphylla*) forests with heath (*Pyrola, Vaccinium vitis-idaea*); pine (*P. sylvestris*) forest with *Rhododendron dauricum*
	More or less steppe forests: larch (*L. sibirica*), larch-birch or larch-birch-pine and pine forests; birch (*B. platyphylla*) and poplar (*Populus tremula*) forests
Taiga	Boreal taiga forests: Siberian cedar (*Pinus sibirica*) sometimes with *Abies sibirica*; *Picea obovata* and *Picea-Abies*; mixed *Larix sibirica-Pinus sibirica* mountain forests
Cool conifer forest	Pine forests with *Ulmus pumila* at the river terraces

Table 2.6. Continued.

Biome name	Vegetation type
Steppe	Meadow-steppes and steppe-meadows of low-, middle- and high mountains
	Dry graminoid steppes of plains and low mountains
	Desert steppes (with *Artemisia* and grasses) of plains and foothills
Desert	Dwarf shrub and shrub deserts of plains and foothills: *Anabasis brevifolia, Salsola passerina, Nanophyton mongolicum, Artemisia*-Chenopodiaceae, *Zygophyllum xanthoxylon, Ephedra przewalskii, Haloxylon ammodendron* (sometimes with *Nitraria sphaerocarpa* and *Salsola passerina*), *Sympegma regelii, Reaumuria songarica, Kalidium gracile-K. foliatum* and *Nitraria sibirica* deserts
	High-altitude rocky deserts

Table 2.7. Numerical comparison for each site between biomes derived from modern surface samples (indexed by a «p») and observed biomes (indexed by an «a»). (TAIG = taiga, CLDE = cold deciduous forest, STEP = steppe, DESE = desert).

	TAIGp	CLDEp	STEPp	DESEp
TAIGa	3	2	0	0
CLDEa	1	5	1	0
STEPa	4	1	76	0
DESEa	0	0	8	1

A visual comparison of the pollen-derived biome map (Figure 2.13a) with a map of the actual biome distribution (Figure 2.13b) shows a good agreement. Results of a numerical comparison (Table 2.7) also demonstrate the success of pollen-based biome reconstruction in Mongolia. Biomes were correctly predicted 85 times (83% of cases). However, a few discrepancies should be mentioned between actual and reconstructed biomes in Table 2.7.

In 9% of the cases (8 times) steppe was reconstructed where the vegetation map shows desert. All of these samples were collected in the Great Lakes Basin where vegetation maps (Lavrenko, 1979) show a mosaic of vegetation representing both desert and dry steppe. Moreover, in all cases the modern surface samples were collected either in a large river valley or close to big lakes. There vegetation in fact is closer to steppe than to desert as shown on a large-scale vegetation map. This, however, indicates that underestimation of desert is a problem of the data, not of the method of biomization.

A cold deciduous forest was reconstructed two times at places where taiga dominates the vegetation cover. Such a situation occurred when surface samples were collected in the Dood-Nur basin at low and waste lacustrine terraces currently occupied by larch forest. Picea appears there in a few local habitats but does not produce enough pollen for analysis.

Application to the fossil pollen data

The biomization method was applied to fossil data sets, including 25 core pollen records from Mongolia. Maps of reconstructed biomes were produced at 500-yr interval since 15,000 yr B.P. to identify changes in biomes distribution in late Quaternary. Here, we present maps only of those time slices that demonstrate the most pronounced vegetation changes. When more than one record was available from the same site, the sample with better dating control defined as in Webb (1985) for each time slice was used for mapping. A visual comparison of the reconstructed biomes (Figures 2.14 & 2.15) with actual biome distribution (Figure 2.13b) shows strong broad-scale and regional coherent patterns of vegetation change in late Quaternary.

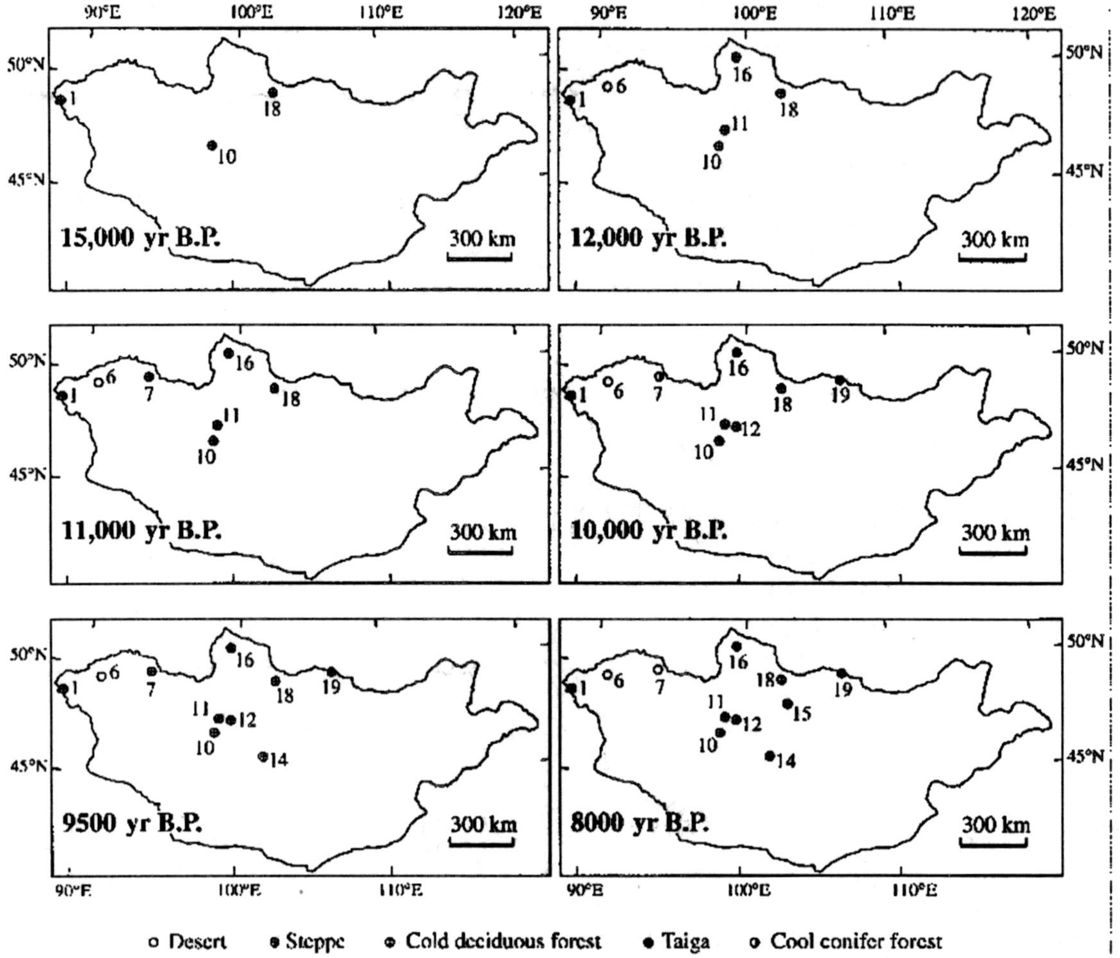

Figure 2.14. Pollen-derived biomes at 15,000; 12,000; 11,000; 10,000; 9500, and 8000 yr B.P. For Figs. 2.14 and 2.15 when more than one core is available from the same site the core with better dating control has been used for mapping purposes.

The late Glacial pollen records are confined to the north-western quarter of Mongolia. Maps of reconstructed biomes (Figure 2.14) demonstrate that between 15,000 and 11,000 yr B.P. vegetation was steppe at most sites, including those sites in northern Mongolia where forest vegetation dominates today. Since 12,000 yr B.P., desert reconstructed at the Achit-Nur site where modern vegetation is dry steppe.

At 10,000 yr B.P. taiga and cool conifer forest replaced steppe at three sites in northern Mongolia (Figure 2.14). There was no change in the reconstructed biomes at sites in western and central Mongolia.

In the early Holocene (e.g. 9500 and 8000 yr B.P.; Figure 2.14) taiga and cold deciduous forest were reconstructed at a very limited area in the northernmost part of the country. Steppe dominated at most sites and desert still existed in the large depressions of western Mongolia.

A pronounced change in biome distributions appeared in middle Holocene (e.g. 7000 and especially 6000 yr B.P.; Figure 2.15), when taiga and cold deciduous forests became very important in the vegetation of western and

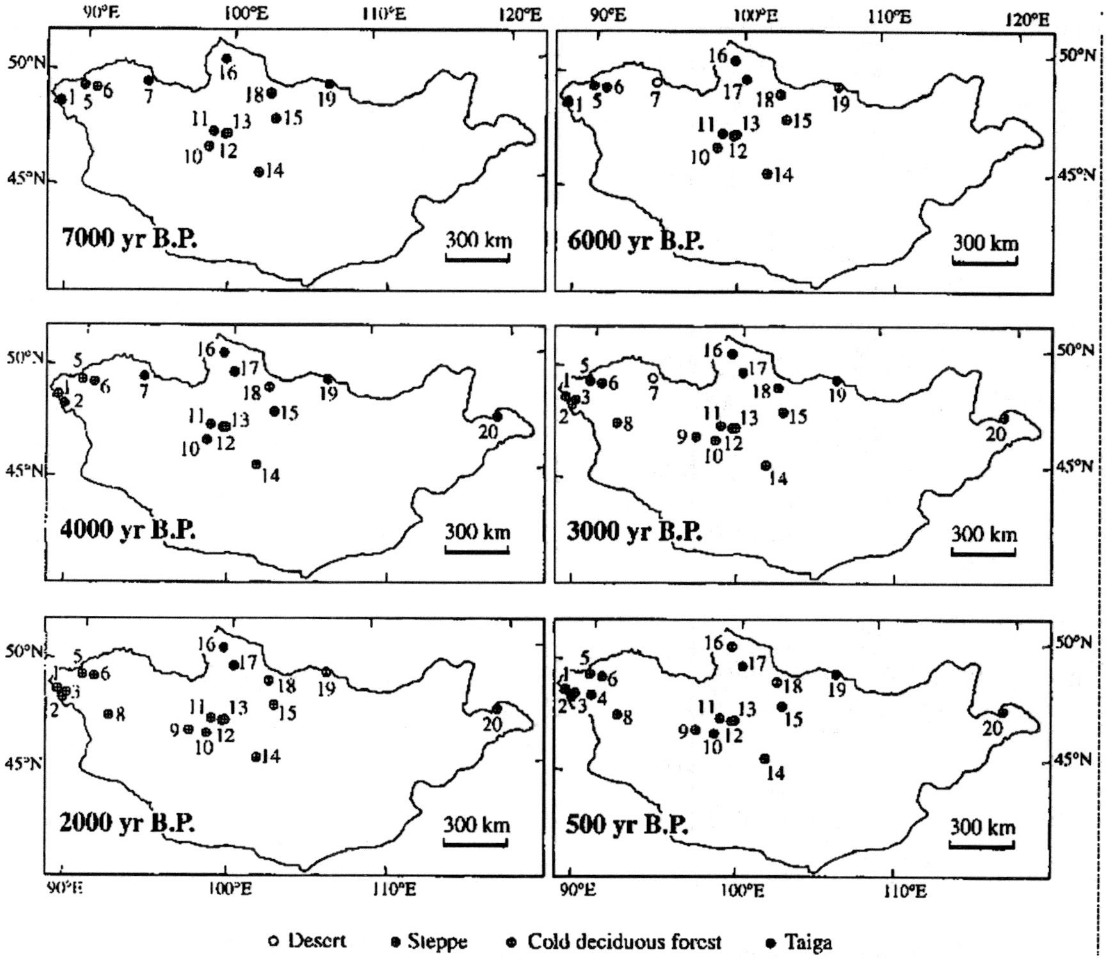

Figure 2.15. Pollen-derived biomes at 7000; 6000; 4000; 3000; 2000, and 500 yr B.P.

northern Mongolia. A mid-Holocene appearance of taiga at sites where forest-steppe or cold deciduous forest dominate today suggests that boreal evergreen conifers (*Picea* and *Abies*) occupied larger area in Mongolia than they do today. Desert was not reconstructed anymore, being replaced by steppe. Pollen-derived biomes for sites situated in the central part of Mongolia were steppe. Steppe is a dominant vegetation type there today; also patchy forests dominated by *Larix* and *Betula* are a typical landscape characteristic. Individual pollen records from central Mongolia show increases in arboreal pollen in middle Holocene. However, our biome reconstruction does not provide evidence that a mid-Holocene aforestation of central Mongolia was large enough to change the dominance of steppe vegetation there.

At 4000 yr B.P. taiga was replaced by steppe in western Mongolia, but retained its position in northern Mongolia (Figure 2.15). This retreat of forests, started in late Holocene, became more pronounced at 3000 yr B.P. (Figure 2.15). At that time taiga was reconstructed only at three sites in the northernmost part of Mongolia, suggesting conditions became less favourable for the growth of boreal conifers than in middle Holocene. Desert re-appeared then at Hoit-Gol, where modern conditions also are very dry. Our reconstruction suggests that vegetation had become similar to todays by 2000 yr B.P. There were no significant changes in biomes distribution during the last 2000 years (Figure 2.15). Hilbig (1995) after Savina et al. (1981) suggested that fires, wood utilization and grazing caused a forest retreat in the forest steppe zone of Mongolia during the last several centuries. Significant forest cuttings in the most populated regions of Mongolia has been documented only since the last quarter of the 19th century (Hilbig, 1995). A comparison of pollen-derived biome distribution at present (Figure 2.13a) with that at 500 yr B.P. (Figure 2.15), when human impact was unimportant (and compared with that during earlier parts of the records), show good similarity. This suggests that vegetation changes in Mongolia likely have natural origins up to the most recent time and can be explained mostly by climate changes.

2.7 GENERAL DISCUSSION AND CONCLUSIONS

The main features of these late Quaternary pollen-derived biome distributions for Mongolia are in good agreement with more recent qualitative interpretations of palaeovegetation records both at regional (Dinesman et al., 1989; Vipper et al., 1989) and local (Sevastyanov et al., 1993) scales. However, none of the previous reconstructions presented data from the whole country at late Quaternary. The palaeoclimatic significance of the patterns described here needs to be discussed.

An extension of steppe and desert north of their present position in Mongolia during late Glacial times and early Holocene suggest conditions were generally drier than at present. Lakes in Mongolia were lower than today between 10,000 and 9,000 yr B.P. (Harrison et al., 1996; Tarasov et al., 1996), which is consistent with this interpretation. However two lakes from Mongolian Altai were higher at 12,000 yr B.P. Such disagreements with vegetation patterns can be explained by local effects of melting water from mountain glaciers in the Altai Mountains at Glacial times as suggested by geomorphological and sedimentary records from there (Malaeva, 1989b; Devyatkin, 1993). A direct effect of higher (than present) summer insolation on a temperature increase at the end of late Glacial likely was strong enough to melt glaciers, causing a lake-level rise. However, scantiness of glacial water sources associated with the highest (for the last 18,000 years) summer insolation anomalously caused low lake levels at the beginning of Holocene. Pollen records from China synthesised by Winkler & Wang (1993) show patterns similar to those found in Mongolia. Arid steppe occupied areas larger than today in northern China around Mongolia's border through late Glacial and early Holocene, suggesting drier climates than the present. A weaker than present summer monsoon was cited as the most probable reason for these drier patterns suggested both by lake level (Harrison et al., 1996) and vegetation (Winkler & Wang, 1993) records. Furthermore a possible decrease in mean temperature for the coldest month below –35°C (ca 5 to 10°C lower than today) could be another limiting factor for boreal evergreen conifer growth in late Glacial-early Holocene, when winter insolation was 10 to 12% lower than present values at 50° N latitude (Berger & Loutre, 1991). This reason also was used to explain the disappearance of boreal conifers at Ozerki in eastern Kazakhstan, where *Betula* continuously grew through the last 13,000 years (Tarasov et al., 1997). Furthermore, late Glacial temperatures, inferred from biological evidence, were 8–10°C lower than today at 45°N in Northeast China (Winkler & Wang, 1993). We are not able to explain a short-term appearance of forest vegetation reconstructed at Hoit-Gol and Gun-Nur in northern Mongolia at about 10,000 yr B.P. Both sites are situated at low elevations close to a river. Even today forests dominated by evergreen conifer and deciduous trees («urema») grow in river valleys in Mongolia in more favourable moisture conditions, although watersheds outside those valleys are covered by steppe. Probably at 10,000 yr B.P. a combination of temperature and moisture coming from a local source e.g., melting of permafrost, was favourable enough to support tree growth. Numerous freshwater ponds, indicated by ostracods, developed as permafrost melted in the Beijing lowlands (North China) at 12,000–8000 yr B.P. (Winkler & Wang, 1993), are consistent with our interpretation. A short-term climate oscillation may provide another explanation of this reconstructed pattern. However, this hypothesis must be supported by evidence.

The expansion of boreal coniferous trees (e.g. *Picea* and *Abies*) in the mountains of northern and western Mongolia (where they either are not present or are restricted to local habitats today) between 8000 and 4000 yr B.P. indicates wetter than present conditions. This result is in good agreement with that suggested by other palaeoproxy data from the region. Lake level records in Mongolia reflect more positive moisture balance during the mid-Holocene with a maximum around 7000 yr B.P. (Harrison et al., 1996; Tarasov & Harrison, 1998). Studies of buried soil horizons from the eastern Mongolia plains currently occupied by dry steppe suggest meadow steppe vegetation and wetter than present conditions between 6150 and 4700 yr B.P. (Dinesman et al., 1989). At 6000 yr B.P. temperate forest and forest-grassland vegetation dominated in China north of 42°N latitude largely replacing arid steppe (Winkler & Wang, 1993). Reconstructed vegetation changes likely reflect greater than present precipitation in the mid-Holocene. Yu & Qin (1997), using pollen data from eastern China, estimated that annual precipitation was ca 10–280 mm greater than present between 8000 and 5000 yr B.P. These values are comparable with these we estimated for Gobi Altai. Major environmental changes in Central Asia can be explained by strengthened mid-Holocene monsoon activity in the region, caused by higher than present summer insolation and consequently greater than present land-ocean temperature contrasts (Wright et al., 1993; Harrison et al., 1996). The palaeodata from north China, including biological, archaeological, and geomorphic evidences show that the limit of monsoon activity was at least 500 km further north at 6000 than at 9000 yr B.P. (Winkler & Wang, 1993).

A retreat of boreal conifers and a reduction of forest vegetation in Mongolia during last 4000 yr B.P. suggest moisture conditions then became more or less similar to the present. This pattern parallels that shown by lake levels in Mongolia during late Holocene (Harrison et al., 1996; Tarasov & Harrison, 1997), following a gradual attenuation of summer insolation anomalies. However, processes of desiccation apparently were not gradual as evidenced by pollen and lake data. Thus Mongolian lakes were the same or drier than present conditions at 3500 yr B.P.; but wetter than present at 1800 yr B.P. (Tarasov & Harrison, 1997). Qualitative climate reconstructions based on pollen and lake-level records from China (Winkler & Wang, 1993) shows conditions drier than today in regions that bound Mongolia at 3000 yr B.P. Our study results prove once more that broad environmental changes (e.g. vegetation or lake levels) at late Quaternary times can be explained by orbitally-induced insolation changes. However, these gradual insolation changes apparently are not able to explain a shorter-term (millennial to century-scale) fluctuation. More data still are needed to do that.

These recently compiled pollen and macrofossil data from Mongolia demonstrate strong regional coherent patterns of vegetation changes through late Quaternary. Results obtained both with traditional (qualitative) and new quantitative methods of vegetation reconstruction show excellent compatibility. However the method of biomization provides the best possibility for a more objective interpretation of these data.

CHAPTER 3

ASSESSING PRESENT-DAY PLANT COVER DYNAMICS

3.1 INTRODUCTION. MODERN METHODS FOR STUDYING AND MONITORING PLANT COVER

The dynamics of plant cover has long received much attention and is addressed in numerous publications. Methods for the investigation of vegetation dynamics were treated in detail in the monograph Field Geobotany (Aleksandrova, 1964). Today, in addition to these conventional methods, remote-sensing (Vostokova et al., 1988a, b) GIS and ecological databases are used extensively (Trofimova & Novikova, 1994).

Our investigations of Mongolia's vegetation cover dynamics involved several methods. The most important were:

– extrapolation of spatial vegetation changes to temporal changes;

– comparison of geobotanical descriptions at different times of the same area;

– comparison of vegetation maps compiled at certain timed intervals of the same territory;

– comparison of aerial and satellite photographs made at certain timed intervals;

– historical comparisons using archive materials.

One widely used method was to extrapolate spatial changes in vegetation cover to their temporal changes. In that work both stationary and enroute studies were conducted in which ecological profiles were laid and described thoroughly. Normally, these profiles were laid across the relief, i.e., across the slope of an elevation or from periphery to the center of a lowland or a bound depression, etc. In describing a relief much attention was given not only to the relief but also to the ecological conditions of habitats and to natural or human-associated factors affecting the vegetation. Sampling sites were chosen along the profile where the vegetation, soils, moisture conditions and the nature of surface deposits were thoroughly described. Such samples served to characterize a specific plot of a particular plant community. Numerous ecological profiles provided data to document vegetation ecology and to show changes caused by specific human-associated factors, i.e., the landscape-ecological factors of plant cover dynamics.

Such vegetation series were obtained for mountains and plains, depending on their zonal position. These ecological vegetation series were treated separately for mountain versus plain areas. That made it possible to elucidate vegetation dynamics for taiga mountain zones and also for steppe and desert zones. On the plains, vegetation dynamics were studied in steppe and in desert zones. True enough, it was impossible to obtain ecological series for every landscape and ecological regions in a big country like Mongolia. But the major ecological communities succession patterns were studied. For more completeness additional methods also were used. Specifically, we used data from previous observations that are published in the Proceedings of the Joint Russian (Soviet)-Mongolian Complex Biological Expedition and in various journals (Meteltseva, 1986; Hilbig, 1991; Gubanov & Hilbig, 1993).

The use of personal observations made over long time periods (from 5 to 10 years), plus the findings in earlier geobotanical studies provided us with vegetation descriptions of the same areas at earlier times. Those studies were essentially all in territories that conducted stationary or semi-stationary integrated studies. The location of these stations is presented in Figure 3.1. The integrated nature of observations at these stations enabled us to analyze different factors affecting plant cover. We can distinguish major factors in the dynamics of plant cover, i.e., long term changes of plant communities or replacement by other plants. In most cases, we can not account for seasonal fluctuations of vegetation. For some stations it was possible to monitor human-associated vegetation dynamics in communities on the basis of site descriptions and also by using heterotemporal cartographic materials, i.e., using maps of the same plot compiled at times of 5 to 7 years.

The use of satellite photographs made at time intervals of up to 10 years also provided some parameters of vegetation dynamics. Thus, it was possible to monitor the dynamics of forested areas and the distribution of ploughed lands, including the transfer of previously cultivated lands into fallow lands. For that purpose, photographs of the same plot were used. Normally, such a plot also was examined, using land enroute observations. Heterotemporal photos were deciphered separately and their patterns compared on the same scales.

Most satellite photos were obtained from Kosmos satellites. The resolution of these photos was from 2 to 5 to 10 m (Kravtsova, 1992). In handling remote-sensing information, methods developed earlier were used (Vostokova, 1980; Vostokova et al., 1988a, b; Knizhnikov & Kravtsova, 1991).

To study the present-day status of Mongolian ecosystems, a map was drawn to reflect modern ecosystem distributions and the extent of their disturbance by anthropogenic factors (Gunin & Vostokova, 1995b). The map scale was 1:1000000. Using such maps for

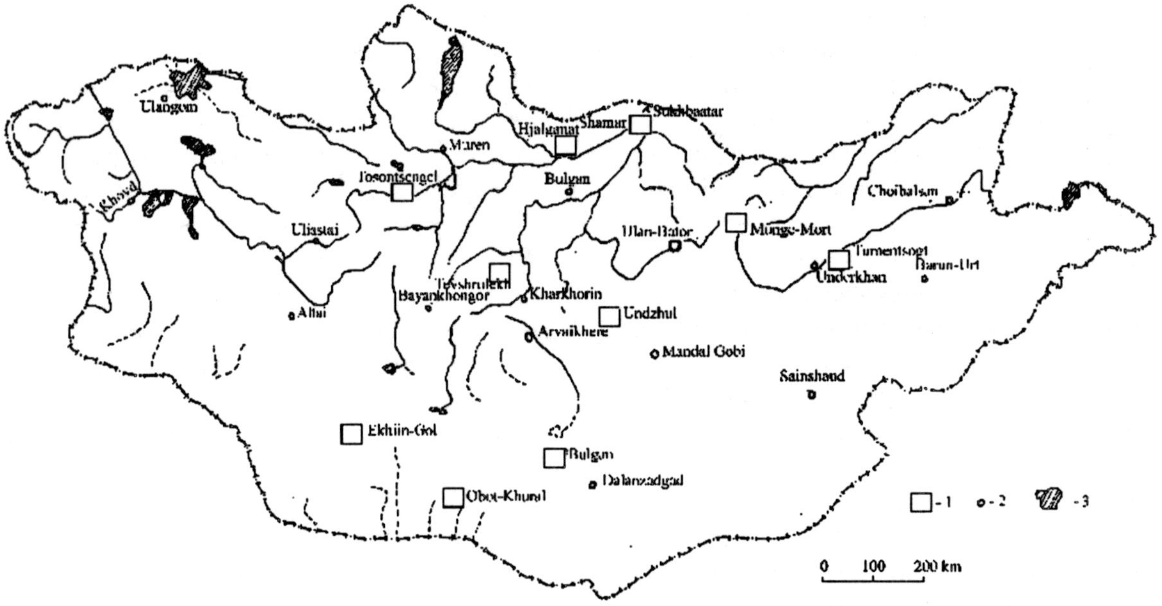

Figure 3.1. Schematic map locating experimental sites of Joint Russian-Mongolian Complex Biological Expedition.

1 – stationary observation sites; 2 – large settlements; 3 – lakes.

different natural and altitudinal zones, we obtained a generalized ecological series reflecting modern human-associated vegetation dynamics.

3.2 MOUNTAIN PLANT COMMUNITY DYNAMICS

Each mountain altitudinal zone is characterized by a specific ecological series reflecting its plant communities dynamics. Thus each altitudinal zone reflects its own factors for changes in plant cover. For high-altitudinal vegetation, natural factors of plant successions are the most important. For forest and forest-steppe zones, the most important are anthropogenic factors (felling, grazing of livestock) or human-stimulated natural factors, fires for instance.

The dynamics of high-mountain vegetation is associated mainly with the variation of natural factors, principally precipitation. But such natural features are very gradual and are not revealed by enroute observations. However, by extrapolation of spatial ecological series to temporal changes, as affected by a particular factor such as precipitation, we uncovered the dynamics of high-mountain communities.

The fact that Mongolian Altai stretches roughly 600 km from north to south determines the position of its top surfaces in different physiographic zones, which are characterized by different precipitation levels. The differences in moisture conditions plus an increase in aridity from north to the south both help show the dynamics of high mountain vegetation in this mountain range. In fact, on the mountaintops of Mongolian Altai, with precipitation of over 300 mm per year, moderately moist *Kobresia* and sedge communities are common. For example, lichen-cryophyte-forb sedge communities (*Carex rupestris*, *Bistorta vivipara*, *Schultzia crinita*, *Oxytropis alpina*, *Cetraria nivalis* Ach.) in combination with cryophyte-forb *Kobresia* communities and *Dryas* communities (*Dryas oxyodonta*), sedge *Kobresia* communities (*Kobresia myosuroides*, *Carex obtusata*, *C. melanantha*, *Festuca kryloviana*, *Lagotis integrifolia*) in combination with swamp sedge communities (*Carex sempervirens*, *C. melanocephala*, *C. orbicularis*), and occasionally high-mountain swamps with *Eriophorum polystachyon* and *Ptilagrostis mongholica*. But when precipitation declines (less than 300 mm per year), the high-mountain vegetation consists of somewhat steppe-like communities of cryophyte forb-sedge *Kobresia* communities. These communities are dominated by *Kobresia myosuroides*, *K. smirnovii*, *Carex sempervirens*, *C. rupestris*, and also fairly numerous forbs (*Potentilla nivea*, *Lagotis intergrifolia*, *Thalictrum alpinum*, *Oxytropis strobilacea*, *O. oligantha*, *Bistorta vivipara*, *Aconogonon alpinum*). Occasionally, these are combined with willow groves (*Salix berberifolia*, *S. arctica*). At still lower atmospheric moisture, willows are replaced by cryophyte-herbaceous *Kobresia* communities (*Kobresia humilis*, *K. myosuroides*, *Dryas oxyodonta*, *Bistorta vivipara*) in combination with cryopetrophyte communities (*Rhodiola quadrifida*, *Saxifraga hirculus*, *Stellaria petraea*, *Waldheimia tridactylites*), which occupy the top surfaces of Gobi Altai (Volkova, 1994). These vegetation series in the Mongolian and Gobi Altai likely reflect successions of high-mountain vegetation with an increase in aridity.

The changes in high-mountain vegetation ecological profiles are more pronounced in different mountain areas situated within a particular physiographic zone. For example, in northern Mongolian Altai, lichen cryophyte-forb *Kobresia* communities and sedge communities, involving forbs, are replaced downslope by steppe-like grass-cryophyte-herbaceous sedge communities and *Kobresia* communities (*Carex rupestris*, *C. sempervirens*, *Kobresia myosuroides*, *K. smirnovii*, *Poa attentuata*, *Festuca lenensis*,

Arenaria meyeri, Potentilla nivea, Oxytropis strobilacea, etc.). In the central part of Mongolian Altai the cryophyte-forb-sedge *Kobresia* communities on top surfaces are replaced downslope by grass-pulvinario-forb sedge communities (*Carex rupestris, Stellaria pulvinata, Saussurea leucophylla, Oxytropis oligantha, Festuca brachyphylla, Poa altaica*). In Gobi Altai, the *Dryas* and cryophyto-forb *Kobresia* communities are replaced downslope by steppe-like cryophyte-grass sedge communities and *Kobresia* communities (*Kobresia humilis, K. myosuroides, Carex rupestris, Festuca lenensis, Poa attenuata, Potentilla nivea, Oxytropis oligantha, Rhodiola quadrifida,* etc.).

A comparison of the main species in these communities with the precipitation level on the top surfaces and at each altitude reveals their similarity in terms of species community compositions. Hence, we conclude that these ecological series illustrate probable successions of high-mountain vegetation with a gradual increase in aridity.

However, under existing harsh climatic conditions, particularly in the southern mountains (Gobi Altai and Gobi Tyan-Shan), high-mountain cryoxerophyte vegetation can be in direct contact with steppe communities. In Gobi Altai, for example, cryophyte-forb-grass sedge communities and *Kobresia* communities are replaced downside by moderately dry and dry steppes – forb-sheep's fescue or wheatgrass-sheep's fescue communities (*Festuca lenensis, Agropyron cristatum, Arenaria meyeri, Amblynotus rupestris, Dracocephalum origanoides, Oxytropis trichophysa* etc.). Therefore, with an increase in aridity and a deterioration of moisture conditions both in Gobi Altai and in Gobi Tyan-Shan and on southern ranges slopes of Mongolian Altai and Khankhukhiin, we believe high-mountain communities with a predominance of *Kobresia* and sedges will be replaced by cryophyte steppes (cryophyte-forb-sheep's fescue) in combination with *Kobresia* communities) and by forb-sheep's fescue communities.

Human-associated effects on high mountain vegetation is only minor so far, but on high-mountain meadow steppes large changes in plant cover successions are caused by anthropogenic (pastoral) effects. Most human-induced changes in high-mountain plant cover are a result of overgrazing. In fact, with cryptophyte scale communities, great pastoral pressure primarily removes grasses (*Festuca lenensis, Poa attenuata,* and *Koeleria altaica*), and under still greater pressures, scales are destroyed. Thus, cryophilous scales are replaced by an impoverished plant community with a sparse cover (the cover changes from 50 – 60% to 10 – 15%).

A characteristic feature of vegetation dynamics in mountain regions is a change in the boundaries of vegetation zones. This is well illustrated in considering the boundaries of a forest zone that can be found from remote-sensing information. Changes in high-altitude zones of meadow-steppe and mountain-desert vegetation are less readily observable – they can only be observed from direct observations and by comparison with previously established altitude positions of specific plant communities. For example, the disturbance of under-bald-mountain open woodlands results in a lower upper border of the forest, replaced by shrub thickets of *Betula exilis, B. fruticosa, B. rotundifolia, Potentilla fruticosa,* and, occasionally, forest-willow open woodlands.

When dry years cycles are prolonged, changes occur in the lower borders of the forest zones with a replacement of pine or larch forests by steppe communities. As result, the entire forest zone is reduced. However, during moist years, if drought duration is less than 4 to 5 years, a reverse process may occur, i.e., an advancement of forest vegetation onto steppes (Korotkov, 1978; Bannikova, 1983). Cases of forest repopulation in a meadow forb-feather grass steppe occurred in eastern Khangai, to a less extent in central Khangai, and also in eastern Khentii. An undergrowth of larch, that roots in moist periods, subsequently serves as an advanced position for forest vegetation regrowth on steppes (Figure 3.2).

Figure 3.2. Advanced posts of *Larix* on steppe slopes of North-West Khangai (above) (photo by A.V.Prishchepa) and Central Khankhukhiin Nuruu (below) (photo E.N. Matyushkin).

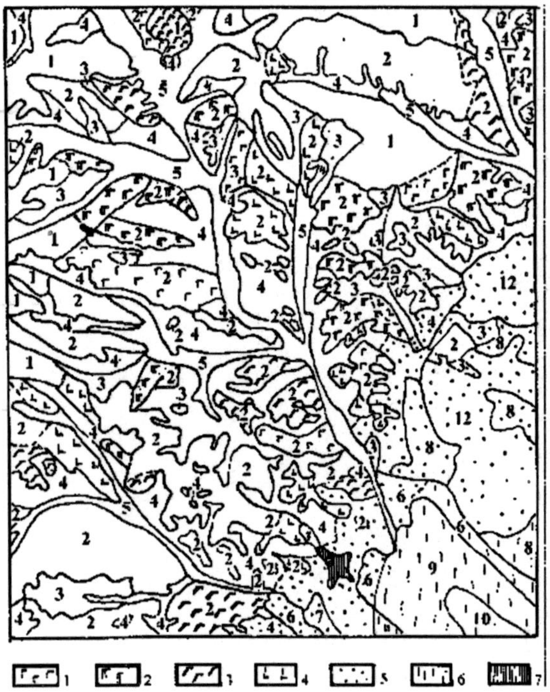

Figure 3.3. Main factors of forest disturbance in Eastern Khentii (according to the materials of Mungen-Mort stationary observation site).

1–3. Disturbances caused by forest fires: 1 – less than 20% of trees are damaged; 2 – 20 – 50%; and 3 – more than 50% of forest are damaged. 4. Disturbances caused by forest clearing (more than 70% of the total area). 5 – 6. Damage to grass cover: 5 – weak; 6 – heavy. 7. Settlement.

Figures show: 1 – 4. Vegetation of low and middle height mountains: 1 – taiga-type larch stands; 2 – steppized larch stands; 3 – steppized larch stands with meadow steppes; 4 – petrophyte – forb-meadow steppes. 5 – 12. Vegetation of Kerulen valley and its tributaries: 5 – yernik thickets; 6 – forb – sedge meadow; 7 – sedge meadow and sedge-forb water-logged meadow; 8 – willow thickets; 9 – rich forb – feathergrass steppe in combination with sedge swamps and participation of willows and *Potentilla fruticosa*; 10 – rich forb – feathergrass steppe; 11 – *Leymus* – forb steppe (formed as a result of hay cutting over 50 years); 12 – grass – sedge meadow.

With the formation of forest communities, an abundance of the main steppe community-forming species is reduced, including feather grass and meadow-steppe forbs *(Stipa krylovii, Aster alpinus, Stellera chamaejasme*, etc.). At the same time, under the forest canopy, increases occur in the abundance of forest forbs and grasses, e.g., an increased abundance of *Sanguisorba officinalis, Artemisia tanacetifolia, Carex lanceolata* and *Thalictrum minus*.

The human-associated dynamics of forest vegetation is strongly influenced by clear-fellings, by heavy forest fires that result in a high mortality of primary species, and by intensive livestock grazing that also result in undergrowth mortality and changes in forest regeneration conditions (Figure 3.3).

The dynamics of forest vegetation is largely a function of the species and the extent of human-associated effects; but in addition the primary forest community plus existing climate conditions must be considered.

On the upper forest boundary, which is formed mainly by Siberian pine or Siberian pine-larch open woodlands, a tree stand destruction often leads to an expansion of the *Kobresia*-sedge subalpine communities.

Siberian pine and fir-Siberian pine forests account for a very small area, but their economic importance is very high. Hence their regeneration after fires and other disturbances has been receiving much attention (Semechkin, 1980). The largest concentration of Siberian pine forests are recorded in central Khangai and Khentii. Normally, Siberian pine appears as an undergrowth in larch forests where there are favorable forest growing conditions (moist and rich soils). Because Siberian pine takes up mostly north-facing slopes that are exceptionally steep, heavy disturbances or mortality of tree stands, irrespective of the reasons, often cause erosion processes to develop. In the worst case, the entire fertile layer can be washed off and the denuded cliff rocks remain forestless for a long time. In such areas, the succession of forest vegetation begins along rock streams with very protracted regrowth stages, or the forest simply may not regenerate.

The main forest zone is formed by larch forests, both pure or with other species including the Siberian pine, pine, and birch.

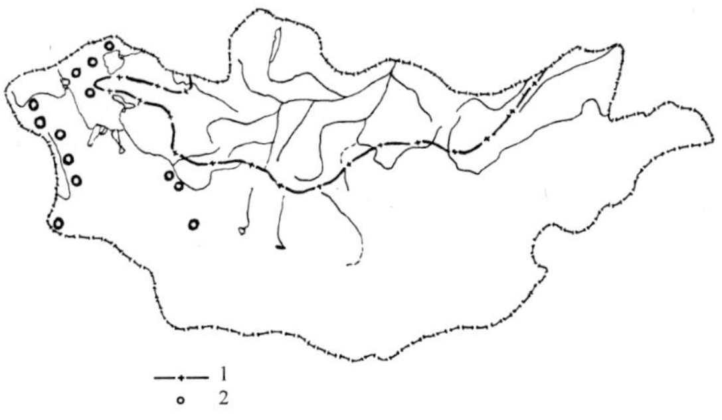

Figure 3.4. Distribution of larch (according to E.A.Volkova).

1 – southern boundary of larch forests; 2 – separate larch stands.

Larch belongs in open woodlands at the upper boundary of forests and, at the same time, it forms stands along the lower forest boundary that reach to north-facing slopes, far beyond the main forest zone (Figure 3.4). Hence, in Mongolia, the most diversified and best-studied successions are of larch forests as caused by combinations of both natural and anthropogenic factors.

Particularly sharp forest ecosystem successions are caused by forest fires. Often forest fires are a continuation of human-induced steppe fires. After fires, the burns develop intensive erosion processes, particularly on slopes with thin cryomorphic taiga soils. The entire soil layer may be washed off and at larch forests and Siberian pine forests sites, only rock streams will remain. Such sites will not be overgrown with forests for an indefinitely long time (Korotkov & Dorzhsuren, 1988; Cherednikova et al., 1991). On the other hand as a result of fires, forest soils are enriched by nutrients, which promotes an active development of herbaceous vegetation. Pine forests are very common in the western Khentii spurs of the eastern part of the Selenga River. In Khentii, between 1982 and 1988, 172 fires were recorded that involved an area of over 250 ha (Cherednikova et al., 1991). Particularly destructive fires occurred in the spring of 1996, when virtually all the forests at lower parts of the mountains along the Selenga river and western Khentii were damaged to varying extents. The worst affected were pine forests and small pine groves of the mountains forest-steppe zone (Figure 3.5). The most hazardous fire period is April through May, which accounts for up to 95% of all forest fires.

Figure 3.5. Pine tree forests damaged by low-level fire (photo by A.V.Prishchepa).

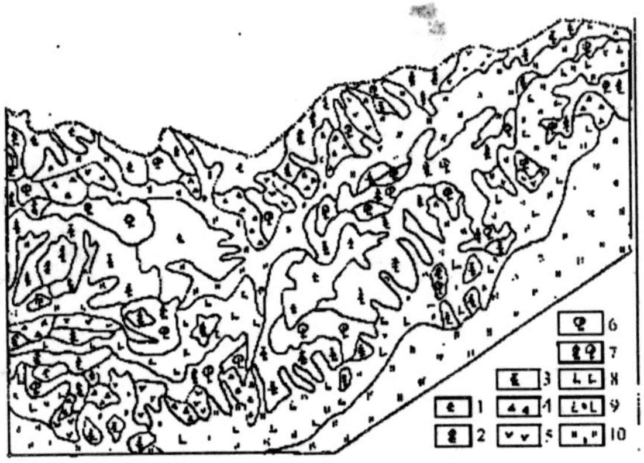

Figure 3.6. Forest successions in Buteiliin Nuruu (according to materials of Khyalganat observational station).

1 – 3. Non-disturbed forests: 1 – sub-bald larch open woodlands; 2 – Siberian stone pine – larch and larch dwarf bush – green moss forests; 3 – pine – larch forests and pine stands with participation of birches. 4 – 7. Pyrogenic successions: 4 – substitution of forests on stony places with lone petrophytes; 5 – substitution of forests by grass communities, sometimes with participation of larch; 6 – substitution of Siberian stone pine – larch forests by birch stands, sometimes with aspen; 7 – substitution of pine and larch-pine forests by birch or larch-birch forests. 8 – 9. Anthropogenic successions: 8 – forest clearings with renewal of forests (previously existed or long-derivative ones); 9 – forest clearings with no forest renewal, and forests substituted by mountain meadow steppes. 10. Mountain meadow rich forb steppes.

After-fire successions vary widely in different mountain zones and with different forest types. On bald-mountains, open woodlands often are replaced by shrub thickets of *Betula exilis, B. fruticosa* and *B. rotundifolia* after fires; a landscape that is close to mountain shrub tundras. Forests may not regenerate here for an indefinitely protracted time.

After strong fires or clear fellings, taiga forests comprised of larch-whortleberry-green moss tall-herb and larch-forb-whortleberry that are found mainly on north-facing slopes of middle-height mountains are replaced by Siberian pine-larch or larch-Siberian pine forests. Such replacements were observed at mountains along the Selenga River and Khangai (Savin & Dugurzhav, 1980).

In reality, however, the replacement of larch by Siberian pine is extremely rare, since recurrent fires prevent the development of Siberian pine, which suffers from fires to a much greater extent when compared with larch.

In the mountains along the Selenga river, both degradation and regenerative successions are present (Figure 3.6). Frequently, the denuding processes for soils and vegetation do not coincide (Krasnoshchekov et al., 1990; Gunin & Vostokova, 1995a). When areas that previously were occupied by Siberian pine or Siberian pine-larch forests are burned or felled, Siberian pine rarely plays a pioneering role. Normally it is regenerated under larch canopy, or more rarely a birch canopy (Semechkin, 1980).

Siberian pine forests occurring in small plots are exceptionally hard to regenerate after anthropogenic changes. In northwestern Khentii the mountain-taiga fir-Siberian pine forests, which grow mostly on steep slopes (up to 10 – 15°), regenerates very poorly after strong fires and Siberian pine often is replaced by fir. But on Khangai, Siberian pine forests only occupy the upper forest zone, frequently on the boundary with subalpine and there Siberian pine regenerates fairly successfully after disturbances (Semechkin, 1980). In the absence of repeated fires, different-age Siberian pine forests form normally, occasionally with larch, or a larch-whortleberry-shrub-moss forest develops with scattered Siberian pines.

After-fire vegetation successions frequently start with an active development of *Chamaenerion angustifolium* thickets that have a distinct dark-rosy color during flowering. Under its canopy, larch seedlings develop well, giving rise to a consecutive succession series. In the absence of recurrent fires, a mixed coniferous-small-leaved forest develops first, with birch and occasionally aspen dominating. In the case of subsequent low fires, the larch or pine understory can be destroyed. At the site of former coniferous forests, birch forests will develop. Occasionally, repeated ground fires

Figure 3.7. Ulmus pumila in the Selenga valley (photo by A.V.Prishchepa).

also badly affect birch undergrowth and promotes the formation of steppe ecosystems at the forest site. Such fires may promote a tendency to develop patchy forest communities that are separated by patches of steppe vegetation.

Siberian pine taiga larch forests can regenerate after fire either without replacement of species or after short successions of birch forests. Pine forests associated with low mountains or lower zones of middle mountains regenerate after fires via mixed pine-birch forests. But with recurrent fires at the site of pine forests, herbaceous communities can develop. Indeed, at a site of former pine forest growth on the sandy terraces of Selenga, meadow steppes have developed. Intensive grazing in these areas also was conducive to the formation of plots of poorly fixed wind driven sands, occasionally with group of *Thymus* sp. Subsequent stages of overgrowth on these sands are associated with the development of sparse *Ulmus pumila* communities (Figure 3.7).

At the southern forest boundary, forest fires result in an irreversible replacement of pine forests and steppe larch forests by steppe communities, mainly forb-sheep's fescue. Hence a reduction of forest-covered areas and an increase in hazardous erosion areas occurs.

Clear felling is the most devastating human factor in forest ecosystems dynamics. In clear fellings not only are the tree phytomass removed, but also the undergrowth and shrub-herbaceous layers are strongly disturbed. Indeed many important environmental/ecological conditions are changed drastically. In felled areas, for example, illumination increases and surface runoff is markedly increased. Even negligible precipitation on clear-cut mountain slopes brings about sheet wash and linear erosion (see Section 1.3).

An augmentation of erosion also is promoted by the use of heavy equipment for timber skidding and transportation. The productive and accessible larch forests of Khentii and Khangai are clear-cut extensively.

In clear-felled areas vegetation dynamics are largely dependent on the technical conditions of felling, on the species presence prior to the felling, on the growth stage of the primary undergrowth species, and on the number and fertility of mature trees left.

In eastern Khentii the regeneration dynamics in clear-felled mesophyte larch forests are largely dependent on the felling technique, the number of seeds trees left, and the felling area. In felled areas that exceed 5 ha, natural regeneration is very slow in plots distant from the un-felled forest. The initial aspect of herbaceous cover regeneration in felled area occurs only when the crowns of young trees join, which happens in 15 to 20 years under favorable conditions (Korotkov, 1978). During the first years after felling, the herbaceous cover is drastically transformed with respect to changes in species ecology. The species of forest forbs most abundant under the forest canopy (*Vicia venosa, Lathyrus humilis,* and *Aegopodium alpestre*) almost completely disappear. In contrast, other species increase in abundance to play a greater role in herbaceous cover. Primarily these include meadow-steppe species (*Festuca ovina, Vicia cracca,* and *Trisetum sibiricum*). Still, the general number of species remains the same and the similarity of herbaceous cover in felled areas versus the forest canopy understory reaches 83 to 87% (Table 3.1).

But even under favorable conditions of larch natural regeneration after clear fellings, mixed tree stands normally develop and birch becomes important in the regenerated forest. Thus a short-lived community is formed since the lifespan of birch is shorter than larch. Under fairly unfavorable natural regeneration conditions a birch forest develops in the felled area that can exist fairly long particularly with recurrent disturbances of the ecosystem (e.g., forest fires). A wide distribution of birch forests is favored by only felling healthy larch trees, leaving sick larches and birch in the felled area.

In Khangai, after intensive fellings, larch forests of the forest-steppe zone often are replaced by meadow-steppe communities. The sod cover that forms in a felled area hampers the development of larch undergrowth and often eliminates it completely (Korotkov, 1978). Such plant cover changes also occur in steppe herbaceous larch forests. There, the abundance of forest species (*Lathyrus humilis, Cypripedium guttatum, Fragaria orientalis,* and *Viola biflora*) also declines. Meadow species appear in the herbaceous cover during the first year, including *Sanguisorba officinalis, Adenophora stenanthina, Galium verum* and *Astragalus mongholicus* while the abundance of *Festuca ovina, Bromopsis sibirica* and *Vicia megalotropis* increases considerably (Table 3.2).

Table 3.1. Grass cover changes in a grass-forb larch stand in forest clearings (according to Korotkov & Dorzhsuren, 1988).

Main species	Cover percentage, %				
	Under crown cover	At clearings			
		1 year	2 years	4 years	5 years
Vicia venosa	20.0	2.5	2.7	3.7	0.0
Lathyrus humilis	21.0	0.3	0.4	0.3	0.0
Poa sibirica	4.0	5.9	3.9	0.3	1.0
Calamagrostis obtusata	<0.1	1.1	1.3	3.5	2.0
Bromopsis sibirica	<0.1	0.9	2.4	2.6	2.3

Table 3.1. Continued.

Main species	Cover percentage, %				
	Under crown cover	At clearings			
		1 year	2 years	4 years	5 years
Viola biflora	2.0	–	–	–	–
V. uniflora	3.0	1.9	2.0	2.5	3.0
Artemisia tanacetifolia	1.5	0.2	0.2	0.5	0.5
Moehringia lateriflora	<0.1	1.4	1.8	0.7	0.9
Aegopodium alpestre	3.3	0.9	0.9	1.0	0.8
Carex amgunensis	1.5	<0.1	1.6	1.5	0.3
Geranium pseudosibiricum	2.0	1.4	1.4	1.4	1.7
G. eriostemon	3.0	1.9	2.0	2.5	3.0
Bistorta vivipara	<0.1	0.8	0.5	0.3	<0.1
Fragaria orientalis	2.0	3.3	3.5	6.1	5.0
Chamaenerion angustifolium	2.0	0.2	0.3	0.5	0.7
Festuca ovina	6.0	17.4	18.6	22.5	19.9
Ranunculus propinquus	<1.0	3.6	3.7	1.6	<1.0
Galium boreale	1.0	0.5	0.6	0.1	0.3
Trollius asiaticus	0.5	0.8	0.8	1.5	1.0
Carex lanceolata	<1.0	0.5	0.5	0.7	1.0
Equisetum pratense	1.0	–	–	–	–
Artemisia sericea	0.5	1.1	0.7	1.2	1.1
Dendranthema zawadskii	<0.1	0.5	0.3	1.0	0.4
Polemonium chinense	<0.1	0.8	0.9	0.8	0.7
Vicia cracca	<0.1	0.9	0.6	0.8	1.5
Aconitum barbatum	<0.1	0.4	0.4	1.0	0.2
Trisetum sibiricum	<0.1	0.7	1.4	1.4	1.8
Larch regrowth, trees/ha	–	400	1900	160000	33300
Similarity to forest, %:					
according to specific composition	100	87.6	85.3	83.0	85.0
According to cover structure	100	39.7	42.2	36.4	36.7

Table 3.2. Grass cover changes with steppized larch stands in forest clearings (according to Korotkov & Dorzhsuren, 1988).

Main species	Cover percentage, %		
	Under crown cover	At cuttings	
		After 1 year	After 12 years
Lathyrus humilis	15	5	1
Pulsatilla multifida	10	10	13
Fragaria orientalis	10	5	3
Poa attenuata	<1	<1	1
Artemisia tanacetifolia	5	5	5
A. sericea	5	3	2
Vicia unijuga	5	5	4
Galium boreale	<1	<1	1
Bromopsis sibirica	4	4	5
Dendranthema zawadskii	<1	5	1
Saussurea elongata	2	<1	<1

Table 3.2. Continued

Main species	Cover percentage, %		
	Under crown cover	At cuttings	
		After 1 year	After 12 years
Festuca ovina	2	5	13
Scorzonera radiata	<1	<1	1
Carex lanceolata	<1	2	5
Anemone crinita	<1	1	1
Vicia megalotropis	<1	3	5
Sanguisorba officinalis	–	<1	1
Similarity with forest, %:			
specific composition	100	82	84
cenotic structure	100	62	55.4

If a final felling coincides with an arid year cycle, a meadow steppe may develop in the absence of young larch trees. The replacement of forest vegetation by meadow vegetation is common with steppe larch forests of lower mountain zones. In such plots forest herbs rapidly lose their ecological importance and occasionally are completely eliminated from the herb stand. By contrast, the abundance of small-sod grasses, primary *Festuca ovina* and *Carex lanceolata*, rapidly increases. This brings about problems in the regeneration of larch – either a forest-steppe complex develops, or the original forest-covered area is occupied by steppe vegetation. In marked contrast, selective fellings fail to cause substantial changes in forest vegetation and virtually do not affect natural or anthropogenic dynamics of forests.

In the eastern Selenga basin pine forests are influenced mainly by fires or haphazard fellings (field, selective, or occasionally clear). And to a great extent they also suffer from grazing by domestic animals, so characteristic of forests near built-up areas. In these cases a large number of weeds also

Figure 3.8. Zoogenic erosion on forested slopes (photo by A.V.Prishchepa).

Figure 3.9. Grass cover of pine forest after fire (photo by A.V.Prishchepa).

appear *Urtica cannabina* being particularly abundant. Weedy vegetation and numerous livestock paths markedly change forest-growing conditions. The surface soil horizons are compacted substantially, the moisture supply deteriorates, the litter is disturbed, and erosion processes develop (Figure 3.8). Forest regeneration is considerably hampered, occasionally completely absent, and seedlings are trampled.

In clear fellings pine forests often are replaced by small-leaved forests or mixed pine-birch forests. In steppe pine forests, disturbances lead to further acquisition of steppe features and to the development of steppe communities at forest sites. At the same time, on north and north-east-facing slopes in the northeastern Khentii at sites where pine forests develop with an undergrowth of *Rhododendron* sp. after fellings, the primary species regenerate satisfactorily (Savin & Dugarzhav, 1980).

Degradation of forest ecosystems becomes acute upon strong disturbances of tree stands, e.g., mortality of over 50% after felling or fires. But the degree of degradation for forest community successions depends not only on the species and the extent of disturbance, but also on the stand of trees. Forest disturbances that are not associated with a high tree mortality, for instance, running ground fires or restrictive selective felling, only result in plant cover changes and changes in the forest type or subtype (Figure 3.9).

However, the regeneration of forest community successions proceeds fairly slowly. Even without replacement of the primary arboreous species, the recovery of all forest vegetation functions takes 60 to 100 years (Table 3.3). And, according to Yu.N. Krasnoshchekov et al. (1990), the regeneration of vegetation versus soil cover is not synchronized. A still greater difference occurs in the slow regeneration of all forests

Table 3.3. Period of renewal for forest community successions and their water-regulating functions (Krasnoshchekov et al., 1990).

Factors of dynamics	Direction of phytocenotic changes	Renewal Period:		
		Phytocenoses	Soils	Water-regulating function
		Years		
Extensive fires	Without changes of tree species	60–100	100	100
Forest clearings	Without changes of tree species	100	100	100
	Through coniferous-leaved plantations	130	80	150
	Through short-period derivative birch stands	160	40	200
	Through long-period derivative birch stands	300	40	300

protective properties, including its regulation of water.

The present-day condition of Mongolia's forest is strongly determined by the effects of human activities over the last 100 years (human-associated forest fires, final felling, grazing, transport vehicle pressures, etc.). Destabilizing effects in the disturbed forest communities also are caused by outbreaks of forest insect pests (Zhukov & Savin, 1980). Presently disturbed forests account for about 40% of the entire forested area of Mongolia. Their destruction as a result of aggressive vegetation successions is shown in Table 3.4.

The least disturbed are open woodlands under bald mountains, which are most distant from built-up areas and hard to access. At the same time, thanks to moisture, they have little fire hazardous.

Pine forests and larch forests in the lower forest zone and forest steppe are the most affected. They are located near built-up areas, frequently neighboring areas of intensive

Table 3.4. Changes of forest areas due to degrading alterations (according to Korotkov & Krasnoshchekov, 1992; Krasnoshchekov et al., 1990, 1992).

The state of disturbed forest areas	Areas of changed forest communities, th. ha						
	Cedar and larch open woodlands	Cedar, cedar-larch communities		Larch, dry moss commu-nities	Larch, birch-larch grass-moss communities	Pine, larch, birch grass communities	Total
		Grass - green moss	Dwarf shrub - green moss				
Burns with no plant renewal	6.4	194.7	61.7	22.2	87.4	311.9	684.3
Cuttings with no plant renewal	–	–	12.5	36.4	27.9	172.5	249.3
Young growth of *Populus tremula*	–	–	18.7	–	–	180.9	199.6
Young growth of *Larix dahurica*	9.6	282.9	27.8	43.8	390.2	311.0	1065.3
Young growth of *Pinus sibirica*	9.8	111.2	44.2	–	–	–	165.2
Birch forests instead of *Pinus sylvestris*	–	–	–	–	–	686.7	686.7
Birch forests instead of *Larix dahurica*	–	132.0	1.9	–	54.6	546.2	734.7
Birch forests instead of *Pinus sibirica*	–	60.0	55.7	–	–	–	115.7
Stony placers	1.7	5.6	1.7	12.2	12.2	23.2	56.0
Steppe communities instead of forests	0.2	–	–	41.3	41.3	59.4	103.7
Total	27.7	786.4	224.2	55.9	574.5	2291.8	4060.5

livestock breeding and farming. These forests have a particular fire-hazard.

Burns and felled areas that were not regenerated account for about 20% of the original forest area, i.e., about 930 thousand ha. Consequently, we believe that only one third of these areas has a potential for recovery as primary forests. About one half will be taken up by secondary birch forests for a long time. In the remaining area, meadow and steppe communities and rock streams will develop on steep slopes. A recovery of primary forest communities (without a succession of forest-forming species) is observed roughly in only 30% of the forest area. Apparently they will fully regenerate in the absence of recurrent fires.

Secondary small-leaved birch and aspen forests are widely distributed – they account for about 40% of the forest area. They emerge at taiga larch forest sites, and may turn into primary. Larch forests sites with steppe elements can exist indefinitely. This is promoted by recurrent fires and gypsy moth outbreaks that destroyed the undergrowth of coniferous forests. In an area of about 160 thousand ha, original-forested areas have been replaced by rock streams, steppes, meadows and shrub thickets. Natural forest regeneration in such former forest communities is impossible because natural forest breeding is not feasible.

Human-associated disturbances of forest communities essentially destroys their protective functions, particularly, water conservation and water regulation. Whereas under an undisturbed forest canopy, the proportion of liquid surface runoff accounts for only 2 – 6% of the total precipitation, in felled areas and in burns runoff increases to 60 – 80%. Estimates show that with a 10% decline of forested area, the magnitude of mean annual surface liquid runoff increases by 15 – 25 mm. When large forest areas are disturbed, this results in the formation of heavy floods in rivers and to the drying up of small streams and rivers during low-water periods (Krasnoshchekov, 1992; Krasnoshchekov & Korotkov, 1992). The destruction of upper sandy horizons caused by mechanical skidding and intense ground or tree-crown fires, results in the removal of plant materials for soil formation in the surface runoff that causes a strong degradation of soils and soil cover plus a disasterous deterioration of their forest-growing properties (Krasnoshchekov, 1980). This undoubtedly affects the processes of forest regeneration by prolonging them and by lowering forest productivity. Over the last 20 to 25 years over 684 thousand ha of non-regenerated burns and over 200 thousand ha of felled areas have accumulated. Evenso the current rates of anthropogenic forest degradation is not declining!

Therefore, natural-anthropogenic and anthropogenic factors cause different forest successions, amongst which we have noted:

- replacement of forest communities by petrophile plants on rock streams or unfixed sands;
- replacement of forest communities by herbaceous (meadow, steppe) or shrub thickets;
- replacement of one forest type by other semi-weed relatives;
- replacement of primary coniferous species by secondary leaved (birch and aspen) forests plus changes in the herbaceous cover;
- regenerative successions without much change in plant structure.

Along with forest communities, mountain slopes also are inhabited by steppes and forest-steppes. For herbaceous vegetation in these zones the most substantial changes are those caused by increased grazing pressure from livestock breeding.

In north-eastern Khangai in the forest-steppe zone of middle mountains, grazing on steppe plots reduces the species composition and the contribution of wormwoods increases. The development of communities dominated by *Artemisia frigida* is particularly characteristic.

Mountain steppes are exposed to a still greater extent to human-induced effects, since these are the main summer livestock grazing ranges. Intensive grazing causes changes in the

Table 3.5. Degrading alterations of forb-fescue mountain steppe in Mongolian Altai due to grazing intensification (according to Gunin & Vostokova, 1993).

Impact	Plant communities	Total percentage of coverage, %	Participation in the cenosis (%) of:	
			Grasses	Warm-woods
Increase in pasture load	*Festuca lenensis* + herb-sedgegrass	80	60	–
	Festuca lenensis + herb-sedgegrass + *Artemisia frigida*; *Artemisia frigida* + *Carex duriuscula*	60	40	15
	Festuca lenensis + herb-sedgegrass + *Festuca lenensis* + *Artemisia frigida* + herb-sedgegrass; *Artemisia frigida* + *Festuca lenensis*	60	5–10	40
	Artemisia frigida + *Potentilla* sp.	40	<5	50
	Artemisia frigida, A. sericea, A. annua + *Chenopodium album*; *Artemisia frigida, A. annua* + *Leymus chinensis*	20	<5	>50

rich forb feather grass steppes dominated by *Stipa krylovii* plus increasing the contributions of rock inhabiting and high-mountain forbs (*Pedicularis achilleifolia, Silene repens, Lagotis integrifolia, Bistorta vivipara, Schultzia crinita, Minuartia verna, Leontopodium ochroleucum*) and encourages the development of impoverished wormwood and cryophilic wormwood-cinquefoil communities. At permanent animal holding sites bare fields develop almost free from vegetation. Such an aggressive replacement of steppe vegetation was observed in Mongolian Altai (Table 3.5). A large replacement of forb-wheat grass (*Agropyron cristatum*) communities by wormwoods (*Artemisia frigida, A. dracunculus,* and *A. santolinifolia*) also was recorded by E.A. Volkova (1994) for the Gobi Altai.

Similar changes in meadow-steppe communities were recorded in Khentii. In the upper Kerulen basin with mountain meadow steppe vegetation, an increase in grazing pressure resulted in a gradual decline of the cover and at the same time an increasing role of aggressive plant species (Figure 3.10). However, one indice of initial stages of such successions is a penetration into the meadow steppe communities by *Artemisia frigida*, which is regarded as one indicator of pastor degradation (Gunin & Vostokova, 1989).

The lower zones of mountain slopes, normally taken up by steppe and dry stand communities, are characterized by a replacement of mesophilic vegetation with xerophilic as determined by cold aridization (Bannikova, 1983). In these forb-large-sod-grass steppes dominated by *Stipa baicalensis* and forbs (*Serratula centauroides, Echinops latifolius, Bupleurum scorzonerifolium, Heteropappus hispidus, Potentilla conferta*) when under intensive grazing, wormwoods appear and feathergrass is replaced by more grazing-resistant *Cleistogenes squarrosa, Leymus chinensis,* and *Carex duriuscula*. At permanent camps and in enclosures, wormwood-weed communities form, including *Artemisia scoparia, A. macrocephala, Chenopodium album, Hyoscyamus niger*, and *Urtica cannabina*. Along side many dirt roads forb-feathergrass steppes are replaced by grasses-wormwood with a considerable contribution from semi-weed species (*Potentilla multifida, P. supina, Rheum undulatum, Salsola collina, Sphallelocarpus gracilis, Chenopodium acuminatum, Axyris*

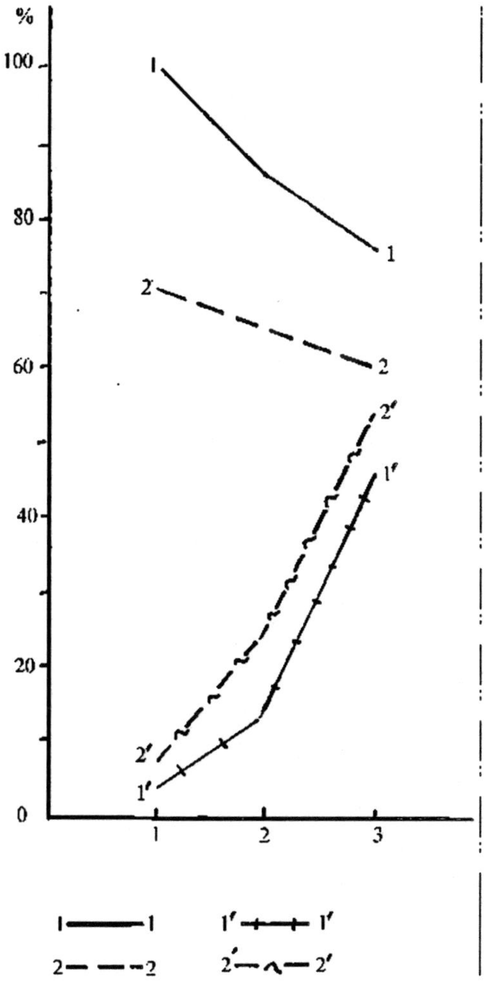

Figure 3.10. Changes in cover percentage of mountain-meadow steppes as a result of grazing intensification (according to the materials of Mungen-Mort station).

1–2 Total coverage of: 1 – mountain – meadow steppe (*Poa attenuata, Festuca lenensis, Koeleria cristata, Agrostis trinii, Helictotrichon schellianum*), 2 – petrophyte – mountain meadow steppe. $1^1 - 2^1$. Coverage by digressive species: 1^1 – at mountain – meadow steppe; 2^1 – at petrophyte mountain meadow steppe.

hybrida, Fallopia convolvulus, Urtica cannabina and others).

With the same type of anthropogenic effects in mountain steppe zones the relationship between the replacement of plant communities and soil-ground conditions is very pronounced. I.A. Bannikova (1983) distinguished the following succession series under such increased human-associated pressures:

– On rocky-sandy soils under human-induced removal of fine earth, an increase in petrophyte abundance was observed (*Poa attenuata, Thymus gobicus*).

– On sandy deposits occupied by grass-forb steppes involving *Caragana microphylla*, intensive grazing induced wind erosion of sands; initially *Caragana stenophylla* appeared and subsequently, feathergrasses were replaced by *Leymus racemosus, Calamagrostis epigeios*; and on semi-broken sands *Ulmus pumila* appeared with some individual grasses and forbs (*Leymus racemosus, Calamagrostis epigeios, Thalictrum squarrosum, Thymus* sp.).

– In depression soils with a heavy surface, salinization increased and *Stipa krylovii, Cleistogenes squarrosa*, and *Koeleria cristata* were replaced by other species - also *Leymus chinensis* and *Carex duriuscula* appeared plus numerous semi-weed forbs (*Serratula centauroides, Thermopsis lanceolata, Artemisia scoparia, A. adamsii, Cymbaria daurica, Medicago ruthenica, Chenopodium aristatum*).

In the northeastern Khangai forb-large-sod grass steppes dominated by *Stipa baicalensis* and forbs (*Serratula centauroides, Echinops latifolius, Bupleurum scorzonerifolium, Heteropappus hispidus, Potentilla conferta* and others), intensive grazing induces an emergence of wormwoods; and feathergrass was replaced by more grazing resistant *Cleistogenes squarrosa, Leymus chinensis*, and *Carex duriuscula.* At camps and near enclosures wormwood-weed groupings form, including *Artemisia scoparia, A. macro-cephala, Chenopodium album, Hyoscyamus niger*, and *Urtica cannabina*. Along many dirt roads forb-feathergrass steppes wide bands were replaced, by vostrets-wormwood, involving to a great extent semi-weed wormwood (*Potentilla multifida, P. supina, Rheum undulatum, Salsola collina,*

Sphallerocarpus gracilis, Chenopodium acuminatum, Axyris hybrida, Fallopia convolvulus, Urtica cannabina and others).

Similar plant community successions also have been observed in mountain forb-tyrsa meadow petrophyte steppes. But dry sod-grass steppes especially are exposed to anthropogenic disturbances mainly associated with the lower zone of middle mountains. Here the most affected community is wheatgrass (*Agropyron cristatum*). Under moderate and heavy grazing they are replaced by communities with a dominance of *Artemisia frigida, Potentilla acaulis, Iris tenuifolia, Oxytropis aciphylla* and other unpalatable semi-weed species. Under intensive grazing the small-sod-grass-forb steppes of northern-eastern Khangai foothills are replaced by more xeromorphic communities dominated by *Koeleria cristata, Cleistogenes squarrosa* and *Carex duriuscula*. New abundant forb species also emerge naturally (*Thermopsis lanceolata, Aconogonon angustifolium, Erysimum flavum,* and *Artemisia frigida*).

Overall, different subzones of mountain forests are characterized by an elimination of grass species even under a minor increase in rangeland pressures. In the next stage, new communities are formed with a dominance of wormwoods and subsequently weed species. Normally the abundance of xeromorphic species increases and mostly semishrubs dominate.

Such regressive successions of vegetation in mountain lower zones are found virtually ubiquitously in Mongolia. Anthropogenic effects, which aggravate climate changes to promote dryness and dryland features, cause dry-steppe complexes to development. As a result species develop that are less valuable as livestock fodder and the altitudinal boundaries of dry steppe zones expand.

Thus, anthropogenic successions of steppe vegetation are essentially regressive. However, in a number of cases one can observe an expansion of steppe communities involving original forest habitats upon strong disturbances of the trees.

In addition artificial irrigation of mountain ranges, as practiced in certain Mongolian Altai valleys, entails gradual changes in the plant cover. Irrigation is mainly by a redistribution of river water. With such plots an increase in plant growth by 3 to 4 times is recorded even the first year of additional moisture. In this case new modifications of plant communities emerge. For instance with *Stipa glareosa* steppes of the Saksai Depression in Mongolian Altai under the effect of irrigation, *Agropyron cristatum* appeared but the number of *Stipa glareosa* declined. Subsequently under both irrigation and grazing the grass-forb steppe was succeeded by a forb-sedge community dominated by *Carex stenophylloides* and *C. duriuscula*. With poorly controlled irrigation, salinization begins and dry steppe communities are replaced on saline soil meadows with *Puccinellia tenuiflora* and others.

In lower mountain zones, steppe vegetation successions are directly associated with anthropogenic effects in areas adjacent to intermountain depressions and are similar to steppe plains community successions.

3.3 PLANT COMMUNITY DYNAMICS IN PLAINS AND ROCKY AREAS

The vegetation of plains and rocky topography is exposed to modifications by man to the greatest extent. These territories not only are used intensively as ranges, but here non-irrigated farming is very common, plus many residential areas, and a multitude of unpaved dirt rut roads and a few paved highways. Few people live in deserts, nevertheless, human-induced successions are almost ubiquitous, though they are mostly of a local and focused nature. All of this cause replacement of certain plant communities by others, i.e., predetermines the trend of plant community dynamics. But along with human activities these areas have characteristic natural factors of plant cover successions. Such natural factors include the formation of soil and plant cover at cliff rock outcrops, overgrowth of sand sediments, and the effects of burrowing animals.

The steppe vegetation of Mongolia has a number of specific features, which must be taken into account in developing general approaches for investigations on the natural state and dynamics of plant communities and their response to human modifications. Also there are major climatic and geological factors in steppe vegetation development (Gunin & Vostokova, 1993). The most important climatic factor in plant cover dynamics is a cyclic alternation of moist and arid years. Such cycles result in large external and internal changes in soils and in the structure and composition of plant communities. Amongst the geological factors, which largely determine the peculiarity of soils and vegetation is a poor development of the soil mantle. Thus bedrock occurs close to the surface, plus salinization and carbonate patterns and rock debris patterns occur in surface soil horizons.

Such geological features are indicative of the relative youth of Mongolian steppe communities. Even in Holocene, steppe development was mostly associated with geological processes, e.g., the development of waste (soil) mantle, washing, erosion, and the redistribution of silts (Dinesman et al., 1989). Most of these factors prevent a deep penetration of plant root systems into soil and relates to the specific physiology of its vegetation.

Natural vegetation successions determined by erosion and soil formation processes are observed at cliff rock outcrops and their eluvium. At the first succession stage polydominant groups of petrophytes are established, and at the final stage representative plant communities are established. The particular vegetation species composition of each succession series in particular subzones are to a huge extent affected by the geochemical and lithological composition of the soil-forming bedrock in each subzone (Volkova, 1976). Similar to all natural successions they proceed at a very slow rate, and the substitution of one member of the series by another can only be established via extrapolation of spatial series to temporal changes. These successions have been examined in greatest detail in desertified steppes in steppe-like and in true deserts with rocky topography areas where plant

Table 3.6. Final stages of a calciphyte-psammo-petrophyte succession sequence (according to Volkova, 1976).

Sub-zones	Plant communities at eluvial substratum of:		
	Basalts	Tuffs and lavas of basal composition	Marbleized limestones
Semidesert steppe (northern desert)	*Stipa gobica* + *Lagochilus ilicifolius*, *Limonium tenellum*, *Krascheninnikovia ceratoides*		*Stipa gobica* + *Ajania fruticulosa*, *Krascheninnikovia ceratoides*
Steppized desert (medial desert)		*Anabasis brevifolia* + *Stipa glareosa* + *Krascheninnikovia ceratoides*, *Oxytropis aciphylla*, *Allium mongolicum*, *Zygophyllum pterocarpum*	*Salsola laricifolia* + *Stipa glareosa*, *S. gobica*, *Caryopteris mongholica*, *Lagochilus ilicifolius*, *Scorzonera capito* (Alashan Gobi, Dornod Gobi)
True desert (southern desert) (Transaltai Gobi, Alashan Gobi)	*Sympegma regelii* + *Anabasis brevifolia*, *Rheum nanum*, *Ajania achilleoides*, *Krascheninnikovia ceratoides*	*Sympegma regelii* + *Nitraria sphaerocarpa*, *Ephedra przewalskii*, *Oxytropis aciphylla*, *Convolvulus gortschakovii*	*Sympegma regelii* + *Salsola arbuscula* + *Stipa glareosa*, *Ptilagrostis pellioti*, *Ajania achilleoides*, *Ephedra sinica*, *Caryopteris mongholica*

Table 3.7. Final stages of plant succession sequenes at shales eluviums and salinized sandstones in deserts (according toVolkova, 1976).

Sub-zones	Plant communities at eluvium of:	
	Shales	Salinized sandstones
Semidesert steppe (northern desert)	*Stipa gobica* + *Limonium tenellum, Ajania fruticulosa, Allium polyrrhizum* + *Anabasis brevifolia, Reaumuria songarica, Salsola passerina*	*Potaninia mongolica* + *Zygophyllum xanthoxylon, Asterothamnus centrali-asiaticus* (Alashan Gobi, Dornod Gobi)
Steppized desert (medial desert)	*Anabasis brevifolia, Salsola passerina, Reaumuria songarica* + *Stipa glareosa, Rheum nanum, Arnebia guttata, Allium polyrrhizum*	*Potaninia mongolica* + *Salsola passerina, Anabasis brevifolia, Reaumuria songarica* (Alashan Gobi, Dornod Gobi) *Sympegma regelii* + *Anabasis brevifolia, Salsola passerina* (Transaltai Gobi)
True desert (southern desert)	*Reaumuria songarica, Salsola passerina, Anabasis brevifolia, Asterothamnus molluscus, Allium polyrrhizum, Limonium tenellum*	*Salsola passerina* + *Potaninia mongolica* (Alashan Gobi, Dornod Gobi)
Extra arid desert	*Iljinia regelii (Nitraria sphaerocarpa)*	

community successions are the most dynamic. In rocky topography of the steppe zone, natural successions have been examined with special reference to the processes of erosion and the formation of soils on the rock eluvium (Volkova, 1976). For example in desertified steppes the cliff crevices of large-grain granites are populated by *Artemisia pamirica*, *A. xanthochroa* and *Caragana leucophloea*. The formation of brown soils is associated with the development of a feathergrass-petrophyte-forb community, including *Stipa gobica, Haplophyllum davuricum, Astragalus variabilis, Oxytropis tragacanthoides, Clematis fruticosa* and others. On limestones and marbleized limestones other features distinguish the species composition (Table 3.6).

E.A. Volkova (1976) suggested some plant indicators, such as of carbonate rocks, are species as *Zygophyllum pterocarpum, Tugarinovia mongolica*; or a steppe-like desert zone by *Ajania achilleoides, Ephedra przewalskii, Sympegma regelii, Salsola arbuscula* and *Ephedra sinica*. With the development of a pale-brown soil cover on steppe-like deserts, feathergrass-saltwort communities develop (*Salsola laricifolia, Stipa glareosa, S. gobica*); and on gray-brown desert soils, grass-saltwort-*Sympegma* communities develop (*Sympegma regelii, Salsola arbuscula, Stipa glareosa*). In eastern Gobi regions winter-fat – bean-caper communities develop on such rocks (*Zygophyllum xanthoxylon, Krascheninnikovia ceratoides*). Still more substantial differences are found with halophilous – petrophilous vegetation developing on shale deposits of gypsum clays (Table 3.7). On the eluvium of these rocks the succession series opens with gypsophyte groups. South of desertified steppes *Salsola passerina, Reaumuria songarica* and *Anabasis brevifolia*, and more rarely only in Gzungarian Gobi *Anabasis truncata* and *A. eriopoda* are common. Developing ubiquitously on these habitats are communities and groups of more xeromorphic plants than those characteristic of the zone. The final communities in these succession series are: on brown soils of desertified steppes, perennial saltwort-petrophyte-forb-feathergrass (*Stipa gobica, Limonium tenellum, Ajania fruticulosa, Allium polyrrhizum, Anabasis brevifolia, Reaumuria songarica, Salsola passerina*); on steppe-like deserts, *Anabasis* and perennial saltworts dominate to form a community including

Anabasis brevifolia, Salsola passerina, Reaumuria songarica, and *Stipa glareosa*; in true deserts on gray-brown soils on the eluvium of shales, the community only retains perennial saltworts, including *Asterothamnus molluscus, Limonium tenellum* and *Allium polyrrhizum*; and in extra-arid gray-brown soils, the final community in the succession series on the eluvium of gypsyferous rocks are sparse groupings of *Iljinia regelii*, occasionally including *Nitraria sphaerocarpa* (Volkova, 1976).

On a salinized sandstone eluvium the final succession stage is characterized by intermediately mixed communities of halo-psammophyte vegetation (Table 3.7), for which regional differences are very common (Volkova, 1976). On desertified steppe brown soils in the Alashan and eastern Gobi, the community is *Potaninia mongolica*, plus *Zygophyllum xanthoxylon* and *Asterothamnus centrali-asiaticus*. On steppe-like desert pale-brown soils in the same sector, perennial saltwort-*Potaninia* communities develop (*Potaninia mongolica, Salsola passerina, Anabasis brevifolia, Reaumuria songarica*), and in similar habitats of Transaltai Gobi the final succession stage is borbudurgan-baglur-*Sympegma* communities (*Sympegma regelii, Anabasis brevifolia, Salsola passerina*). The final stages of these successions are listed in Tables 3.6 and 3.7.

Despite some species differences, similar to the entire vegetation succession series that develop on salinized rock eluviums, these communities are characterized by sparseness (the total plant cover does not exceed 10%), a presence of typical halo- and gypsophytes, xeromorphic plants, and a minor presence of other forbs and grasses.

The dynamics of vegetation evolution on rock eluviums is strongly affected by soil-formation processes from the vegetation itself, i.e., vegetation itself contributes to the endogenous dynamics of plant cover formation.

On steppe plains burrower mammals play a large role in plant cover successions (Dmitriev & Guricheva, 1978; Dmitriev et al., 1990; Kozhemyakin, 1978). Burrower colonies occupy an area ranging from 3 to 900 m^2 (Tables 1.2 & 1.3). Brandt's vole burrows frequently grow *Caragana microphylla, Leymus chinensis, Chenopodium album, C. acuminatum* and *Artemisia macrocephala* on forb-grass-peas-shrub steppes (Dmitriev & Guricheva, 1978). In pika burrows of feathergrass steppes, these authors report a dominance of *Artemisia frigida, A. glauca, Leymus chinensis, Echinops latifolius, Axyris amaranthoides* and *Chenopodium album*. Such burrows give steppes a peculiar micro-complex pattern, particularly when the animal colony numbers are high.

The plant cover micro-complex pattern mostly is caused by burrowing, trophic level and other metabolic activities of plant eating mammals and active burrowers (Dmitriev & Guricheva, 1978, 1983a, b; Dmitriev & Shauer, 1987; Dmitriev & Khudyakov, 1989; Dmitriev et al., 1990). On the burrows, succession is triggered by lowering the threshold for a limiting factor as a result of zoological induced bedrock erosion. Burrowers bringing rock debris material onto the soil surface promotes its rapid breakdown and a leaching of carbonates. Thus burrows change the general ecological conditions and increase aeration of plant habitats causing the development of specific mico-ecosystems. For example, maps (Figure 3.11) show these microcomplexes that changed the original structure of the tansy-forb and forb-feathergrass steppes of Eastern Mongolia (Dmitriev et al., 1990).

Therefore these animal activities lead to different plant succession stages. Dmitriev et al. (1990) believed that burrowers intensify the processes of bedrock erosion, forming thick non-lithic soils. In Eastern Mongolia, on a sheep's fescue -petrophyte-forb steppe with thin soils (*Festuca lenensis, Cleistogenes squarrosa, Poa attenuata + Carex duriuscula + Orostachys malacophylla, Potentilla leucophylla, P. verticillaris* and others), removal of rock debris from the burrows reached 2 kg/m^2. On fresh burrows *Leymus chinensis* dominates; and on abandoned burrow

Figure 3.11. Zoogenic microcomplexes of forb – feathergrass steppe (according to Dmitriev & Guricheva, 1983; Dmitriev et al., 1989 and others).

1–2. Unchanged sites: 1 – feathergrass (*Stipa grandis, S. sibirica*); 2 – forb – feathergrass (*Stipa grandis, S. sibirica + Serratula centauroides*). 3–6. Sites changed by fossorial animals: 4 – with pea shrubs (*Caragana microphylla*), 5 – with feathergrass (*Stipa grandis*) and participation of *Caragana microphylla*; 6 – with feathergrass-forb cover (*Stipa grandis, S. sibirica + Serratula centauroides, Potentilla tanacetifolia, Artemisia frigida* with participation of *Leymus chinensis*).

soils, *Caragana stenophylla, C. microphylla, Artemisia frigida, Kochia prostrata* and *Stipa krylovii* dominate. Apparently there is a direct transfer from a petrophyte steppe to sod-grass. Similar zoogenic vegetation successions have been observed on arid tansy sod-grass steppes with petrophyte forbs on mealy-carbonate soils; and also on forb – feathergrass steppes with pea-shrub soils (Table 3.8). Similar zoogenic successions have been recorded for other steppe regions in Mongolia, particularly on the undulating plains northeast of Khangai (Figure 3.12).

The tarbagan marmot is big burrower with still greater effects. The material removed from its burrows often are distinctive due to the dark green color of pea-shrubs and weeds. A great abundance of animals (*Ellobius talpinus, Lasiopodium brandtii*) in a colony promotes not only a microcomplex pattern of plant communities showing different succession stages on the mounds or removed material, but hillocks also are formed that make the complex pattern of the vegetation still more showy (Figure 3.13).

Observations made at the Tumentsogt Station in Eastern Mongolia (Dmitriev et al., 1990) show these vegetation dynamics trends. Extensive activity of burrowing mammals there resulted in the formation of feather-grass-pea-shrub communities on thick non-lithic non-carbonate soils. This generalized succession series is given in Table 3.9 (after Dmitriev et al., 1990).

On desert plains where human pressure is not so great, vegetation dynamics as determined by natural factors can be seen clearly; but, on the other hand, the conditions for water consumption by plants are very difficult. On these plains under protracted droughts (lasting 4–5 years running or more) the plant cover shows degradation processes. At the beginning a decrease in vegetative mass occurs and generative shoots are absence on

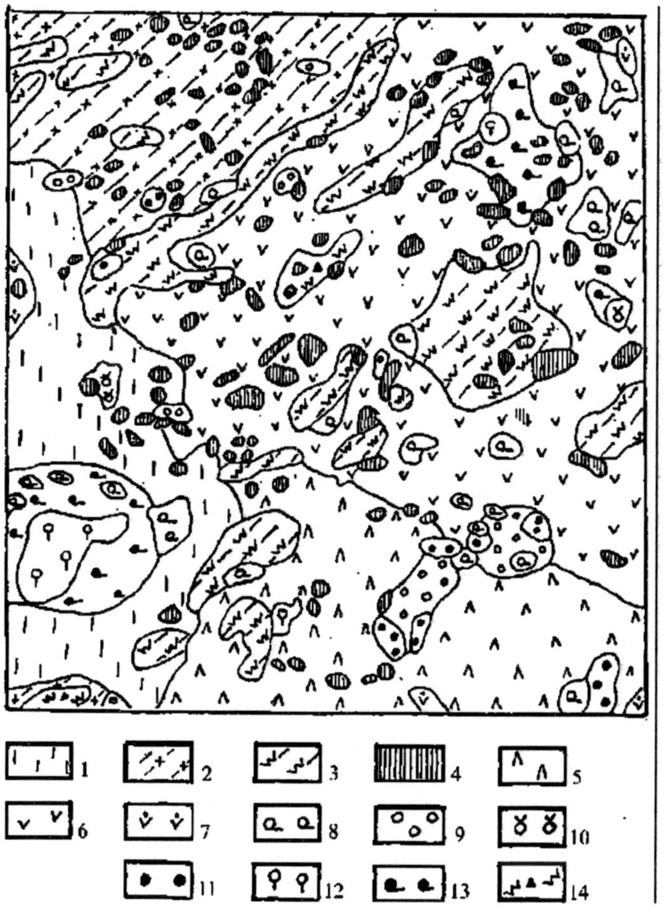

Figure 3.12. Zoogenic microcomplexes of forb – feathergrass steppe in Eastern Khangai (according to Guricheva & Buevich, 1986)

Plant communities: 1–4. Unchanged sites covered by: 1 – *Stipa baicalensis, Festuca kryloviana* + *Carex pediformis* + *Potentilla tanacetifolia, Dendranthema zawadskii, Oxytropis microphylla* with participation of *Scabiosa comosa, Bupleurum scorzonerifolium, Halenia corniculata*, 2 – *Stipa krylovii, S. baicalensis* + *Festuca kryloviana, F. sibirica* + *Potentilla tanacetifolia, Potentilla fruticosa*, 3 – *Festuca lenensis* + *Orostachys spinosa, Arenaria capillaris, Androsace incana, Potentilla sericea, Artemisia commutata, Halenia corniculata*, 4 – *Potentilla fruticosa* + *Potentilla tanacetifolia, Galium verum, Silene jeniseensis*. 5–8. Sites with shrew activity (initial stages of zoogenic successions): 5 – *Potentilla tanacetifolia* + *Stipa baicalensis* + *Scabiosa comosa, Aconogonon angustifolium, Trifolium lupinaster, Vicia multicaulis, Echinops latifolius*, 6 – *Stipa baicalensis, Festuca kryloviana* + *Potentilla tanacetifolia, Delphinium dissectum, Sanguisorba officinalis, Valeriana alternifolia, Thalictrum minus, Trifolium lupinaster*, 7 – *Carex pediformis* + *Oxytropis microphylla, Bupleurum scorzonerifolium* with participation of *Stellera chamaejasme, Halenia corniculata, Stipa baicalensis*, 8 – *Potentilla fruticosa* + *Echinops latifolius, Elymus gmelinii* + *Schizonepeta multifida, Scabiosa comosa*. 9–13. Sites changed by activity of fossorial animals (mature stages of zoogenic successions): 9 – *Galium verum, Thalictrum petaloideum* + *Stipa baicalensis* with participation of *Potentilla tanacetifolia, Echinops latifolius*, 10 – *Echinops latifolius, Adenophora stenanthina, Schizonepeta multifida, Delphinium dissectum* + *Carex pediformis*, 11 – *Schizonepeta multifida, Adenophora stenanthina, Delphinium dissectum* + *Elymus gmelinii, Bromopsis inermis*, 12 – *Echinops latifolius* + *Elymus gmelinii, Bromopsis inermis* + *Delphinium dissectum, Valeriana alternifolia, Sanguisorba officinalis, Rheum undulatum, Artemisia changaica*, 13 – *Sanguisorba officinalis, Potentilla tanacetifolia* + *Stipa baicalensis* with participation of *Echinops latifolius, Schizonepeta multifida*, 14 – covered stones.

Table 3.8. Plant Succession sequence on Eastern Mongolia steppes depending on soil transformation by fossorial animals (*Lasiopodomus brandtii* Radde, *Ochotona daurica* Pall., *Microtus gregalis* Pall.) (according to materials of Dmitriev et al, 1990).

Stage	Soils	Vegetation	
		Cover, %	Specific composition
I	Stony – detrital powder-calcareous	18–35	*Filifolium sibiricum, Poa attenuata, Festuca lenensis, Koeleria cristata, Medicago ruthenica, Potentilla acaulis, P. leucophylla, P. verticillaris, Arenaria capillaris, Orostachys fimbriata, O. malacophylla*
II	Detrital, shallow dark chestnut	45–70	*Stipa grandis, S. krylovii, S. sibirica, Leymus chinensis, Agropyron cristatum, Poa attenuata, Potentilla tanacetifolia, Galium verum, Cymbaria daurica, Serratula centauroides*
III	Sceletonless, non-carbonate, deep dark chestnut	52–70	*Caragana microphylla* at fresh colonies: *Artemisia scoparia, A. dracunculus;* at elder colonies: *Stipa grandis, S. sibirica, Agropyron cristatum, Potentilla tanacetifolia*

Table 3.9. Generalized steppe vegetation successions and soil types occurring due to fossorial animal activities (according to materials of Dmitriev et al., 1992; Khramtsov et al., 1992 and others)

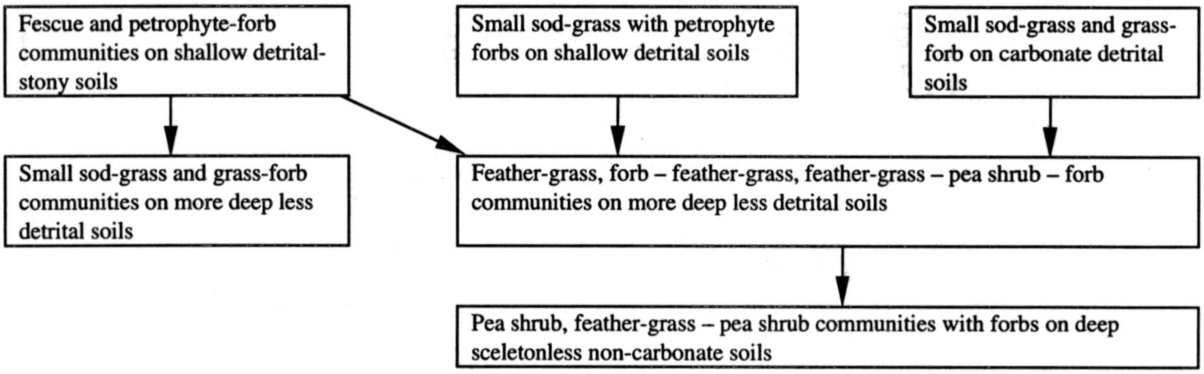

Table 3.10. Haloxylon Dynamics in deserts over 12 years (Kazantseva et al., 1992).

Sub-zone	Species	Relief	1978		1989	
			Pieces/are	Cover, %	Pieces/are	Cover, %
True desert	*Haloxylon ammodendron*	Watershed	0.77	0.419	0.0	0.0
		Depression	3.72	2.45	1.0	1.95
Extra-arid desert	*Haloxylon ammodendron,*	Saira	2.45	2.65	2.2	2.52
	Ephedra przewalskii,		0.67	0.5	0.68	0.73
	Calligonum mongolicum,		0.02	0.03	0.03	0.02
	Ajania fruticulosa,		0.12	0.02	0.2	0.02
	Sympegma regelii		0.35	0.15	0.22	0.12

Figure 3.13. Zoogenic micro-hills in steppe (photo by A.V.Prishchepa).

the main community formers. Thus, according to station observations from 1971 to 1984 during arid years the total phytomass diminished considerably (Lavrenko & Bannikova, 1986; Sokolov & Gunin, 1986). Over 12 years (1978–1989) the increment of saxaul growth in true deserts decreased from 2.5 to 0.5 centners/ha; the increment of saxaul annual shoots in extra-arid deserts was still lower: from 32 to 3 kg/ha (Kazantseva et al., 1992). However, according to the same authors, *Ephedra przewalskii* over the same period somewhat increased the number of individuals per unit area and its total cover (Table 3.10). Hence we believe that with the current stable trend towards a precipitation decrease in the Transaltai Gobi (Table 1.1), communities will lose mesoxerophilous species, and their niche taken by xeromorphic plants.

Indeed in 1991 in the lower zone of the low mountains of Gobi Tyan-Shan, P.D. Gunin observed a range expansion by *Ephedra sinica* which, prior to that, was only recorded in low abundance on stony cliffs (Gunin et al, 1993). Earlier the contribution of *Ephedra sinica* to plant communities did not exceed 1% (Yunatov, 1954; Volkova, 1976; Banzragch et al., 1978). In 1991 we observed onion-*Ephedra*-feathergrass communities (*Ephedra sinica* + *Stipa gobica* + *Allium polyrrhizum*) and *Ephedra*-feathergrass-baglur communities (*Anabasis brevifolia* + *Stipa gobica* + *Ephedra sinica*). *Ephedra* reproduces vegetatively and grows through drought-weakened feathergrass sods (*Stipa gobica, Stipa glareosa*) and onion (*Allium polyrrhizum*). Excavations demonstrated that *Ephedra sinica* roots reach a depth of over 2 m and its rhizomes reach out several meters. *Epherdra*'s morphophysiological fea-

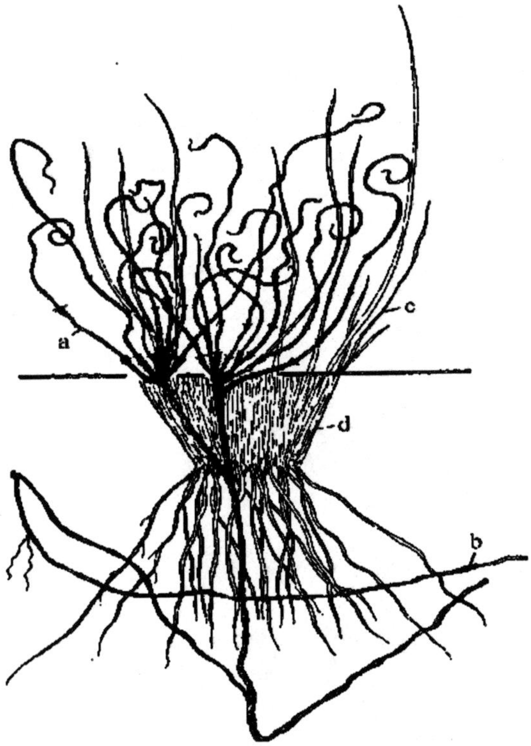

Figure 3.14. Growth of *Ephedra sinica* (a, b) through *Stipa gobica* sod cover (c, d).

Figure 3.15. Distribution of *Ephedra sinica* communities in eastern outskirts of Gobi Tyan-Shan Mountains (Tost Uul) (according to Gunin et al., 1993).

1 – *Ephedra sinica* + *Stipa gobica* + *Allium polyrrhizum*, 2 – *Anabasis brevifolia* + *Stipa gobica* + *Ephedra sinica*, 3 – same as 2 but with participation of *Artemisia xanthochroa, Amygdalus mongolica*, 4 – *Stipa gobica* + *Anabasis brevifolia* + *Ephedra equisetina* with participation of *Caragana leucophloea, Amygdalus mongolica*, 5 – *Anabasis brevivolia* + *Stipa gobica*, 6 – *Arenaria meyeri, Scorzonera capito* + *Stipa gobica, S. krylovii* + *Anabasis brevifolia* with participation of *Caragana leucophloea, Amygdalus mongolica*, 7 – sairas.

tures appears to be conducive not only to its survival, but also its wider distribution under deteriorating moisture conditions (Figure 3.14 & 3.15).

Natural plant community dynamics can be seen well with successions on sand deposits. The hillock sands in northern Mongolia are occupied by pine forests plus herbs, and an occasional dog rose undershrub. The plant cover contains some very sparse *Calamagrostis epigeios, Oxytropis microphylla, Polygonatum humile, Thalictrum squarrosum, Allium ramosum, Leontopodium leontopodioides, Lespedeza davurica* and others. In the zone of moderately wet steppes, more to the south on hillock sands amongst forb-feathergrass steppes, the contribution of shrubs is great, *Caragana microphylla* and *Prunus pedunculata*; but *Ulmus pumila* trees are especially important, which make these steppes unique. The herbaceous cover normally is very sparse and common steppe species are present. In the subzone of true and dry steppes, mounds of hillock sands occur mostly in river valleys. Elm (*Ulmus pumila*) also is common there in a community with *Salix* sp. and *Caragana* sp. In sparse herbaceous cover, meadow-steppe forbs often occur (*Stipa krylovii, Agropyron cristatum, Leymus racemosus, Krascheninnikovia ceratoides, Rhaponticum uniflorum, Thalictrum squarrosum* and others). Occasionally birch grows in such sandy mounds plus pea-shrubs and *Ribes rubrum.*

However quite distinct plant communities are characteristic of hillock sands in the desert zone where the sandy desert mounds often contain accumulations of barkhans and large areas devoid of vegetation. Only occasionally in depressions individuals do occur of *Leymus racemosus, Iris tenuifolia* and *Caragana bungei*. Such is the case on Mongol-Els sands situated southwest of Khangai.

In hillocky sands the vegetation is better established. (Figure 3.16). The main plant structure there is saxaul (*Haloxylon ammodendron*) that occasionally forms tall stands (Buyan-Orshikh, 1985). The shrub – herbaceous cover in such saxaul stands is very sparse. Occasionally *Calligonum mongolicum* and *Corispermum mongolicum* occur on the hillocks. The vegetation here also is more diversified in depressions. In hillock sand depressions in subzones of desertified steppes and steppe-like deserts one commonly finds *Asparagus gobicus* and *Agriophyllum pungens* (Yunatov, 1974). In true desert depressions of hillock sands, in addition to the saxaul, one occasionally finds *Artemisia xerophytica, Allium mongolicum* and *Salsola arbuscula*; still less abundant are *Asterothamnus centrali-asiaticus, Oxytropis aciphylla, Stipa glareosa, Chesneya mongolica, Echinops gmelinii, Ptilotrichum canescens, Bassia dasyphylla* and *Micropeplis arachnoidea*. But overall in hillocky sands *Corispermum mongolicum* is the most abundant (Rachkovskaya & Volkova, 1977; Rachkovskaya, 1993).

This ecological series of sand-loving vegetation that we just examined meridianally, from north to south, can be regarded as a conventional succession! Hence, as the aridity

Figure 3.16. Mounded sands at the margin of Mongol-Els massif (photo by A.V.Prishchepa).

increases to the south, pine forests on hillock sands are replaced by elm forests and, subsequently, by saxaul stands.

On piedmont erosion plains one can see successions associated with the wind transfer of sand materials (Vostokova et al., 1994). On stony areas with coarse rock debris *Sympegma regelii* individuals grow; and on small rock debris-rubble eluvium in deserts *Krascheninnikovia ceratoides, Stipa glareosa, Zygophyllum xanthoxylon* and *Amygdalus mongolica* grow. During wind transfer the air currents are intercepted by shrubs and large sods. In their wind shade peculiar shaped sand accumulations develop near such plants that are reminiscent of a drop when viewed from top. *Ajania fruticulosa* and *Ephedra przewalskii* grow in these sand drop drifts. In still thicker sand capes, groups of true sand-loving plants form, e.g., *Calligonum mongolicum* and occasionally *Caragana leucophloea* and *Atraphaxis frutescens.*

All of these sand-loving succession series are characterized by a dominance of arboreous-shrub vegetation that presumably associate with the improved moisture conditions in sands, where even in deserts, lenses of fresh ground water form that are accessible to tree roots. But when the tree vegetation here is disturbed by felling, fires, or other factors, some specific regressions of plant community succession occurs in each subzone. For example, pine forests on sands are devastated by forest fires. Even small ground fires that affect only the lower parts of the trunks may prove fatal for undergrowth. At the same time herbaceous vegetation, normally recovers very rapidly, and afterwards it develops intensively. Recurrent fires can bring about irreversible replacement of pine forests by meadow-steppe vegetation.

On the sand mounds of dry-steppe zones vegetation regression is mostly determined by grazing related disturbances. The destruction of herbaceous cover results in the formation of numerous wind-blown areas, whose refixation occurs very slowly, since the typical strong winds of Mongolia handicap the overgrowth of sands. On such regressing plots, only *Leymus racemosus* and *Agriophyllum pungens* grow.

Figure 3.17. Shifting barchans near Davst settlement, Transaltai Gobi (photo by A.V.Prishchepa).

Tall saxaul stands on hillock sands of deserts are exposed to human effects not only with grazing, but most drastically by the direct use of saxaul for firewood. Saxaul is normally uprooted for wood and the remaining hole becomes a source for wind erosion. When animals are grazed on saxaul stands, not only is herbaceous cover disturbed, but young saxaul undergrowth tops are browsed too. As a result, at the site of tall saxaul stands, high hillock sand sediments forms that are almost free of plant cover (Figure 3.17).

Such age-old grazing practices in the dry-steppe zone, in desertified steppes, and in steppe-like deserts results in wind erosion of sands deposits, their transfer and accumulation in «wind shades» (Bold, 1989). Mobility of the sand substrate promotes the formation of relatively sparse plant communities, where only shrubs and herbaceous sand-loving plants occur (*Agriophyllum pungens, Ceratocarpus arenarius, Corispermum patelliforme, C. mongolicum*, etc.). Upon further sand burial these are replaced by sparse groupings of *Leymus racemosus, Agriophyllum pungens* and others. In the presence of a water table lense under the sand mounds, plant community complexes can develop further. On such hillock sands plant communities develop as groups whereas depressions are taken up by ground water utilizing communities (Vostokova, 1980). On sand mounds of desertified steppes and steppe-like deserts of Mongolia meadow-shrub communities are common in depressions, occasionally including *Ulmus pumila, Populus diversifolia* or *Tamarix* sp., *Pinus* sp. occasionally appears on sand mounds of dry steppes. On the hillock sands of Obon-Els in the Dariganga region of Eastern Mongolia we found: on hillocks, groups of *Artemisia* sp. and *Caragana* sp., including *Leymus racemosus*; on less fixed plots, trees of *Ulmus pumila* including *Caragana microphylla*; and in depressions, dense meadow communities develop with *Artemisia xanthochroa, Aconogonon divaricatum, Hedysarum fruticosum, Oxytropis hailarensis* and others.

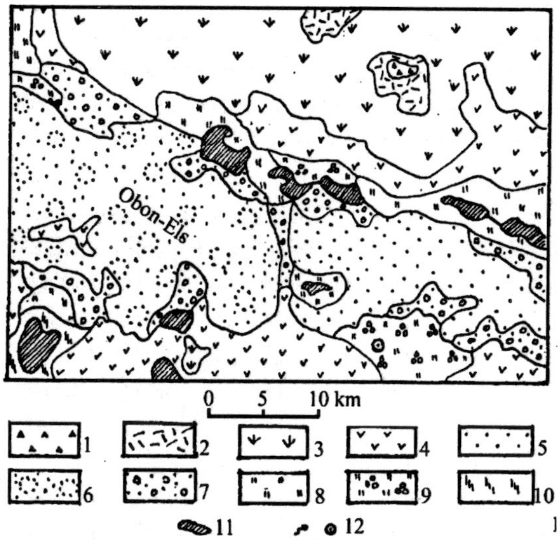

Figure 3.18. Alternation of vegetation communities in sands of Obon-Els and eluvium of Dariganga basalts (space image interpretation).

1–4. Plant successions at basalt eluvium: 1 – stony volcano cone with lone petrophytes, 2 – stony-detrital slopes with the community *Koeleria cristata, Cleistogenes squarrosa, Agropyron cristatum + Arenaria capillaris, Orostachys spinosa, Stellaria dichotoma*; 3 – gently sloped hilly plane with the community of *Artemisia frigida + Stipa krylovii, Cleistogenes squarrosa* with participation of *Caragana stenophylla, C. pygmaea*; 4 – gently sloped plane with the community of *Stipa krylovii, S. baicalensis, Poa attenuata + Artemisia frigida* with participation of *Carex korshinskyi*. 5–8. Vegetation of the sand massif: 5 – low mound sands fixed by the community of *Psammochloa villosa, Calamagrostis epigeios, Bromopsis korotkiji + Artemisia xanthochroa, A. halodendron* with participation of *Caragana microphylla, Oxytropis gracillima, Thalictrum squarrosum*, 6 – mound-ridge sands with groupings of *Leymus racemosus, Artemisia xanthochroa, Hedysarum fruticosum* with participation of *Prunus pedinculata*; 7 – weakly mounded massif margin with the complex community of *Leymus racemosus, Calamagrostis epigeios + Artemisia halodendron + Clematis aethusifolia* and bush thickets (*Tamarix ramosissima, Salix microstachya*) with considerable participation of *Ulmus pumila, Caragana leucophloea*. 8–10. Hydropmorphic vegetation of depressions with shallow aquifers: 8 – *Achnatherum splendens + Carex* sp. + *Puccinellia mongolica*, 9 – *Puccinellia mongolica, Psathyrostachys lanuginosa + Carex* sp. with participation of *Salix* sp., 10 – *Achnatherum splendens + Hordeum brevisubulatum* together with *Kalidium foliatum*, 11 – solonetz, 12 – springs.

Occasionally shrub forms of *Ulmus pumila* were found (Figure 3.18).

On hillock sands in the Cis-Khyngan Region a complex developed including groups of *Leymus racemosus* on hillocks and dense rich forb-grass meadows with *Salix* sp., *Prunus* sp. and *Pinus* sp. in depressions.

It is noteworthy that the formation of large mounds of unfixed sands is mostly the result of natural wind processes, which are only accelerated by human modifications! Essentially the greatest human-induced successions of plant cover on steppe and deserts plains of Mongolia are caused by long and intensive domestic livestock grazing practices. These regressive vegetation successions involve different plant species responding differently to the same pasture pressures. Thus, in a rich forb-feathergrass steppe *Stipa baicalensis, Festuca lenensis, Helictotrichon schellianum, Agrostis trinii* and *Poa attenuata* are not resistant to grazing (Chognii, 1988); and grazing removes numerous forb species (*Lilium pumilum, Linum baicalense, Dactylorhiza salina, Papaver nudicaule*, etc.). Simultaneously the abundance of *Koeleria cristata, Leymus chinensis* and *Carex duriuscula* increase considerably, and semi-weed forbs appear (*Thermopsis lanceolata, Artemisia frigida, A. changaica, Veronica incana,* and *Potentilla acaulis*). Therefore, another plant community develops that has a lower fodder value.

At the Tevshrulekh Station, O. Chognii (1988) studied such changes in species composition and productivity of green mass in feathergrass and low-grass steppes (Tables 3.11 & 3.12). He observed an initial abundance of mesoxerophilous grasses (*Agropyron cristatum, Stipa krylovii, S. baicalensis*, etc.). At the same time, the number of xeromorphic species increased including weeds. Under an excessive pastoral effect, plant communities completely degrade, all of the ecosystem links are disturbed, and, at the original site of a complex plant community, primitive groupings of weeds and unpalatable plants dominate. Wormwoods (*Artemisia frigida, A. adamsii*) or groups of *Urtica cannabina* and other weeds often become dominant.

Table 3.11. Changes in coverage (mean percentage for 1970–1974) and productivity (dry weight g/m^2, average for 1971–1976) in a *Stipa baicalensis + Carex pediformis* community with intensive grazing (according to Chognii, 1988).

Species	Cover percentage and productivity of species at pasture degrading:					
	Weak		Medium		Strong	
	Cover percentage	Dry weight	Cover percentage	Dry weight	Cover percentage	Dry weight
Artemisia changaica	1.0	–	1.0	2.4	32.0	123.2
A. frigida	1.0	0.8	1.0	2.1	6.0	7.3
Carex duriuscula	1.0	–	1.0	3.0	9.0	11.5
C. pediformis	10.0	39.3	20.0	28.2	1.0	2.8
Dendranthema zawadskii	–	7.2	–	8.8	–	0.1
Halenia corniculata	–	3.0	–	16.8	–	2.6
Heteropappus altaicus	–	1.1	–	1.8	–	6.0
Koeleria cristata	1.0	1.6	1.0	5.6	7.0	11.3
Leymus chinensis	1.0	1.6	1.0	21.2	7.0	54.1
Potentilla tanacetifolia	4.0	20.8	15.0	29.1	1.0	6.6
Scabiosa comosa	–	10.4	–	15.5	–	6.5
Stellera chamaejasme	–	11.1	–	10.7	–	7.0
Stipa baicalensis	39.0	75.2	22.0	29.4	1.0	4.1
Veronica incana	1.0	0.8	1.0	1.2	30.0	2.8
Thermopsis lanceolata	1.0	–	1.0	–	4.3	–
Sanguisorba officinalis	1.0	–	2.0	–	5.0	–

Table 3.12. Changes in coverage (mean percentage for 1970–1974) and productivity (dry weight, g/m^2, average for 1971–1976) in a *Koeleria cristata* + *Cleistogenes squarrosa, Poa attenuata* + herb sedgegrass community with intensive grazing (according to Chognii, 1988).

Species	Stages of pasture degrading					
	1 – weak		2 – medium		3 – strong	
	Cover percentage	Dry weight	Cover percentage	Dry weight	Cover percentage	Dry weight
Agropyron cristatum	2.0	3.2	1.0	0.9	1.0	2.4
Androsace incana	–	13.1	–	16.8	–	9.8
Artemisia frigida	<1.0	7.0	–	–	13.0	19.9
Bupleurum bicaule	–	3.5	–	3.1	–	0.7
Carex duriuscula	1.0	43.0	5.0	10.9	19.0	19.5
Cleistogenes squarrosa	10.0	13.8	6.0	13.3	5.0	10.7
Arenaria capillaris	2.0	4.5	3.0	4.2	3.0	4.5
Heteropappus altaicus	–	5.5	–	3.3	–	46.0
Koeleria cristata	8.0	7.3	4.0	2.9	5.0	4.4
Leymus chinensis	<1.0	0.5	<1.3	2.2	<9.0	6.6
Leontopodium ochroleucum	3.0	6.5	1.0	0.4	<1.0	–
Phlojodicarpus sibiricus	9.0	4.2	4.0	4.3	1.0	–
Poa attenuata	9.0	53.7	14.0	10.6	2.0	2.9
Potentilla acaulis	1.0	5.1	16.0	22.1	3.0	4.1
P. sericea	2.0	6.4	1.0	2.4	<3.0	4.6
Stipa krylovii	1.0	6.3	1.0	3.9	<1.0	3.2
Veronica pinnata	6.0	2.1	4.0	2.7	1.0	0.2

Generally this human-induced pastoral succession series on the steppes of Central Mongolia consists of a four-chain step replacement as follows:

feathergrass steppes replaced with mesophylous forbs;

small-sod-grass-hard sedge replaced with mezoxerophilous forbs;

hard-sedge-wormwood replaced with weed forbs;

and finally weed groups only.

A similar succession series is present in Eastern Mongolia where changes in the number of species, in their morphological appearance, and in total phytomass were found (Table 3.13; Figure 3.19).

Table 3.13. Changes in above-ground biomass (%) on a sandy desert due to grazing intensification in the Kerulen valley (according to Z.V. Karamysheva and V.N. Khramtsov, personal communication).

Biomorphs	Biomass (%) in the succession sequence due to intensification of grazing →			
	Stipa baicalensis, S. krylovii + *Cleistogenes squarrosa*	*Carex duriuscula* + *Cleistogenes squarrosa*	*Carex duriuscula* + *Artemisia frigida*	*Carex duriuscula* + *Agropyron cristatum, Thymus* sp.
Dwarf semi-shrubs	7.3	12.5	26.6	20.5
Perennial grasses	79.0	79.2	66.7	74.0
Annual and biennial grasses	13.7	8.3	6.7	5.5
Total biomass, kg/ha	1108.5	1056.8	659.0	1108.3
% of the initial community	100	95.2	59.4	99.8

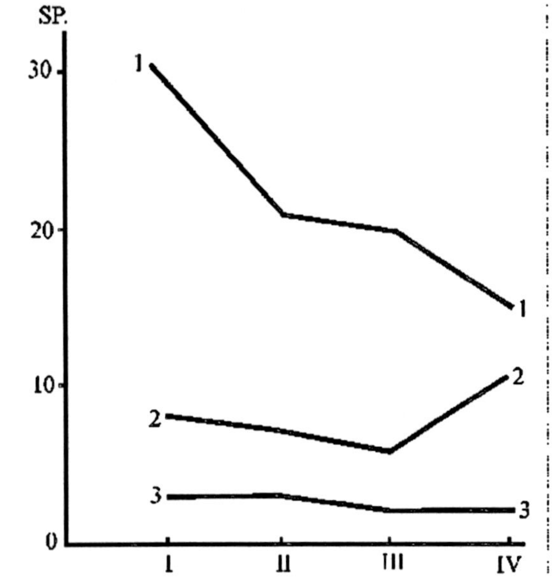

Figure 3.19. Changes in species number according to biomorphs in pasture succession community catena in Kerulen valley (according to materials of Z.V.Karamysheva and V.N.Khramtsov).

I–IV. Communities: I – *Stipa krylovii, Cleistogenes squarrosa* + forbs; II – *Cleistogenes squarrosa* + *Carex duriuscula*, III – *Carex duriuscula* + *Agropyron cristatum* + *Thymus* sp., IV – *Artemisia frigida* + *Carex duriuscula*. 1–3. Biomorphs: 1– dwarf semishrubs, 2 – perennial grasses, 3 – annual and biennial grasses.

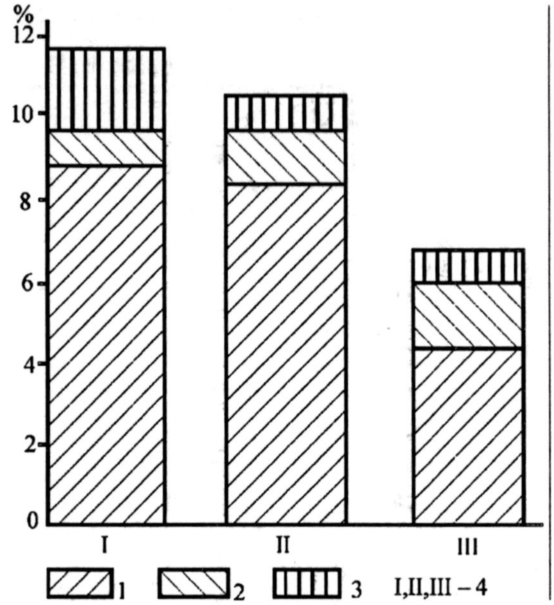

Figure 3.20. Dynamics of above-ground biomass of pasture community catena of sandy steppes and changes in their structure due to grazing intensification in the Kerulen Valley (according to materials of V.N.Khramtsov).

I–III. Communities: I – *Stipa krylovii* + *Cleistogenes squarrosa* + forbs; II – *Carex duriuscula* + *Cleistogenes squarrosa* + *Leymus chinensis*, III – *Carex duriuscula* + *Artemisia frigida* + *Leymus chinensis*. 1–3. Biomorphs: perennial grasses, 2 – dwarf semishrubs, 3 – annual and biennial grasses, 4 – number of plant communities.

Morphologically the number of shoots on different plants is variable when grazed: with shrubs the shoot number increases more than 5-fold; with perennial plants with long rhizomes 3 to 7-fold increases; and with annuals, 3-fold. Concurrently, loose-sod plant numbers decline 4-fold; with compact-sod plants the numbers decline until these plants are completely removed; and shoot numbers of biennials decline 3-fold. A general trend toward a weight reduction in vegetative shoots by 2 – 3 times is observed. According to changes in the number and weight of the shoots, the aboveground biomass in certain species and communities as a whole change differently. An average level of grass stand grazing disturbance corresponds to a forb-feathergrass-bistort community; a high level corresponds to a hard-sedge-bistort community; and plots with heavy overgrazing develop a weed-hard-sedge community. Whereas in the first two stages of pastoral succession, productivity remains the same, the third stage has a marked reduction – by 1.6–1.7 times.

The wormwood-hard-sedge community on light texture soils in areas of excess grazing is replaced by groups of *Thymus* sp. with forbs. In terms of numbers and productivity of species, the community structure of these groups differs sharply from all other communities (Figure 3.20). Further investigations of grazing effects with rich-forb

Table 3.14. Plant cover succession changes on bunch-grass rich forb steppes as a result of grazing (according to materials of Tumentsogt station)

Vegetation	Soils					
	Dark chestnut stony-detrital soils	Dark chestnut loamy-sand and loam soils		Dark chestnut loamy-sand and sand soils		
Initial communities	*Stipa* spp. + *Festuca* spp. + petrophytic forbs	*Stipa krylovii* + forbs with *Stipa sibirica* + *Caragana* spp.	Rich forb-grass with forbs + *Stipa sibirica* + *Caragana* spp.	–	Rich forb-grass with forbs + *Caragana* spp. + *Stipa* spp.	Rich forb-grass and *Stipa* spp. + *Caragana* spp. + *Aconogonon divaricatum*
Succession changes due to intensification of grazing	Petrophyte forb+*Festuca* spp. with *Stipa* spp. + *Caragana* spp.	*Stipa krylovii*+ *Leymus chinensis* + forbs with *Stipa sibirica, S. grandis* + *Caragana* spp.	–	*Filifolium sibiricum* + *Festuca* spp. + forbs with grass + *Caragana* spp. and *Aconogonon divaricatum*	Rich forbs + *Stipa* spp. + grass steppe with *Caragana spp.* + *Stipa* spp. + *Aconogonon divaricatum*	–
	Festuca spp. + *Caragana* spp. + ruderal herbs	*Stipa krylovii, Cleistogenes squarrosa* + forbs and *Caragana* spp. + *Stipa* spp. + ruderal herbs	*Leymus chinensis* + *Stipa sibirica* + forbs and *Stipa* spp. + *Caragana* spp. + ruderal herbs	*Filifolium sibiricum* + *Lespedeza davurica* + forbs and grass steppe + *Caragana* spp. + *Aconogonon divaricatum*	Losse-bunch grass + rich forbs + *Stipa* spp. with *Caragana* spp. + *Stipa* spp. + *Aconogonon divaricatum*	*Cleistogenes squarrosa, Stipa* spp., *Agropyron cristatum* + forbs with *Caragana* spp. + *Stipa sibirica* and *Aconogonon divaricatum*
		Cleistogenes squarrosa, Leymus chinensis +forbs with *Caragana* spp. + *Leymus chinensis* and ruderal herbs	Loose-bunch grass + forbs with *Artemisia dracunculus* + *Stipa* spp. + *Caragana* spp. and ruderal herbs	*Lespedeza davurica* + forbs with *Caragana* spp. + *Stipa* spp. + *Aconogonon divaricatum*	–	*Cleistogenes squarrosa* + *Carex duriuscula* + forbs with *Caragana* spp. + *Stipa* spp.
		Leymus chinensis + *Carex duriuscula* + forbs with *Carex duriuscula* + ruderal herbs and *Caragana* spp.	*Carex duriuscula* + *Leymus chinensis* + *Stipa* spp. + ruderal herbs with *Artemisia dracunculus* + ruderal herbs and *Caragana* spp.	*Artemisia dracunculus* + forbs with *Aconogonon divaricatum* and ruderal herbs	*Thymus* spp. + forbs with ruderal herbs + *Caragana* spp. + *Stipa* spp.	*Carex duriuscula* + *Artemisia* spp. + *Cleistogenes squarrosa* with ruderal herbs
		Leymys chinensis + *Carex duriuscula* + ruderal herbs with *Caragana* spp.				

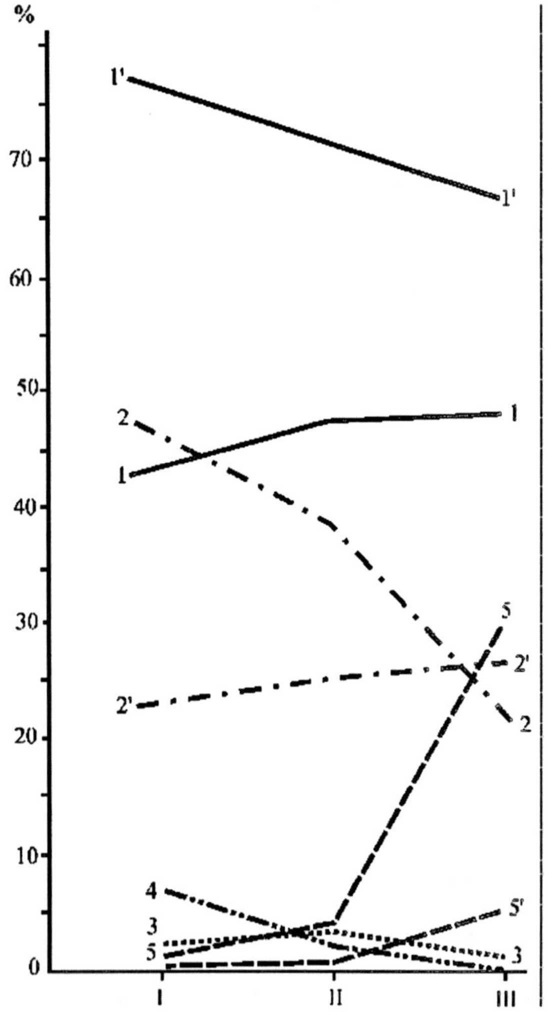

Figure 3.21. Changes in ecological steppes spectra of North-Eastern Khentii (according to Chognii, 1988).

1–5. Ecomorphs of feathergrass steppes: 1 – euxerophytes, 2 – mesoxerophytes and xeromesophytes, 3 – mesophytes; 4 – cryomesophytes; 5 – haloxerophytes and halomesophytes. 1^1, 2^1, 5^1 – ecomorphs of low sod-grass steppes. I–III Degrees of pasture impact: I – weak; II – moderate; III – high and very high.

sod-grass steppes show plant community dynamics on different soil substrates (Table 3.14). The most complete series occurs on forb-feathergrass steppes with pea-shrub (*Stipa krylovii, S. sibirica, S. grandis + Caragana microphylla, C. stenophylla*), associated with loam and sandy loam dark chestnut soils on gently undulating plains. This succession series is completed at the stage of a weed community, developed in a complex with hard-sedge-forb communities, and occasionally with *Caragana* sp.

In undisturbed plots on loam-sandy and sandy soils a complex is found of grass-rich-forb and feathergrass-pea-shrub communities. But intensification of human-related modification results in an early appearance of weeds; and the succession series is completed by hard-sedge-bistort-thyme communities (*Carex duriuscula + Cleistogenes squarrosa + Thymus* sp.). The common weeds are *Thermopsis lanceolata, Artemisia adamsii, A. scoparia, A. frigida* and *Axyris hybrida*.

On the whole, there is a ubiquitous increase in xeromorphic species with increasing levels of human activities resulting in rangeland degradation. This degradation is supported by changes in the ecological spectra studied by O.Chognii (1988) on the feathergrass and small-sod-grass steppes in northeast Khangai (Figure 3.21).

Changes in steppe ecosystems of the Tumentsogt Station region over 20 years (1968–1988) as a result of anthropogenic modifications are illustrated by the diagram in Figure 3.22. Station observations show that the transformation series, correlating to the level of anthropogenic disturbances, have different stages. Xeromesophyte and mesophyte and mesoxerophyte types of communities form a longer series and have a larger number of stages than xerophyte types. These are associated with processes of community xerophytization under pastoral degradation. The number of possible stages in a succession series indicates indirectly the succession resistance level. Mesophyte types under excessive grazing pressures are replaced initially by xeromorphic that rapidly turns into a complete degradation stage.

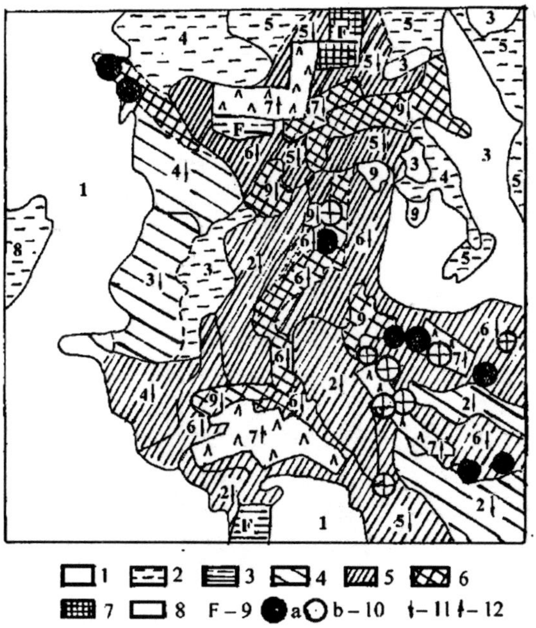

Figure 3.22. Changes of Tumentsogt region steppes over 20 years (1968 – 88) as a result of anthropogenic influences (based on materials of the Tumentsogt station).

1–3. Practically unchanged ecosystems: 1 – climax communities; 2 – weakly disturbed; 3 – heavily disturbed. 4–7. Degrading successions: 4 – undisturbed cenoses transformed into disturbed; 5 – weakly disturbed transformed into moderately disturbed; 6 – moderately disturbed transformed into heavily disturbed; 7 – weakly disturbed transformed into heavily disturbed. 8 – Renewal at idle lands. 9. Arable lands. 10. Sites with destroyed vegetation: a – new sites; b – previously existed. 11–12. Direction of succession development: 11 – degrading; 12 – renewal.

Figures show contemporary plant communities and complexes:

1. Combinations of petrophyte – grass steppes (*Festuca lenensis, Koeleria cristata + Arenaria capillaris, Pulsatilla bungeana, Amygdalus mongolica, Armeniaca sibirica*) on summits and low mountain slopes.

2. Combinations of petrophyte – forb – wormwood – grass steppes (*Festuca lenensis, Koeleria cristata, Stipa krylovii, Cleistogenes squarrosa + Artemisia frigida + Androsace chamaejasme, Orostachys spinosa, Galium verum, Caragana microphylla, C. stenophylla* on highly dissected low hilly terrain.

3. Combinations of wormwood – grass – petrophyte forb steppes (*Pulsatilla bungeana, Filifolium sibiricum + Festuca lenensis, Poa attenuata, Koeleria cristata, Leymus chinensis + Artemisia frigida, A. commutata* with pea shrub – *Caragana microphylla, C. stenophylla* on high undulating plains.

4. Combination of forb – tansy – grass steppes (*Stipa baicalensis, S. krylovii, S. sibirica, Leymus chinensis, Poa attenuata, Cleistogenes squarrosa + Filifolium sibiricum + Haplophyllum davuricum, Potentilla tanacetifolia, Thermopsis lanceolata, Artemisia frigida* with pea shrubs *Caragana microphylla, C. stenophylla*) on low flat plains.

5. Pasture sagebrush – forb – grass steppes (*Stipa krylovii, S. grandis, S. sibirica, Agropyron cristatum, Leymus chinensis + Serratula centauroides, Potentilla tanacetifolia, Cymbaria daurica + Artemisia frigida* with pea shrubs *Caragana microphylla, C. stenophylla*) on low flat plains.

6. Combinations of sedge – wormwood – grass steppes (*Stipa krylovii, Leymus chinensis, Cleistogenes squarrosa + Artemisia frigida, A. adamsii, A. mongolica*) on low flat plains.

7. Cinquefoil – *Leymus* – feathergrass (*Stipa sibirica, S. grandis, S. krylovii + Leymus chinensis + Potentilla tanacetifolia, P. bifurca*) and sedge – wormwood (*Artemisia frigida, A. adamsii, A. mongolica + Carex duriuscula*) communities on old idle lands (30 and 17 years).

8. Pea shrub – sedge – tyrsa – *Leymus* steppes (*Leymus chinensis, Stipa krylovii, Poa attenuata + Carex korschinskyi, C. duriuscula + Artemisia frigida*) of hay meadows on weakly dissected plains.

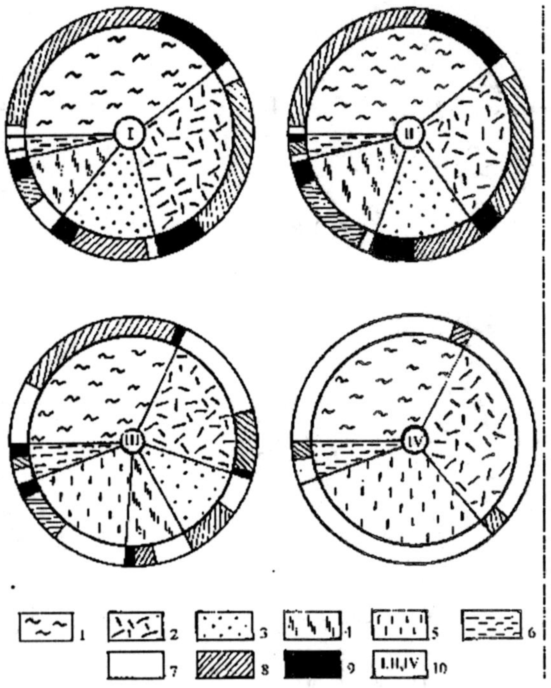

Figure 3.23. Ratio between plant communities with different degrees of distortion by the main edaphic variants in desert sub-zones (according to Evstifeev & Rachkovskaya, 1991).

1 – 6. Edaphic variants of plant communities: 1 – hemipetrophyte and hemipsammophyte; 2 – petrophyte; 3 – psammophyte; 4 – halophyte; 5 – gypsohalophyte; 6 – hydromorphic. 7 – 9. Anthropogenic distortion of edaphic variants of phytocenoses: 7 – very weak or absent; 8 – weak and sometimes moderate; 9 – moderate and sometimes strong. 10. Desert subzones: I – steppized deserts; II – northern deserts; III – southern deserts; IV – extra-arid deserts.

Thus, the main tendency in all types of steppe communities under intense grazing pressures is xerophytization. This tendency involves different features of these communities, e.g., species composition, range of life forms, morphological features of species, species viability, and their changes in numbers and productivity.

In contrast desert vegetation is exposed much less than steppes to the effects of grazing livestock breeding. Considerable grazing pressure mainly is observed in northern, steppe-like deserts, roughly on 14% of their area. About 35% of the desert area is virtually unaffected by grazing. These are extra-arid and true deserts (Gunin & Vostokova, 1995a). Thus there is not an equal level of human related effects in different desert subzones. It should be added, that the level of anthropogenic effects is different even on different desert soil types (Figure 3.23).

Furthermore, disturbances from grazing livestock breeding also are different because of the natural resistance of individual species and communities to grazing pressures. Indeed relatively stable and rapidly recoverable desert plant communities exist on stony-rock-debris or on a screen-rock-debris shield. These land areas are widely distributed in the Gobi in association with plant communities proven to be fairly resistant to grazing. The plants are *Stipa glareosa, S. gobica, Anabasis brevifolia, Ajania achilleoides, Salsola passerina,* and some onions (Evstifeev & Rachkovskaya, 1976, 1991). But on sand or sandy deposits, characterized by an absence of a surface shield, very unstable vegetation exists. So in the case of overgrazing, a loosening and increase in the sandy sheath occurs that leads to the formation of hillocky and hillocky-barkhan unfixed sands and to plant community replacements by primitive groups of sand-loving plants.

An important index to human-caused disturbance is the presence of indicator species for grazing pressure. Almost ubiquitous indicators of grazing pressure are *Chloris virgata, Dontostemon integrifolius, Peganum nigellastrum, P. harmala, Plantago depressa* and *Neopallasia pectinata.* In addition, for the Gobi Deserts, typical plant indices for grazing pressure are *Artemisia anethoides, Eragrostis minor, Lactuca serriola, Polypogon monspeliensis, Lepidium densiflorum* and *Cynanchum sibiricum.* Many of these plant either are poisonous or are not consumed by livestock, hence, they are distributed even wider (Table 3.15).

A diversion to plant community successions also begin with a decline in the viability of dominant species, decreases in percentage

Table 3.15. Indicators – Species for pasture degradation in desert communities (according to Evstifeev et al., 1990).

Distribution	Species – indicators of: Intensive total pasture degradation	Ruderal sites, often with increased nitrogen content
Everywhere	*Chloris virgata, Dontostemon integrifolius, Peganum harmala, P. nigellastrum, Plantago depressa*	*Artemisia annua, Atriplex sibirica, Cardaria pubescens, Enneapogon borealis, Euphorbia humifusa, Halogeton glomeratus, Lappula myosotis, Lepidium obtusum, Salsola tragus, Setaria viridis, Tribulus terrestris*
Lake Valley and Great Lakes Depression	*Artemisia glauca, A. sericea, Astragalus adsurgens, Carex duriuscula, C. stenophylloides, Corispermum mongolicum, Heteropappus hispidus, Iris lactea, I. tenuifolia, Kochia densiflora, Oxytropis aciphylla, O. glabra, Potentilla acaulis, P. anserina, Saussurea amara, Sibbaldianthe sericea, Stilpnolepis intricata, Thermopsis lanceolata, Th. mongolica, Thymus dahuricus*	*Artemisia macrocephala, A. scoparia, A. sieversiana, Axyris hybrida, Bassia dasyphylla, Chenopodium acuminatum, Ch. album, Ch. aristatum, Convolvulus ammanii, Dracocephalum foetidum, Lepidium obtusum, Neopallasia pectinata, Panzerina lanata, Salsola collina, S. paulsenii, Urtica cannabina*
Gobi	*Artemisia anethoides, Eragrostis minor, Lactuca serriola, Polypogon monspeliensis*	*Atriplex laevis, Bromus japonicus, Cynanchum sibiricum, Lepidium densiflorum, Melilotus suaveolens, Micropeplis arachnoidea, Setaria glauca*

cover, and the appearance of indicator plants for grazing pressure (Figure 3.24). Further replacement is associated with the complete elimination of grasses from the community and a wide distribution and formation of communities dominated by *Artemisia frigida* and semi-weed species with a substantial contribution from saltworts *Salsola collina* and *S. tragus*.

At livestock stations the weed vegetation is dominated by nitrogen-loving species primarily *Urtica cannabina* and saltworts (*Atriplex tatarica, Chenopodium album, Peganum nigellastrum*, etc.). So on the whole human-associated effect result in the formation of still more xeromorphic and impoverished vegetation groups when compared with original plant communities.

In steppe plains an important factor in plant cover dynamics is an extensive fallow land farming because ploughing almost entirely destroys the native vegetation. These ploughed lands only are used for several years (3 to 4), whereupon they are abandoned. Such fallow lands show all the regeneration stages of steppe vegetation. Miklyaeva (1996) studied the regeneration of steppe vegetation on fallow lands in Eastern Mongolia with croplands located at the site of forb-sod-grass steppes on dark-chestnut soils (Volkova, 1988). By studying different-age fallows (from 1 to 22 years) it was shown that the entire succession series is characterized by a high constancy of *Artemisia scoparia, A. macrocephala, Leymus chinensis, Linaria acutiloba, Aconogonon divaricatum, Medicago ruthenica* and *Potentilla bifurca*. These species occurred in more than 80% of the sample sites. Annual fallows were the closest to seeded croplands. The most common plants were field weeds with the most frequent being *Chenopodium album, Salsola collina, Setaria viridis* and *Fallopia convolvulus.* Much of the soil remained in stubble with a very poor sod.

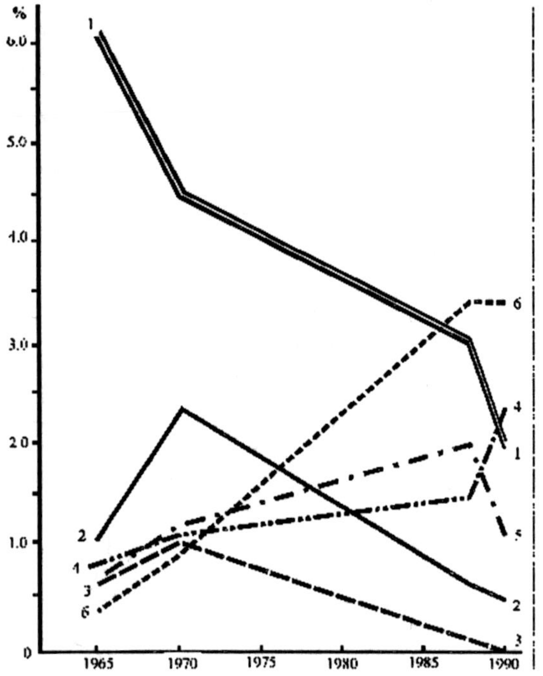

Figure 3.24. Changes in main species coverage in the community *Stipa glareosa, S. gobica + Cleistogenes squarrosa +Artemisia frigida* during 1970 – 1990 due to grazing intensification (according to observations of the Bulgan station).

1 – 6. Main species and groups of species: 1 – *Stipa glareosa, S. gobica*, 2 – *Cleistogenes squarrosa*, 3 – *Caragana leucophloea, Krascheninnikovia ceratoides*, 4 – *Artemisia frigida*, 5 – *Allium polyrrhzum, Gypsophila desertorum*, 6 – annual and biennial semi-weeds (*Aristida heymannii, Eragrostis minor, Corispermum mongolicum, Convolvulus ammanii, Salsola collina, Bassia dasyphylla, Euphorbia humifusa*).

Normally groups of *Leymus chinensis, Artemisia scoparia* and *Linaria acutiloba* dominated. This first regeneration stage of steppe tall-weed vegetation was recorded for about seven years. In the seventh year typical steppe species appear on the original fallow: *Agropyron cristatum, Cleistogenes squarrosa* and feathergrasses (*Stipa krylovii, S. sibirica*). Although a high abundance of typical «fallow» species remain (*Artemisia scoparia, Axyris amaranthoides, Linaria acutiloba*). Together these indicate a second stage in the regeneration of steppe vegetation (Figure 3.25). Finally only 18- and 22-year-old fallows show a grass stand dominance of dense-sod grasses (*Stipa sibirica* and *S. krylovii*). Although *Artemisia scoparia, Linaria acutiloba* and other typically «fallow» species do occur frequently (Miklyaeva, 1996; Hilbig, 1988).

But such a gradual regeneration of steppe vegetation frequently is disturbed by repeated ploughing, or is greatly handicapped by heavy pastoral pressure (Gunin et al., 1996). Such secondary ploughing was observed in 1996 on a large mass of ploughed land near the Orkhon and Khara-Gol rivers. In addition to field weeds, such ploughlands have an abundance of grazing pressure indicator species *Potentilla tanacetifolia, P. bifurca, Carum carvi, Artemisia scoparia* and *Thermopsis lanceolata* The vegetation dynamics of fallow lands was studied by O.A. Klimanova.

The sharpest changes in ecological habitat conditions and successions of plant communities occur as a result of irrigated arable farming. Normally, irrigated plots are small 3 to 10 ha. Irrigation involves either surface or water table water. Arable farming sharply affects vegetation on plots adjacent to the cropland. Due to artificially increased moisture, such plots have a sharp replacement of primary plant communities by secondary ruderal and hydromorphic species.

For instance, in the Obot-Khural region of Alashan Gobi, the irrigated croplands of the Khurshut farm account for an area of 10 ha on a high terrace and a gentle low mountain slope. Spring irrigation waters peter out along a tectonic disturbance (Vostokova et al., 1993, 1994). On non-irrigated plots formed by loam with rock debris, the vegetation is sparse communities of *Reaumuria songarica, Salsola passerina* and *Anabasis brevifolia*; occasionally with *Limonium tenellum, Ajania fruticulosa* and *Stipa glareosa.* This vegetation on the edge of individual vegetable croplands was replaced by patches of weed hydromorphic group of *Malva sylvestris, Atriplex tatarica, Convolvulus chinensis, Lactuca tatarica, Peganum nigellastrum, Inula britannica, Sonchus arvensis* and *Polygonum arenastrum.*

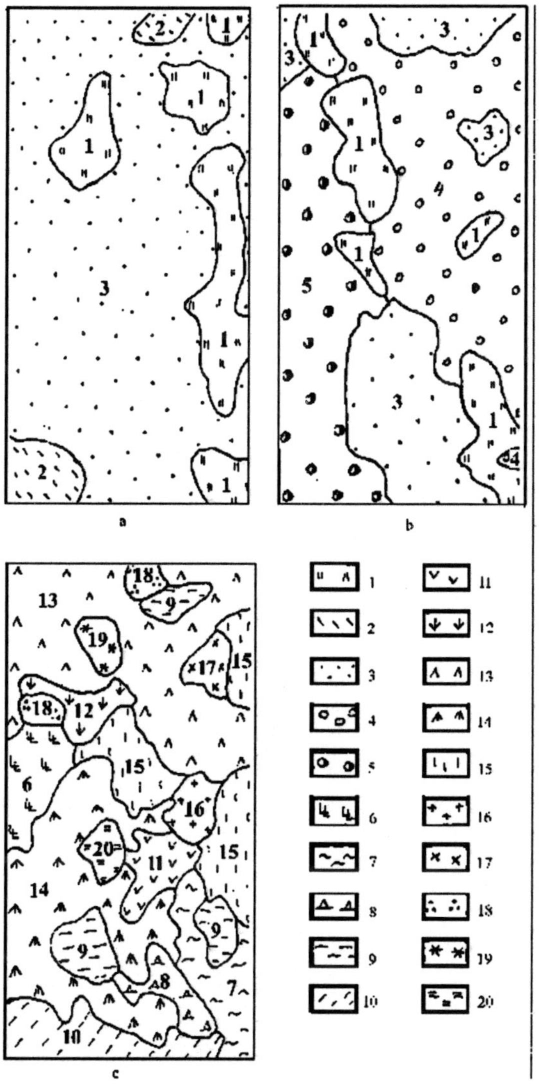

Figure 3.25. The structure of idle dry steppes lands near the Tumentsogt station (Miklyaeva, 1996).

The structure of idle lands: a – one year, b – 7 years, c – 22 – years. 1–20. Plant communities with predominance of: 1 – *Leymus chinensis*, 2 – *Linaria acutiloba*, 3 – *Artemisia scoparia*, 4 – *Artemisia scoparia* + *Aconogonon divaricatum*, 5 – *Artemisia scoparia* + *Poa attenuata* subsp. *botryoides*, 6 – *Artemisia scoparia*, *A. frigida*, 7 – *Artemisia scoparia*, *A. palustris*, 8 – *Artemisia macrocephala*, 9 – *Stipa sibirica* + *Axyris amaranthoides*, 10 – *Stipa sibirica* + *Artemisia scoparia*, 11 – *Stipa sibirica* + *Thermopsis lanceolata*, 12 – *Stipa sibirica* + *Artemisia dracunculus*, 13 – *Stipa sibirica*, *Poa attenuata* subsp. *botryoides*, 14 – *Stipa sibirica*, 15 – *Poa attenuata* subsp. *botryoides*, 16 – *Salsola monoptera*, 17–18. At inhabited fossorial animal colonies: 17 – *Salsola monoptera* + *Artemisia scoparia*, 18 – *Axyris amaranthoides*. 19–20. At abandoned fossorial animal colonies: 19 – *Artemisia palustris*, *A. frigida*, 20 – *Stipa sibirica* + *Salsola monoptera*.

On plot margins a better-formed community developed of *Achnatherum splendens*, where *Sphaerophysa salsula* and *Phragmites australis* are fairly abundant. But where water flow and croplands is not so great, processes of secondary salinization and halohydromorphic communities are established such as deris community replacements by shrubs of *Tamarix hispida*, *T. ramosissima* with *Kalidium foliatum*.

Similar successions of desert and dry-steppe plant communities by halohydromorphic communities are observed commonly at sites of water seepage from irrigated croplands.

For instance, in the Dariganga region (southeastern Mongolia) the community *Reaumuria songarica* + *Salsola laricifolia* + *Anabasis brevifolia* is replaced by group with *Salsola collina*, *S. tragus* and *Saussurea amara*.

When abandoned croplands are no longer irrigated, they are populated by xeromorphic weeds and wormwoods (*Artemisia anethoides*, *A. annua*), and after 10 to 20 years an almost original desert communities recovers. Thus, a sharp change in moisture conditions fairly rapidly results in a replacement of primary desert vegetation by hydromorphic and halohydromorphic. However, regeneration successions on originally non-irrigated croplands proceed very slowly.

In summary, a variety of observations show that modern dynamics of vegetation on steppe and desert plains and hummocky topography areas follow a regression pattern. Successions of plant communites result in the formation of less productive communities with a xerophyte dominances that are impoverished in terms of species composition. When the human-related pressures are removed, regeneration successions proceed at a fairly slow rate over about 20 years.

3.4 DYNAMICS OF WATER-ASSOCIATED VEGETATION

Hydromorphic plant communities are distributed in forest, forest-steppe and steppe regions, mainly along river flood plains and lake shores. In the southern half of Mongolia, such communities are associated either with closed depressions or lake depressions or underground water outcrops.

Natural vegetation successions here are connected with channeling processes in river valleys, variations in lake water levels or their complete drying up, reductions in water flow of springs, and even deeper water tables due to aridization. The major human-associated factors effecting these plant communities are unregulated grazing, tent camps, and traffic pressure, which are quite pronounced in river floodplains. However, hydromorphic vegetation dynamics is affected not only by these direct human economic activities. A strong indirect effect is from operations conducted on adjacent watershed areas. For instance, clear felling on hearby mountain slopes promotes heavy floods and decreases ground water recharging, both of which naturally affect hydromorphic vegetation.

Flood-plain processes are directly responsible for the development of new vegetation habitats (Grubov & Mirkin, 1980). Hence, here the modern vegetation represents a complex mosaic of plant communities, which reflect different stages in succession series.

Sections of mountain river flood plains normally are rubble. There vegetation is comprised of sparse groups of herbaceous plants that frequently are disturbed by sudden rain floods. Only when the river valley expands is it possible to form long-term existing plant communities. In high-mountains only some sedge-Kobresia thickets develop (*Potentilla fruticosa, Betula fusca, Salix bebbiana, Kobresia myosuroides, Carex media, C. dichroa*). But the dynamics of these stream communities is not yet understood.

The lower forest zone and the steppe zone have been studied on expanded sections of river valleys where plant community successions form depending on the alluvial soil composition. In the forest zone on sandy-rubble alluvium in a well-drained near-flood-plain part of the valley, sparse larch forests develop with hydromorphic vegetation cover. Subsequently, these are replaced by a well-developed vegetation, where an arboreous layer is formed by larches and poplar (*Populus suaveolens*), frequently with a birch (*Betula platyphylla*), plus a dense shrub layer of *Padus avium, Salix triandra, S. taraikensis, Swida alba*, etc. On thinly-dispersed alluvium, originally monodominant thickets are formed of *Phragmites australis, Typha laxmannii*, and *T. angustifolia* which gradually are replaced by wetland-forb communities of *Caltha palustris, Leontopodium ochroleucum, Sparganium glomeratum, Alisma plantago-aquatica*, and *Carex* species. Subsequently, they can be replaced by shrub-poplar-urema with a large contribution by willows. In some valleys all three-plant communities occur concurrently, forming a complex mosaic cover.

On alluvial sediments in forest-steppe zone river valleys, combinations of communities are characteristic, forming forest-shrub communities on sod-swamp soils. On newly formed sandy-rubble areas in the Khentii and Khangai forest-steppe zone, pioneering groups include *Myricaria longifolia, Salix schwerinii, S. ledebouriana* and *S. rorida* (Hilbig & Knapp, 1983; Titov et al., 1990a, b). Subsequently, willow forests can be replaced by poplar forests (*Populus laurifolia*). A second layer is formed by *Padus avium, Betula fusca* and *Salix ledebouriana*; occasionally, the undergrowth contains *Hippophaë rhamnoides, Crataegus dahurica, C. sanguinea, Rosa acicularis* and *Ribes rubrum*. Commonly the herbaceous cover includes *Cacalia hastata, Senecio nemorensis, Thalictrum simplex, Poa attenuata, Elymus gmelinii* and *Polygonatum odoratum* (Hilbig & Knapp, 1983; Titov et al., 1990a, b). In moderately wet steppes, often on well-drained plots, sparse communities of

Figure 3.26. Ulmus pumila in the Orkhon floodplain (photo by A.V.Prishchepa).

Ulmus pumila are formed (Figure 3.26) with *Ribes diacantha* and steppe-like grass stands. These communities replace sparse groups of *Ranunculus reptans*; and on silty plots *Bidens tripartita* is replaced.

Large human-related effects on flood-plain vegetation in river valleys results in pressure to change plant communities. With flood-plain forests (uremas) larch and poplars are the first to be eliminated and willow forests become sparse. Herbaceous cover, *Carex duriuscula* and *Potentilla anserina*, become dominant and *Urtica cannabina, Taraxacum leucanthum* and *Plantago depressa* appear which populate meadow plots between willow bushes. Hydromorphic vegetation at this stage is replaced by more xeromorphic.

In steppe zone flood plains, the most thoroughly studied degrading plant cover successions are those which initially differed due to soil-ground conditions. But all succession series are characterized by an appearance of weed and unpalatable species even under moderate grazing pressures (Titov et al., 1990a,b). Particularly characteristic is the expansion of groups of *Iris lactea* and *Carex duriuscula.* Table 3.16 presents the major series of plant degradation with meadow vegetation successions in dry and desertified steppe zones. In the final stages of succession the plant communities characteristic of meadow and steppe-like sod soils are meso- and xeromorphic *species (Carex duriuscula, Artemisia commutata* and *A. scoparia*). And on

Table 3.16. Plant successions in steppe zone river flood plains (compiled according to materials of Titov et al., 1990b).

Load	Plant changes due to increasing pastoral load on soils:					
	Silty humic gley	Meadow	Soddy (under steppization)	At salinized deposits		
				of bog row	meadow	soddy
Weak	*Carex cespitosa, Calamagrostis purpurea*	*Elytrigia repens, Bromopsis inermis, Poa pratensis* subsp. *angustifolia, Leymus chinensis, Thalictrum simplex, Lathyrus pratensis, Valeriana alternifolia*	*Carex duriuscula, Potentilla conferta, P. bifurca, Poa attenuata* subsp. *botryoides*	*Carex enervis, Dactylorhiza salina, Triglochin maritimum, Juncus gerardii*	*Puccinellia tenuiflora, Hordeum brevisubulatum, Poa subfastigiata*	*Achnatherum splendens, Carex duriuscula*
Medium		*Cirsium esculentum, Ranunculus propinquus, Artemisia laciniata, Bromopsis inermis*	*Carex duriuscula, Iris lactea, Dontostemon integrifolius, Artemisia commutata*	*Carex enervis, Triglochin maritimum, Ranunculus repens, Oxytropis salina*	*Puccinellia tenuiflora, Iris lactea, Inula britannica, Oxytropis salina*	*Achnaterum splendens, Iris lactea, Artemisia annua*
Heavy	*Cirsium setosum, Carduus crispus, Carex cespitosa*	*Artemisia laciniata, A. vulgaris, Lactuca sibirica, Cirsium esculentum, Ranunculus propinquus, Bromopsis inermis*	*Iris lactea, Artemisia commutata, A. scoparia, Carex duriuscula*	*Halerpestes salsuginosa, Glaux maritima, Oxytropis salina, Iris lactea, Carex enervis*	*Iris lactea, Oxytropis salina, Glaux maritima, Halerpestes salsuginosa, Saussurea amara*	*Iris lactea, Plantago salsa, Suaeda corniculata*

flood plains of desertified steppes the green sediments in the final succession series are dominated by halophyte and haloxeromorphic species (*Halerpestes salsuginosa, Glaux maritima, Saussurea amara, Plantago salsa, Suaeda corniculata*).

The most pronounced degrading vegetation successions in the flood plains of Mongolia are caused by excessive annual grazing when different mammals graze the same flood-plain range. As a result a majority of the flood-plain vegetation species are modified by 20% or more and even 100% in the middle valley reaches of the Orkhon and Kerulen Rivers (Table 3.17). However only a weak effect on flood plain vegetation is exerted by the winter grazing of horses and cattle (Titov et al., 1990 a,b).

As was found for flood-plain vegetation of Northern Mongolia rivers (Titov et al., 1990a,b), under a very heavy pasture effect, there is a convergence of plant communities. Flat elevations and tops of ridges with sod soils under a weak pasture effect contain hard-sedge communities. With increased grazing they are replaced by wormwood-hard-sedge communities with iris; and further grazing pressure promotes the development of hard-sedge-weed-herb-iris communities. The final community of this degrading succession is a weed-herb-iris group. At inter-ridge depressions and sinks with meadow and meadow-swamp soils, as pasture pressures increase the primary forb or sedge communities are replaced by grass-forb communities with iris or buttercup-iris communities. Under heavier grazing they are replaced by forb-iris; and the final community is a weed-herb-iris group (Table 3.18).

Thus, as a result of general grazing pressure, semi-weed groups of vegetation develop, in which an important role is played by: *Iris lactea, I. sanguinea, I. bungei, Bistorta*

Table 3.17. Human-associated disturbances of floodplain vegetation along principal rivers (Titov et al., 1990b).

Flood plains of rivers	Valley part	Percentage of disturbed communities	Induced by:		
			Grazing	Roads	Others
Muren	–	80	76	3	1
Ider	Upper	50	49	1	–
	Lower	100	92	3	5
Chulut	Upper	70	70	–	–
	Lower	70	69	1	–
Selenga	Upper	80	65	3	12
	Middle	70	67	3	–
	Below Orkhon mouth (till the boundary with Russia)	60	30	5	25
Orkhon	Upper	60	58	2	–
	Middle (from Kharkhorin)	100	30	20	50
	Middle (till the Tuul mouth)	90	40	30	20
	Lower (till the mouth of Khara-Gol)	80	60	15	5
	Lower	100	40	10	50
Ero-Gol	Upper	40	–	10	30
	Middle	80	75	2	3
	Lower	80	35	5	40
Khara-Gol	Upper	80	60	5	15
	Middle	60	50	3	7
	Lower	100	40	10	50
Onon	Upper	10	7	2	1
	Middle (till the boundary with Russia)	20	14	4	2
Kerulen	Upper	0	–	–	–
	Middle (longitudinal)	95	45	10	40
	Middle (latitudinal)	100	45	10	45
	Lower (below the town of Choibalsan)	90	30	15	45

alopecuroides, Aconogonon divaricatum, Rumex crispus, Stellera chamaejasme and *Thermopsis lanceolata* (Figure 3.27).

But river flood plains often are a constant site of summer tent camps. As a result, plots of meadow vegetation undergo drastic changes and groups of weeds and nitrophilous vegetation develop instead of typical grass-forb meadows. Where the earth is compacted by camp tents, groups develop with *Artemisia laciniata, Plantago depressa, P. major, Polygonum arenastrum, Potentilla anserina* and *Taraxacum leucanthum*. Such tent camp sites normally are accompanied by a high soil nitrogen content where these plants develop: *Cirsium setosum, Lactuca sibirica, L. tatarica, Sonchus arvensis, S. oleraceus, Urtica cannabina, Chenopodium acuminatum, Ch. album, Ch. aristatum, Cannabis sativa, Carduus crispus* and *Lepidium obtusum*.

Some flood plain areas in the dry steppe zone show an increased content of readily soluble salts (Grubov & Mirkin, 1980). There typical meadow flood plain herb -forb

Table 3.18. Degradation of flood plain plant communities under grazing pressure (Titov et al., 1990b).

Floodplain Soil Sites:		Flat sites and summits of ridges with soddy soils	Depressions between ridges and slopes with meadow soils	Depressions with boggy soils
Plant communities under grazing influences	Very weak	Rigid sedge	Forb-grass	Forb – sedge
	Medium	Wormwood – sedge with iris	Grass – forb with iris	Buttercup – sedge with iris
	Strong	Sedge – weed – iris	Forb – iris	
	Very strong	Weed – iris		

communities dominated by *Orostachys spinosa, Poa attenuata, Leymus chinensis, Stipa sibirica, Leymus secalinus*, etc., are replaced by wormwood-halophilous, where a large contribution is *Artemisia annua, Puccinellia tenuiflora, Saussurea amara, Triglochin maritimum* and *T. palustre* (Titov et al., 1990 a,b).

Figure 3.27. Iris lactea in the Tuul floodplain (photo by A.V.Prishchepa).

On plain steppes of wide river valleys terraced irrigated croplands are found occasionally. Such plots have been used for the longest time for arable cropping. The largest mass of arable lands in Mongolia is near Kharkhorin in the Orkhon River valley (Figure 3.28); the largest is in the Khara-Gol River valley left tributary. The constant excess moisture due to the discharge of irrigation waters and their seepage from irrigated croplands creates a stable wetland-forb community downstream, dominated by *Phragmites australis* and wetland forbs (*Bidens tripartita, Alisma plantago-aquatica, Butomus umbellatus*). Nearby more elevated areas are dominated by *Gnaphalium baicalense* and small-sod grasses, forming an ecological series, which evidently reflects a succession of plant communities when supplemental watering from arable croplands ceases.

An important role in river flood plains is played by communities of arboreous-shrub species. River flood plains of the forest zone and open woodlands in river flood plains are dominated by shrub thickets of *Betula fruticosa, Potentilla fruticosa* and occasionally with larches. In the steppe zone, the tree vegetation is mainly represented by willow forests, with poplars and occasionally with birches. Vegetation successions here are mainly determined by human-related effects, but partly by fires destroying forest strands and shrubs. The common succession here results in sparse shrub thickets with an important role of semi-weed forbs. But such successions are not pronounced.

The regeneration of hydromorphic vegetation successions on flood plains virtually does not happen because of recurrent human-associated modifications. On the whole, human-associated successions of hydromorphic vegetation on flood plains are essentially degrading, leading to the formation of low nutritional value and low production groups of plants, without strong plant community relationships and with a simplified structure.

On flood plains an indication of succession changes in initial stages is the appearance of indicator species for grazing pressure, salinization, and soil compacting (Table 3.19).

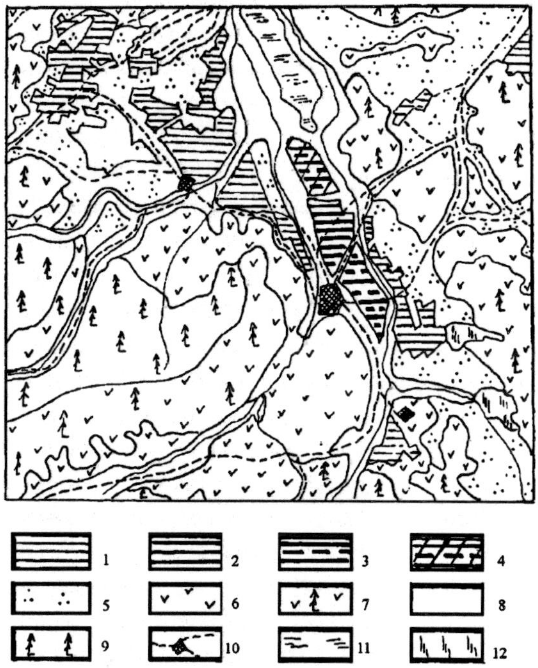

Figure 3.28. Land-use in the Kharkharin region (space image interpretation).

1–10. Land-use: 1–4 – for agriculture (1 – rainfed arable lands, 2 – newly irrigated fields, 3 – old irrigated fields, 4 – idle lands, partially swamped); 5–8 – for grazing stock rising (5 – pastures in the feather-grass – fescue steppes; 6 – mountain pastures of rich forb – feather grass – meadow steppes; 7 – forest – steppe pastures if medium height and low mountains; 8 – meadows and pastures in valleys of rivers and streams); 9 – in forestry : larch stands of medium height mountains; 10 – for settlements and transportation. 11–12 – non-used lands: 11 – swamps; 12 – solonchaks.

Natural climate fluctuations are very important in hydromorphic vegetation dynamics in dry-steppe and desert regions. Multi-year droughts have large effects on vegetation successions because small lakes dry up and large variations occur in the water level of large lakes. Over the last 10 years such variations have been most pronounced in the Lake Valley, and also in Northern Khalkha. Long lasting droughts are the main natural factor in hydromorphic vegetation dynamics on the shores of lakes and large sinks.

Table 3.19. Species indicating pasture ecosystem degradation in flood plains of steppe rivers (according to Titov et al., 1990 b).

Species - indicators of:				
General pasture degrading	Eroded areas	Soil salinization	Soils with high nitrogen content	Soils, strongly compacted at surface
Aconitum barbatum, Acroptilon repens, Heracleum sibiricum, Iris lactea, I. sanguinea, I. bungei, Polygonatum odoratum, Bistorta alopecuroides, Aconogonon divaricatum, Rumex crispus, Stellera chamaejasme, Thermopsis lanceolata, Oxytropis glabra	*Artemisia scoparia, A. dracunculus, Draba nemorosa, Taraxacum officinale, T. bicorne*	*Artemisia annua, Puccinellia tenuiflora, Saussurea amara, Suaeda corniculata, Triglochin maritimum, T. palustre*	*Cirsium setosum, Lactuca sibirica, L. tatarica, Sonchus arvensis, S. oleraceus, Urtica cannabina, Chenopodium acuminatum, Ch. album, Ch. aristatum, Cannabis sativa, Carduus crispus, Descurainia sophia, Lepidium obtusum, L. densiflorum*	*Artemisia laciniata, Chamaerhodos erecta, Plantago depressa, P. major, Polygonum arenastrum, Potentilla anserina, Taraxacum leucanthum*

Hydromorphic vegetation succession changes can be readily observed in steppes when small lakes are drying up and turning into meadow-saline soil depressions. The most pronounced indicator of this vegetation change in the steppe zone are deris communities (*Achnatherum splendens*). Their deris forms communities whose subdominants change with decreased water content and with salinization of soils or ground water near the surface (Vostokova, 1983; Vostokova & Kazantseva, 1995). Yunatov (1974) found that the following main communities form a well-defined succession series from the most drought-resistant to the most hydrophilous series:

– sedge deris stands (*Achnatherum splendens + Carex duriuscula*);

– wormwood-saltwort deris stands (*Achnatherum splendens + Artemisia macrocephala, A. anethifolia + Salsola tragus, S. collina*);

– lyme grass-deris stands (*Achnatherum splendens + Leymus secalinus, L. paboanus*);

– solonchak deris stands (*Achnatherum splendens + Leymus secalinus + Suaeda corniculata, Kalidium foliatum*);

– reed deris stands (*Achnatherum splendens + Phragmites australis*).

However, it is quite evident that sedge and wormwood-saltwort deris stands already are at a degrading succession stage that originated under moderate pastoral pressures and with a lowering of the water table.

In the steppes of Central Khalkha the shores of drying lakes normally contain a succession of lyme grass- deris stands, frequently, including *Hordeum brevisubulatum*, saline soils and, subsequently reed deris stands (Figure 3.29). Under strong soil surface salinization, caused by small salt lakes drying up, deris stands are rapidly replaced by halophilous communities of *Kalidium foliatum* and *K. gracile*, occasionally with suppressed reed individuals.

Figure 3.29. Ecological plant community catena of clay soils in the Choibalsan region.

1 – clay soils without vegetation. 2–4. Microbelts of plant communities: 2 – groupings of *Salsola* sp. 3 – *Achnatherum splendens + Salsola collina*, 4 – *Achnatherum splendens + Hordeum brevisubulatum*, 5 – *Anabasis brevifolia + Reaumuria songarica*.

Table 3.20. Dynamics of productivity (kg/ha of air-dry mass) in *Achnatherum splendens* + *Puccinellia tenuiflora* + *Leymus* chinensis communities (Vostokova & Kazantseva, 1995).

Life form, species	1973	1974	1975	1976	Mean	
					kg/ha	% *
Primitive dwarf semi-shrubs:						
Artemisia adamsii	5	–	–	1	1.5	0.1
Grasses:						
Cereals:						
Achnatherum splendens	1406	991	1394	1052	1052	68.5
Leymus chinensis	182	184	137	199	199	9.9
Puccinellia tenuiflora	230	461	310	383	383	19.5
Perennial forbs:						
Artemisia macrocephala	2	22	–	–	6	0.3
Astragalus adsurgens	–	–	–	1	0.2	0.5
Carex duriuscula	20	27	1	–	12	0.6
Convolvulus ammanii	8	–	–	–	2	0.1
Heteropappus altaicus	1	1	–	1	0.8	0.1
Iris lactea	–	19	–	–	4.8	0.2
Oxytropis myriophylla	–	1	–	–	0.2	0.05
O. selengensis	1	1	–	–	0.5	0.1
Plantago depressa	3	1	–	–	1	0.05
Taraxacum dissectum	4	1	–	1	1.5	0.1
Annual and biennial forbs:						
Artemisia scoparia	–	–	5	6	2.8	0.2
Chenopodium album	4	–	–	–	1	0.1
Salsola collina	1	–	–	–	0.2	0.05
Total above-ground (live) mass	1867	1709	1847	1644	1766.5	100
Dry litter	1260	959	1520	984	1180.8	
Litter	460	371	699	491	505.2	
Total	3587	3039	4066	3119	3452.5	

Note: Dash means absence of production of the given species. Asterisk – means production of the given species as percentage (%) of mean value of the total community above-ground (live) mass.

On dry steppes, deris stand successions also can be caused by the activities of burrowers, in particular, the marmot (Dmitriev et al., 1982, 1989). By bringing up sand onto the surface, marmots promote the formation of wind mounds around deris and cause the replacement of community co-dominants. In fact, herbaceous sand loving plants commonly appear (*Agriophyllum pungens, Ceratocarpus arenarius* and others) and pea-shrubs.

Under strong pastoral grazing pressures deris stands first show a decreased phytomass of grasses (*Achnatherum splendens, Leymus chinensis,* etc.) and other perennials (*Heteropappus altaicus, Artemisia macrocephala, Oxytropis selengensis,* etc.). Subsequently, the community shows a large number of annual-biennual forbs dominated by *Artemisia scoparia* (Table 3.20).

Still greater changes occur with deris in a desert «natural oasis» when they are occupied by humans in the summer time. For instance, in the Alashan Gobi along a flooded fracture in the region of Obot-Khural, a relative abundance of fresh water attracts a great

Figure 3.30. Tamarix ramosissima damaged by goats (photo by A.V.Prishchepa).

number of pastoralists. The concentration of livestock exceeds the natural range carrying capacity resulting in a complete degradation of hydromorphic communities. Almost ubiquitously only disturbed sods remain from deris while meadow forbs (*Glycyrrhiza uralensis, Sphaerophysa salsula, Clematis songarica* and *Cynanchum sibiricum*) are replaced by unpalatable weeds (*Iris* sp., *Urtica cannabina, Salsola collina,* etc.). Even *Tamarix* shrubs, so widely distributed in the zone of flooded fractures (Vostokova et al., 1993), are very strongly affected by excessive animal pressure (Figure 3.30).

Tamarix shrubs frequently mark an original water edge. When lakes dry up, plant communities form around them and regular ecological series develop reflecting community successions. For instance, a small lake drying up in the Lake Valley with a very gentle shore developed a halohydromorphic series (Figure 3.31). Shallow water and very wet habitats along the shore were occupied by a *Phragmites australis* plus *Typha laxmanii* community. With a lower water level plus an intensification of evaporation processes, a salt crust developed on the soil. The glycophilous species *Typha laxmanii* was rapidly eliminated and *Phragmites australis* grew smaller and

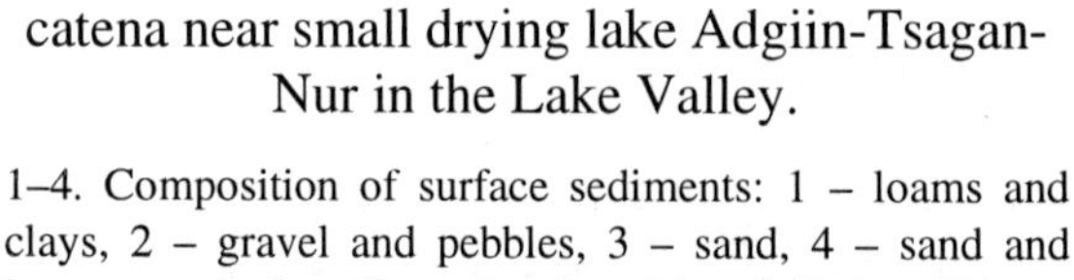

Figure 3.31. Ecological plant community catena near small drying lake Adgiin-Tsagan-Nur in the Lake Valley.

1–4. Composition of surface sediments: 1 – loams and clays, 2 – gravel and pebbles, 3 – sand, 4 – sand and loamy sand. 5 – Groundwater table. 6. Lake. Figures show communities: 1 – *Tamarix ramosissima*, 2 – *Kalidium gracile*, 3 – *Phragmites australis* + *Salsola* sp., 4 – *Phragmites australis* + *Typha laxmannii*.

sparser. In such a newly-formed community an already strong role was played by halophytes (*Salicornia perennis, Suaeda heterophylla* and *S. corniculata*). A community of *Kalidium foliatum* developed at an earlier reclaimed plot around bunches of sandy micro-hillocks. The succession series was completed by *Tamarix hispida* and *T. ramosissima* shrubs forming some fairly large hillocks up to 1.5 m high.

Sandy hillocks around shrubs or *Nitraria* sp. are a fairly typical feature not only near drying up lakes, but also at sites of subterranean water discharge. Such hillocks were observed in the Transaltai and Alashan Gobi (Vostokova et al., 1993). Sand-loving plants appear on these hillocks on a large scale to mark a new plant community succession stage. Hence, in this case, a hydromorphic succession series turns into another hydromorphic.

When this process also is affected by pastoral pressure, sandy hillocks are loosened, and the shrubs, around which the sand accumulates, are destroyed. Wind erosion of loosened sand deposits uncover *Tamarix* roots, which gradually dry up. At such sites, *Tamarix*-psammophyte groups are replaced by sparse individuals of psammophytes *Leymus racemosus* and *Agriophyllum pungens*. Occasionally, the destruction of *Tamarix* sand hillocks is caused by felling for fuel. In this case green vegetative portions are destroyed along with dry branches and trunks. These disturbed hillock areas form sites for wind erosion pressure. As a result, hydromorphic and semi-hydromorphic succession series are replaced by typical xerohydromorphic plants with an impoverished species composition and a simplified structure.

Human-caused degradation plant community successions in the hydromorphic series also are observed on arable croplands in deserts, where subterranean waters are used. Under natural conditions spring water gives rise to a natural oasis development (Lavrenko & Yunatov, 1960; Gunin et al., 1980, Vostokova et al., 1993). When this water is used not only for drinking water by herdsmen and watering of livestock, but also for the irrigation of vegetable crop lands, large vegetation changes occur on their edge. This is mainly associated with the fact that hydromorphic soils are to some extent salinized under arid conditions (Evstifeev & Pankova, 1984). An artificial re-distribution of water often leads to large increases in surface soil salinization (Pankova, 1992). As a consequence of water re-distribution and subsequent salinization, large changes in hydromorphic vegetation develop at irrigation springs.

The greatest attention has been paid to vegetation dynamics of arable croplands in the Ekhiin-Gol area of Transaltai Gobi. Outcrops of subterranean waters are associated with a tectonic dislocation stretching from west to east. The springs fall into three semi-isolated groups. Around them is a grass-forb meadow, dominated by *Glycyrrhiza uralensis, Pseudosophora alopecuroides, Sphaerophysa salsula* and *Setaria viridis*, forming a thicket-type community. On the edge, this meadow is replaced by a community of *Achnatherum splendens* with *Tamarix*, and, occasionally, poplars. The bottom where it crosses the flooded fracture, the large floodplain is particularly well defined thanks to its *Tamarix-Populus* community, with a typically floodplain pattern, similar to the arboreous-shrub vegetation of Central Asian flood plain forests. The herbaceous cover is formed by *Phragmites australis* and *Pseudosophora alopecuroides*. At a distance from the fresh-water source, the glycohydromorphic vegetation is replaced by halomorphic with the appearance of *Halogeton glomeratus* and subsequently, *Kalidium foliatum*.

This ecological succession series under a deteriorating water supply also may be regarded as a vegetation succession manifested under a deep water table and a large increase in water salinization. In Ekhiin-Gol the expansion of halophytes extended to a nearby settlement where common weeds were replaced by groups of *Salsola tragus* plus *Kalidium foliatum*. But

Figure 3.32. Drying species of *Populus diversifolia* in the natural oasis of Transaltai Gobi (photo by A.V.Prishchepa).

small areas between plants are covered with a white salt crust and free from vegetation.

Under lasting droughts, with no ground water recharge, a natural desert oasis will show a gradual degradation of hydromorphic vegetation, then replacement by typical xeromorphic vegetation. Typically, poplars (*Populus diversifolia, P. pilosa*, etc.) dry up and die (Figure 3.32); communities of *Achnatherum splendens* become sparse; shrubs appear, such as *Lycium ruthenicum*; and, subsequently, *Haloxylon ammodendron* appears.

Next xeromorphic communities develop with very sparse gypsophytes (*Reaumuria songarica* and *Salsola passerina*), and only occasionally weak shrubs of *Nitraria* or *Lycium ruthenicum* remain. Such a replacement of hydromorphic vegetation by xerogypsomorphic occurs on oasis edges where the discharge of springs is low without additional recharge.

Thus, studies on natural and human-modified communities both in flood plains and in the oases of deserts show a well-defined trend toward the formation of xeromorphic and salt-resistant plant communities.

3.5 CONCLUSIONS

Modern dynamics of plant communities in Mongolia are determined by sharp climatic fluctuations, by long-term droughts, and by the imposition of numerous human-associated factors. The most important human-associated factors are the use of vegetation for domestic livestock grazing and concentrating activities near water sources. The most important factors for forest vegetation dynamics are forest fires and clear fellings of trees. Arable and irrigated agricultural cropping are less importance since these occupy a small land area.

An intensification of human-related effects against a background of increasing climatic aridization across all of Mongolia leads to the development of degraded vegetation successions and to the origin of more xero-morphic and salt-resistant plant communities. A strong role is played amongst these changing communities by alien weedy plant communities. Today, the plant cover across Mongolia is largely a complex combination of drastically modified plant communities with domestic livestock grazing applying relentless pressure on the land and vegetation.

CHAPTER 4

ANALYSIS OF PRESENT-DAY VEGETATION DYNAMICS

4.1 BASIC CHANGES IN VEGETATION

An analysis of modern changes in plant communities reveals that the most important active factors today are climate and human-associated modifications. Indeed major detrimental effects are exerted by both climate cycles and human activities.

Climate effects are associated with cyclic patterns of precipitation and drought durations. In the available data short climatic cycles can be distinguished lasting 2, 5–6, 11, 22, 44 and 60 years. Longer lasting cycles of 80–90, 500–600 and 1000–1200 years are known but their influences on the characteristic of modern vegetation dynamics are not well understood. Such long cycles are not detectable by current local monitoring and likely can only be analyzed by biogeographical studies.

The effects of climate on nature and on the agrarian Mongolian society are huge. The productivity of ranges and crops; the health vigor and mortality of livestock; the activities of water erosion and winds; and outbreaks of episodic and other natural and social features are largely determined by the level and cyclic distribution of precipitation. In particular, intimate relationships between the 2-year cycle of precipitation and several natural phenomena are well known. During dry years: decreases occur in the productivity of ranges, cereal and vegetable crops, berries, mushrooms, and Siberian pine nuts; the thickness of annual tree rings growing in the lower zones of mountain areas change; the discharge of river water and lake water levels decline; at the same time, the numbers of rodents increase, including *Lasiopodomys brandtii* Radde, thereby more plague outbreaks occur; and wind-associated processes become increasingly active.

The 2-year cycle is evident throughout the Northern Hemisphere temperate zone, with an increase in air currents, both latitudinal and longitudinal. A more exact duration of this cycle is from 20 to 30 months; on the average, 26 months. In Mongolia, the bulk of its atmospheric moisture is known to arrive with air currents from the Arctic Ocean, i.e., moisture is brought by northwesterly and northerly air currents. Hence, during years of predominant northerly and northwesterly winds the annual precipitation increases sharply. By contrast, in years when latitudinal winds increase the precipitation decreases. The cycle is pronounced and sharply affects the dynamics of natural processes (Zolotokrylin, Gunin, 1986).

Investigations of this 2-year cycle and its correlation with natural and social phenomena are a basis for short-term vegetation dynamics predictions. This feature was noted in early antiquity. In fact, it was thought that crop yields of berries, Siberian pine cones and mushrooms only occurred every other year! With that regularity, graphs of annual precipitation and the natural phenomena depending on them all show a saw-like pattern. However, the 2-year cycle is disturbed by climatic oscillations with longer periods.

The 11-year solar activity cycle plays an important role in climate oscillations, particularly in the arid territories of Mongolia. Because it affects annual precipitation it is conducive to desertification processes by promoting the drying up of semidesert lakes

Table 4.1. Drought distribution between even and odd 11-years solar activity cycles.

	Cycles					
	Even			Odd		
	Ascending part	Descending part	Ascending + descending	Ascending part	Descending part	Ascending + descending
During the period of increased solar activity:						
duration, years	25	36	61	24	41	65
number of years with droughts	6	15	21	1	10	11
probability	24	42	34	4	24	16
During the period of weak solar activity:						
duration, years	34	50	84	34	37	71
number of years with droughts	13	19	32	9	5	14
probability	38	38	38	26	13	19
Total:						
duration, years	59	86	145	58	78	136
number of years with droughts	19	34	53	10	15	25

and determining the intensity of plant community regressive dynamics. This cycle also was noted by arats (native Mongolians). Livestock herders of the Lake Valley (Bayankhongor Aimak) at the beginning of the 20th century noticed that once in ten years, a reduction in the Tuin-Gol River water discharge reduced the level of Lake Orog-Nur basin to such an extent that cows and horses could walk on its bottom (Murzaev, 1952). According to available data Orog-Nur dried up in 1890, 1934–1936, 1952–1953 and 1986–1989. Ulan-Nur dried up in 1952–1953. The only reliable information on a large increase in lake levels in the zone of steppe deserts occurred in 1959. During that time the level of Lake Bon-Tsagan-Nur (Lake Valley) rose to such an extent that the nearby village center had to be moved.

A comparison of lake water levels with solar activity periods gives grounds to conclude that lakes dry up with 11-year cycle changes when the Wolf number reaches a minimum or close to a minimum. The lowest Lake Bon-Tsagan-Nur water level occurred in the course of high solar activity, when the annual Wolf of Zurich number was 159.0. The 11-year cycle in 1976–1986 showed the highest Wolf number over the last 300 years. The mean annual Wolf number reached a maximum in 1986, whereupon it declined. Then into the next solar cycle, Lakes Ulan-Nur, Tatsyn-Tsagan-Nur, Orog-Nur, and many small lakes in northern Khalkha dried up. The level of the deep multi-water-feed Bon-Tsagan-Nur Lake declined by two meters. That period of lakes drying up lasted longer than usual, for 4 to 5 years until 1990. This appears to be explained not only by climatic changes, but also by increased human use of water resources. Specifically during those years land improvement irrigation systems were constructed that used the river waters recharging these lakes.

Also it is known that even 11-year cycles of solar activity are distinguished by negative polarity, while uneven cycles are characterized by positive polarity to form a greater cycle that averages 22 years. In different sectors of the 22-year cycle graph and the number of extreme arid weather features in Mongolia are unevenly distributed. Indeed the reoccurrence of droughts is closely associated with the sun's activity (Table 4.1). Droughts occur more

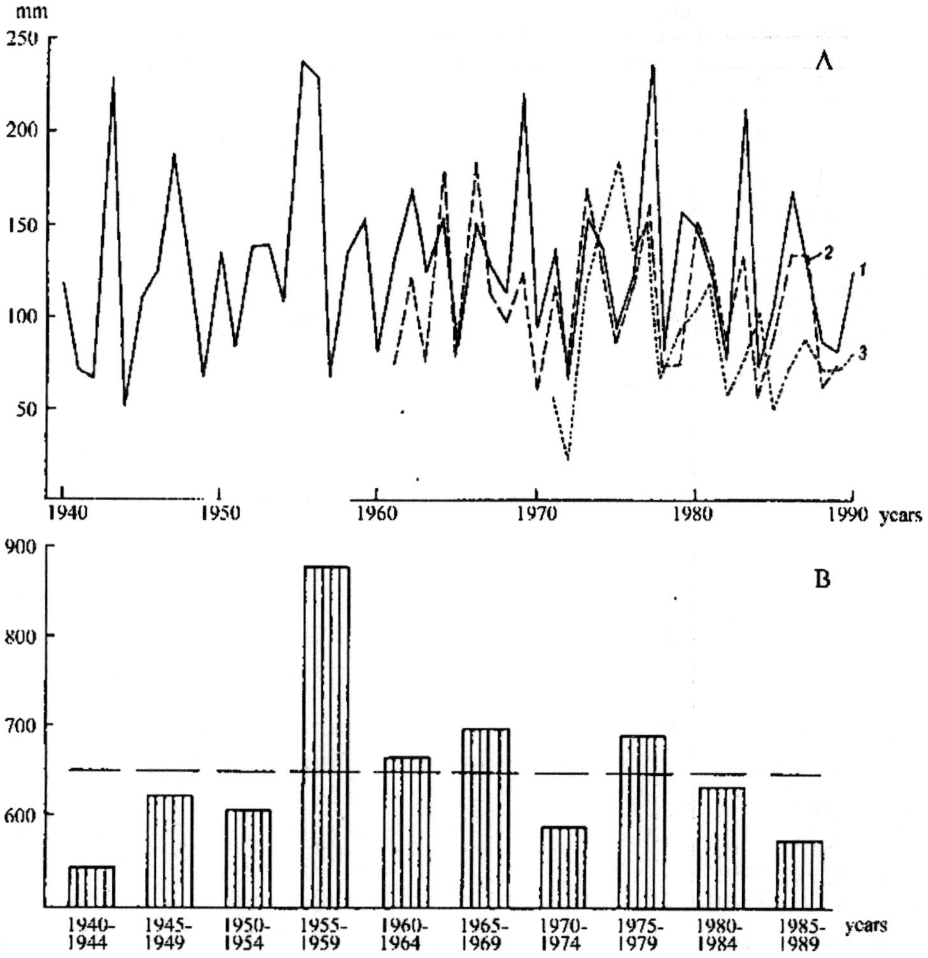

Figure 4.1. Dynamics' of precipitations' sum for one year (A) and five years (B) in Gobi on the data of three meteorological stations (1 – Dalanzadgad, 2 – Bulgan, 3 – Gurvantes) (according to Slemnev, Gunin & Kazantseva, 1994).

frequently in the even cycle, particularly in the descending branch of the annual Wolf number curve (Lovelius, 1998).

Meteorological data analysis over the last 50 years indicate that in addition to the cycles lasting 2, 11, and 22 years, precipitation distribution is substantially affected by a 44-year cycle of solar activity. Integral curves of annual precipitation plotted from meteorological station data with the longest series of observations illustrate this pattern (Figure 4.1). Precipitation reached a minimum during the late 1940s – early 1950s; and a maximum during the late 1960s – early 1970s. By 1990 annual precipitation dropped to a minimum. Subsequently precipitation has gradually increased, apparently to reach a maximum by the late 1990s – early 2000.

These climatic oscillations strongly affect vegetation dynamics.

Detrimental human-associated effects on vegetation, when coinciding with cycles of arid years, result in the most profound changes, e.g., xerophytization, even complete degradation of vegetation, and expansion of desertification processes. Every feature of plant degradation promotes a number of related processes, which profoundly affect the ecological conditions of plant habitats. These considerably handicap the regeneration of plant communities even in climatically favorable years.

Table 4.2. Natural exogenic processes affecting soils.

Type of process	Process action	Results
Eolian	Deflation (blowing out and resedimentation of fine material), sand transport.	Blown out depressions, exposed bush roots.
	Blowing and transport	Low ridges, sandy bars, mounds near plants.
	Accumulation (re-sedimentation)	Barkhans, ridges.
Water erosion	Rill-sheet washout	Small erosion furrows, exposure of subsoil horizons, accumulation of eroded materials downslope.
	Linear erosion	Sinkholes, erosion furrows.
Cryogenic	Water-saturation of soil	Micro-landslides with lines of sod breaks, micro-staircase pattern at slopes.
Hydrogenic	Salinization	Color spots and crusts of salts at the surface.
	Bogging	Predominantly hydromorphic vegetation.
Cryo-Hydrogenic	Bogging in the permafrost regions	Small hillocks at boggy sites with deep fractures between them.

Of these, the most characteristic of Mongolia are erosion processes, both water and wind. Wind erosion dominates in southern Mongolia. On mountain slopes with permafrost and seasonally frozen rocks, cryogenic processes are common. There moist sliding soil is the most frequent, which gives the slopes a small step pattern. But the most widely distributed processes in mountains and hammocky areas are associated with water erosion. Erosion furrows and channels are formed even where the surface is only 5–7° tilted. These especially readily develop in ruts of unpaved roads. On steeper slopes, even after small rains, the ruts on loam soils turn into steep-walled rain channels. When nearby trees and shrubs whose crowns intercept precipitation are destroyed, torrents develop rain channels in the ruts. The degradation of desert vegetation intensifies wind processes that are still more aggravated by human modifications.

Table 4.2 shows the most widely distributed natural exogenic processes. As these processes develop they cause large changes in vegetation, e.g., a reduction of cover and a replacement of compact-sod grasses by annuals or biannuals with poorly developed root systems.

The level of plant habitat disturbance can be assessed from manifestations of exogenic processes (Table 4.3). In Mongolia the strong interaction of natural and human-associated effects gives rise to drastic changes in plant ecosystem successions. Our analysis of modern natural and human-induced plant cover dynamics show that vegetation changes exhibit the following patterns:

– regressive, manifested as a degrading succession by plant communities;

– progressive, determined by sudden changes in climate or human activity that restore plant community successions.

Regressive patterns are the most widely distributed in Mongolia and they are made worst by poor ecosystem management practices.

4.2 REGRESSIVE PLANT COMMUNITY SUCCESSIONS

As a result of climatic oscillations plus human effects, regressive plant community successions are very common in Mongolia. Examples of these successions include:

– replacement of forest vegetation by shrub and herbaceous vegetation and, more rarely, by rock streams free of vegetation;

– rangeland depression of herbaceous vegetation in steppes, deserts, river valleys, and, occasionally, in high mountain areas;

– degradation of desert vegetation, development of desertification processes, and formation of extra-arid lands.

Table 4.3. Intensity of some exogenic processes as criteria for the degree of plant habitat disturbance.

Process	Relief modification	Characteristic	Degree of habitat disturbance		
			Low	Medium	High
Water erosion	Erosion furrows and sinkholes	Quantity per 100 m^3:			
		Ploughed lands	less than 1	1–10	>10
		Forest cuttings and pastures	1–2	2–10	>10
		Depth of incision cm:			
		ploughed lands	<5	5–10	>10
		forest cuttings	<50	50–100	>100
		pastures	<15	15–30	>30
Eolian	Blown out depressions	Depth, cm	<10	10–50	>50
	Mounds, bars	Quantity per 100 m^3, %	<10	10–50	>50
	Sand "mantles"	Thickness, cm	<10	10–50	>50
Hydrogenic	Land subsidence depressions	Quantity per 100 m^3	10–25	25–50	>50
	Moor hillocks	Quantity per 100 m^3, %	<25	25–50	>50

Thus, regressive successions result in the establishment of more xeromorphic vegetation with less cover and lower biological productivity at former sites of more meso- and hydromorphic plants.

A principal reason for such regressive vegetation successions is less precipitation. As precipitation decreases a progressive drying of surface soil horizons occurs and the water table lowers or thin aquifers are depleted, with a decreased ground water recharge. The most obvious consequence is a drying up of rivers and a disappearance of swamps. Such features are essential to prevent vegetation deterioration. The patterns of these regressive ecological changes and plant communities successions are diagramed in Table 4.4. If this process lasts long or indefinitely, plots or areas will develop completely devoid of vegetation.

Table 4.4. Vegetation changes caused by decreases in total precipitation rate.

Forced changes in ecological conditions	Changes in vegetation cover	
	Initial	Final
Desiccation of surface and upper soil horizons.	Decrease of survival ability and bioproductivity.	Dying of ombrophytes.
Fine fracture removal.	Root exposure	Dying of wood-bush vegetation.
Re-sedimentation of sandy material.	Covering of aboriginal vegetation by sand.	Formation of psammophyte groupings.
Decrease in ground water table or depletion of shallow aquifers.	Decrease in survival ability of phreatophytes and in their biomass.	Dying of phreatophytes.
Increase in surface salinization.	Substitution of glycophytes by halophytes.	Dying of halophytes.
Silting of the surface, worsening of aeration and water availability.	Physical elimination of dwarf plants, decrease in survival ability.	Substitution of psammophytes by pelitophytes.

Table 4.5. Generalized scheme of degrading changes in forest vegetation caused by forest clearings and intensive fires.

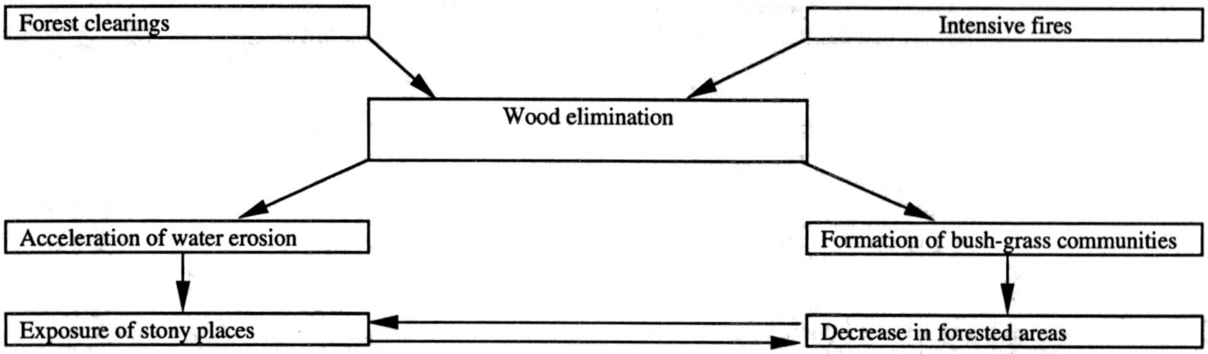

But in addition to direct effects on vegetation caused by a lower water supply, lower precipitation promotes other detrimental exogenous processes (Table 4.2). No doubt, the development of these processes causes great negative changes in ecological conditions for plants. Hence not only is plant water consumption reduced, but their mineral nutrition is deceased without water, and the strong winds blow nutrients away and cover plants with sand, etc.

Also each type of vegetation in a particular altitude zone has some specific features. For example, in forest-mountain and forest-steppe zones of medium-height mountain areas, lower precipitation results in regressive plant community changes. The general deterioration in water consumption handicaps or makes arboreous vegetation seed reproduction impossible. The lack of seeds leads to a turning of forestlands into steppes and to the formation of more xeromorphic vegetation with a smaller biomass; in other words, to a regressive succession. Subsequently in the absence of trees, surface water runoff is increased even when precipitation is less. As a consequence, on steep slopes or in the ruts of timber transportation roads, erosion furrows and rain channels develop whose growth worsens plant habitat ecological conditions to a still greater extent. Increased water drainage promotes the drying of surface soil horizons and also washes off nutrients and litter, which were the sources of nutrition for trees. Therefore herbaceous communities and groups are formed and even vegetation free rock streams can form.

Decreases in precipitation and its absence in spring often promote increased numbers of forest fires. Intensive fires in mountain communities are normally accompanied by other successions: birch forests develop at spruce forest sites and pine forests often are replaced by steppe or meadow communities. These regressive successions are long lasting and clearly reduce forest-covered areas (Tables 4.4 & 4.5). However, despite their ecological and economic importance, forests only account for a fairly small area of Mongolia (Figure 4.2).

Steppe and desert rangelands play a more important role in the country's economy (Kalinina, 1974; Bold, 1989; Yunatov, 1954). Hence, a study of successions in these plant communities under the effects of grazing and an associated increase in transportation pressures, against a background of a deteriorating water supply, are of primary importance for Mongolia.

Intensive rangeland vegetation use and transportation effects with an increased introduction of heavy trucks, all result in a vegetation impoverishment by alien plants and finally to the formation of areas devoid of vegetation (Table 4.6). Lack of moisture, particularly in the spring, effects livestock and rodents (*Lasiopodomys brandtii* Radde, *Ochotona daurica* Pall., *Microtus gregalis* Pall., etc.) which are partially responsible for modern vegetation patterns in the steppes. In eastern Mongolia, I.M. Miklyaeva and O.A. Lysak (1996) found not only a regressive succession series at the level of subassociations, but they

Figure 4.2. Stony outcrops on burn exposed slopes, Hubsugul region's mountains (photo by E.N.Matyuskin).

also found plant groups affected to different extents by grazing (Tables 4.7 & 4.8).

In considering the regression of steppe communities developing on dark-chestnut soils, light-loamy, and moderately rocky debris soils, it has been shown that the regression includes four main stages corresponding to different intensities of pastoral effects (1 –

Table 4.6. Generalized scheme of degrading changes in steppe vegetation due to intensive pastoral and transport loads.

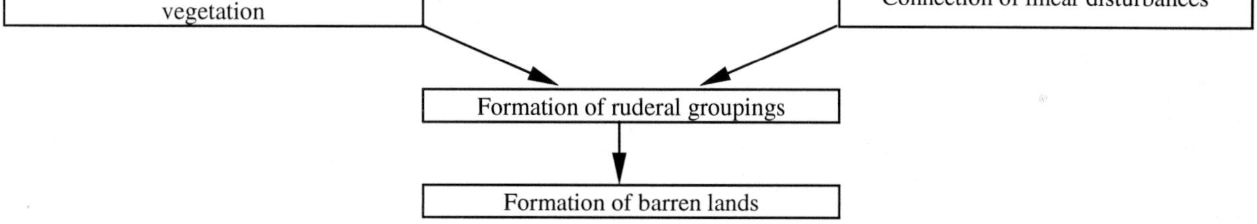

Table 4.7. Sub-association successions on Eastern Mongolian steppes under grazing activity increases (according to Miklyaeva & Lysak, 1996).

Grazing	Degradation Stage	Plant community
Absent	I	*Filifolium sibiricum* + *Thesium refractum*
Very weak	I–II	*Filifolium sibiricum* typicum *Ptilotrichum tenuifolium* typicum
Weak	II	*Limonium bicolor* + *Schizonepeta multifida*
From weak to medium	II–III	*Limonium bicolor* typicum
Medium	III	*Caragana stenophylla* + *Galium verum*
From medium to strong	III–IV	*Caragana stenophylla* typicum *Salsola monoptera* typicum
Strong	IV	*Salsola monoptera* + *Lappula intermedia*

practically with no effect; 2 – weak, 3 – moderate, 4 – strong effect). Three transition stages also are distinguished that floristically unite impoverished plant communities formed under the effect of the main factor with the activities of a massive rodent population. Typically these communities form subassociations. In Eastern Mongolia, the second stage of rangeland depression is the most widely distributed. Communities of the fourth stage only account for a small area and occur in areas of constant tent camps, wintering sites, or wells.

An analysis of community floristic composition and grass community height at different succession stages on steppes show that a weak grazing pressure is most favorable for steppes. At that stage maximal indices of floristic richness were recorded, plus a high percentage of crown density, and the tallest

Table 4.8. Species persistence versus degree of pasture grazing load (according to Miklyaeva & Lysak, 1996).

Not persistent to grazing		Persistent to grazing	
Species name	Load degree	Species name	Load degree
Thesium refractum	I	*Heteropappus biennis*	II–IV
Amblynotus rupestris	I–II	*Serratula centauroides*	II–IV
Polygala sibirica	II	*Dontostemon integrifolius*	II–III–IV
Filifolium sibiricum	II	*Salsola monoptera*	III–IV
Lespedeza juncea	II	*Lappula intermedia*	IV
Pulsatilla bungeana	I–II	*Chenopodium album*	III–IV
Scorzonera austriaca	I–II		
Potentilla leucophylla	II		
P. verticillaris	II		
Melandrium apricum	II		
Stipa grandis	II–III		
Goniolimon speciosum	II		
Pulsatilla turczaninovii	III		
Iris dichotoma	II–III		
Ptilotrichum tenuifolium	II–III		
Koeleria cristata subsp. *mongolica*	III		

Table 4.9. Structure of steppe communities at different pasture degradation stages (Miklyaeva & Lysak, 1996).

Stages of degradation	I	II	III	II	II-III
Succession row of subassociations	*Filifolium sibiricum* + *Thesium refractum*	*Filifolium sibiricum* typicum	*Ptilotrichum tenuifolium* typicum	*Limonium bicolor* + *Schizonepeta multifida*	*Limonium bicolor* typicum
Total number of species	73	44	52	91	70
Mean number of species at the 10 m^2 area	37	26	15	41	40
Mean coverage, %	52	54	56	62	56
Mean height of grass cover, cm:					
maximal	80	80	82	85	87
predominant	18	23	22	23	21

Stages of degradation	III	III-IV		IV
Succession row of subassociations	*Caragana stenophylla* + *Galium verum*	*Caragana stenophylla* typicum	*Salsola monoptera* typicum	*Salsola monoptera* + *Lappula intermedia*
Total number of species	86	64	45	49
Mean number of species at the 10 m^2 area	36	25	16	17
Mean coverage, %	52	44	40	45
Mean height of grass cover, cm:				
maximal	80	35	44	40
predominant	4	6	9	7

stands by a majority of the herbaceous plants. When grazing pressure is weak, the constancy of species which either prefer moderate grazing pressure or that are indifferent to grazing, increases by 1.5 times. Note that these observations were made when the total precipitation level was low.

Where grazing pressure is moderate, the floristic richness remains high but the height of most grass stands decrease more than 4 times; evenso the condition of these ranges is fairly satisfactory on the whole.

Heavy grazing pressure results in a strong vegetation degradation. All plant community indices decline, community structure is simplified, and floristic richness declines. And, compared with communities that experience a weak pastoral pressure, species with a negative response to grazing decrease by 12 times while species with less pressure response decrease by 3.5 times. Individual species in steppe plant communities, at different intensities of grazing pressure, undergo marked changes. The general form of these changes are presented in Table 4.9.

Pastoral regressive successions with steppe plant communities are determined by specific plant traits that change with these ecological conditions. Primary ecological changes are a compaction of loamy soils surface horizons followed by the deterioration of aeration and water conditions for plants. Second, of great importance, is a direct disturbance of sod by animals browsing on new growth, i.e., young and generative shoots. Conversely with sandy soils, extensive grazing cause a loosening of the soil surface, denudation of plant roots and, then followed by wind intensification, formation of wind depressions, removal of sandy materials, and their deposition at other sites.

In both cases the community floristic composition is impoverished, their structure is simplified, and dominant ecological plant groups are replaced. Normally, more primitive communities or groups develop that are

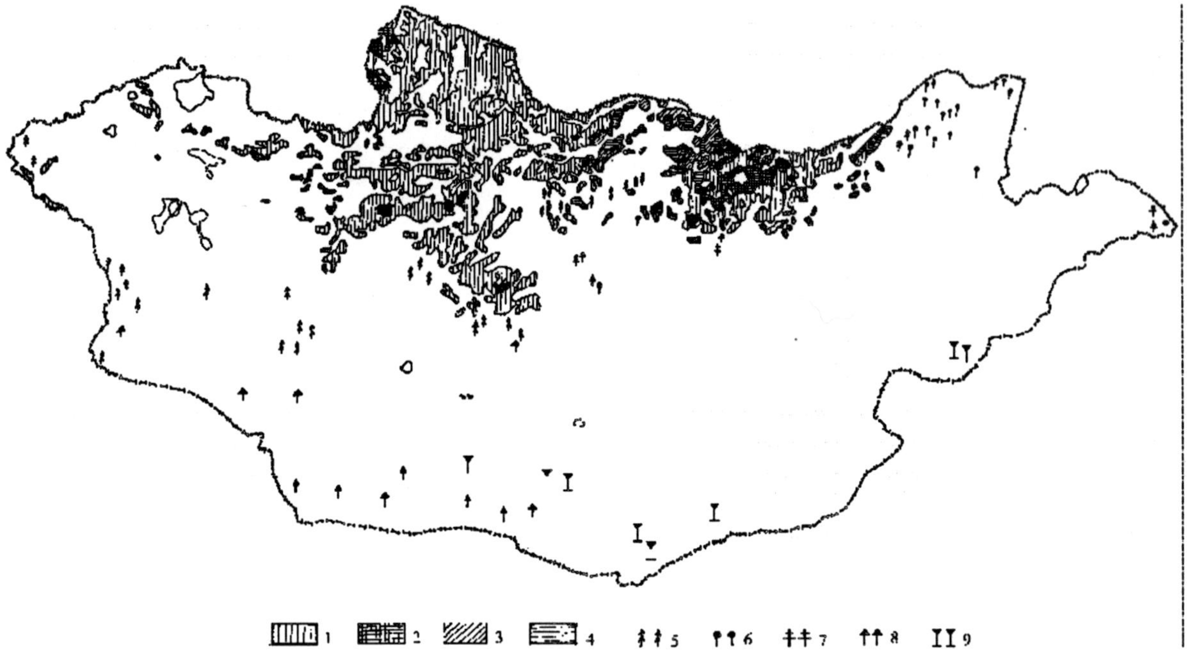

Figure 4.3. Forests of Mongolia

1 – larch forests; 2 – dark coniferous forests (Siberian cedar, fir – cedar, larch – cedar); 3 – pine forests; 4 – secondary birch forests; 5–9 – single habitats: 5 – *Larix sibirica*; 6 – *Pinus sylvestris*; 7 – *Betula platiphylla*; 8 – *Populus diversifolia*; 9 – *Ulmus pumila*.

representative of more xeromorphic species and total phytomass declines considerably. Such vegetation changes can be regarded as desertification.

The term *desertification* is fairly widely applied, sometimes to designate any process leading to ecosystem degradation by an excessive use of their natural capabilities. Hence here more precisely, desertification is defined as the processes of plant ecosystems degradation as a whole in arid and semiarid regions of Mongolia (Khza, 1984; Gunin, 1990a, b; Gunin et al., 1991). At one level desertification is determined by natural climatic factors with lower precipitation being the most important. Long cyclic droughts promote the development of processes such as wind erosion, salinization, etc. These are natural desertification processes. However these effects on plant cover are considerably modified by economic uses of natural land and vegetation resources (e.g., heavy grazing, direct destruction of plant cover on roads in built-up areas, industrial complexes, machine farming, etc.). In the final analysis, simultaneous effects of such natural and human-associated factors can result in the formation of vegetation-free areas (Figure 4.3); which truly is desertification!

Declining precipitation is very detrimental for most plants. For example, with ombrophyte plants whose water consumption is fully supplied by precipitation, an absence or a sharp reduction in precipitation causes immediate plant responses. Numerous ombrophyte plant species are in these steppe and desert communities, including grasses. Attributes of ombrophyte plants can be based on empirical data about their extent of transpiration and dependence on available soil moisture (Lavrenko et al., 1981; Lavrenko & Bannikova, 1986).

Similar conclusions arise from root systems excavations. A majority of steppe and desert

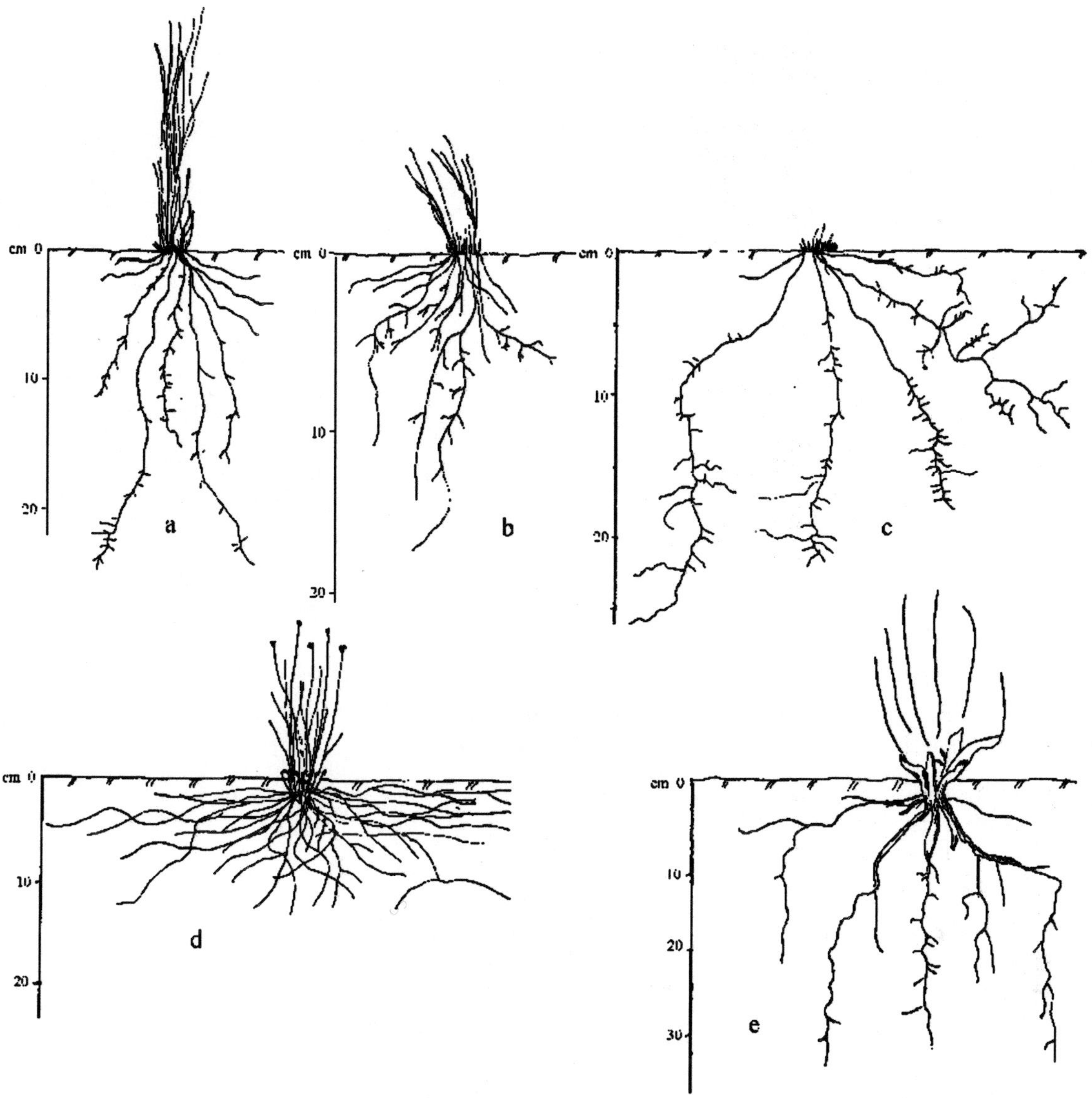

Figure 4.4. Root systems of ombrophytes:

a. *Stipa krylovii*; b. *Stipa glareosa*; c. *Cleistogenes songorica*; d. *Allium polyrrthizum*; e. *Ajania fruticulosa* (Baitulin, 1993).

species roots do not penetrate below 30–60 cm (Figure 4.4). Only certain shrubs, for instance, *Nitraria sibirica, Tamarix ramosissima, T. hispida, Zygophyllum xanthoxylon, Amygdalus mongolica* and *Haloxylon ammodendron* have deeper root systems, that reach a depth of 3 m and more (Figure 4.5; Baitulin, 1993). Such plants form a group of phreatophytes, subgroup of trichohydrophytes, whose water is mainly feed by ground capillary edge water (Beideman, 1983; Meintzer, 1927; Robinson, 1957; Vostokova, 1967, 1980). For instance, with *Haloxylon ammodendron* the most favorable recorded growing conditions have a water table from 3.9 to 10 m, deep, with mineralization of 3–5 g/l (Figures 4.6 & 4.7), and yield the highest productivity within this region (Slemnev et al., 1997).

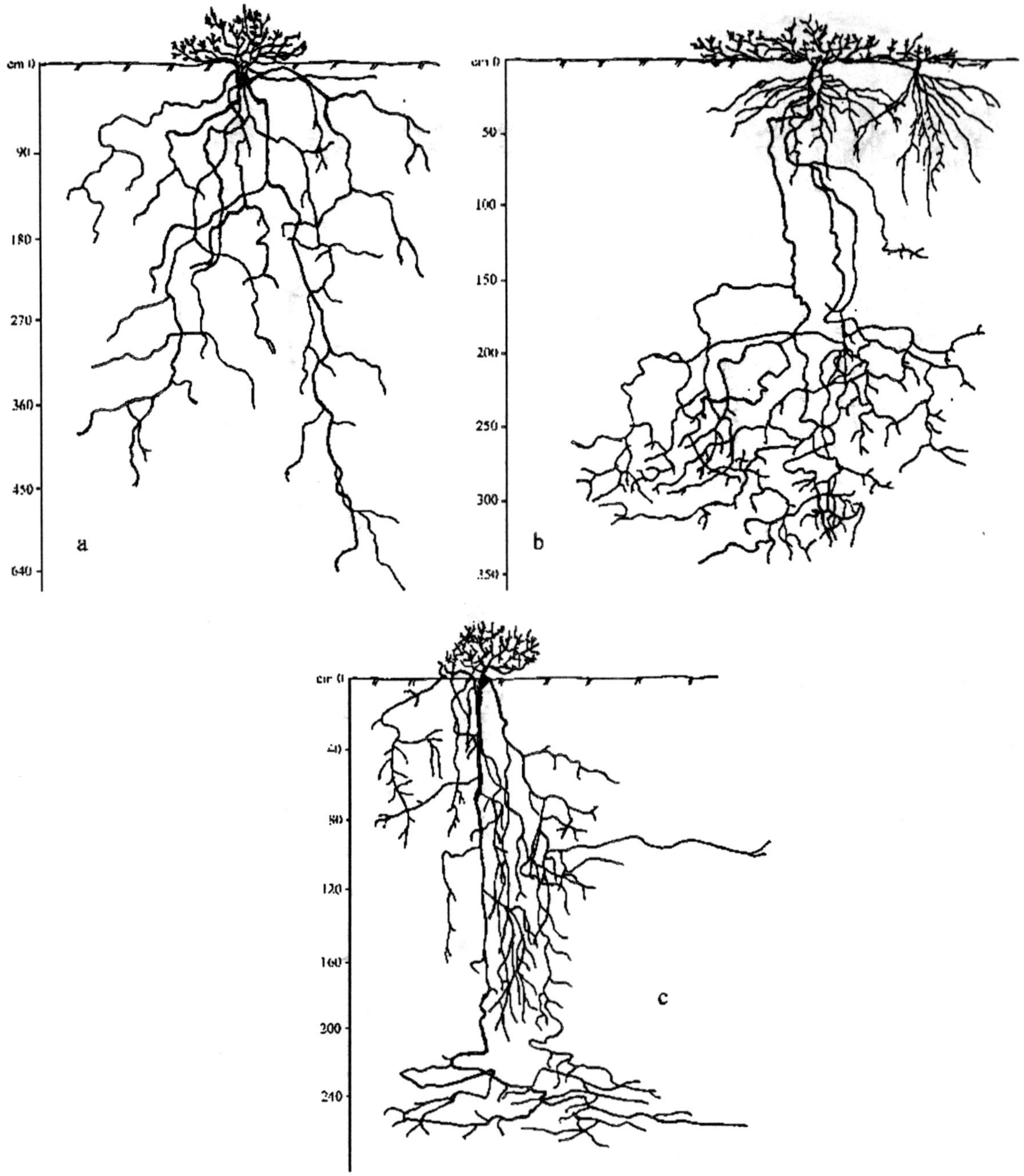

Figure 4.5. Root systems of phreatophytes

a. *Zygophyllum xanthoxylon*; b. *Nitraria sibirica*; c. *Amygdalus mongolica* (Baitulin, 1993)

Figure 4.6. Root system of *Haloxylon ammodendron* exposed for 2 m in the Gurvan-Tes region (photo by A.V.Prishchepa).

A final effect of disastrous decreases in rainfall is a decline in plant viability and subsequently a gradual removal of omborphytes from plant communities. Thus primary communities are replaced by derivative ones that are more drought resistant. These can be either some drought-resistant ruderal species with deeper root systems and reduced water transpiration; or phreatophyte plants that are more independent of precipitation. For instance, the progressive penetration of *Ephedra sinica* into communities of *Stipa gobica – S. glareosa* observed in the Gobi Tyan-Shan piedmont

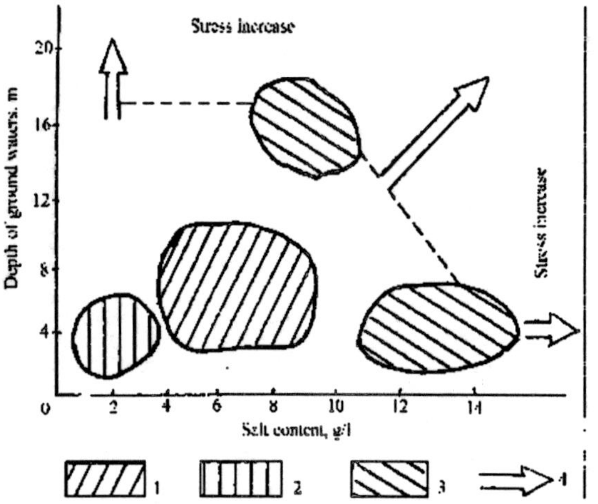

Figure 4.7. Ecological amplitude of *Haloxylon ammodendron* (Gunin, 1991; Slemnev et al., 1997).

1–3 productivity classes: 1 – high; 2 – medium, 3 – low. 4 – directions of changes in ecological conditions causing *Haloxylon ammodendron* community degradation.

(Gunin et al., 1993) can be accounted for by its roots reaching constantly moist soil horizons. Presumably *Ephedra sinica* also can be assigned to the trichohydrophytes group (after Beideman, 1983).

Finally, a long-lasting progressive decline of precipitation causes a lowering of the water table and even depletion of thin aquifers. These lead to lower viability or even mortality of phreatophytes when their roots can not obtain additional water either from ground aquifers or from their capillary edge. Then viability and all other plant community indices decrease.

Such drastic moisture decreases along dry riverbeds, depressions, or flat plateaus of true deserts leads to extra-arid deserts where the aboveground biomass in e.g., *Haloxylon ammodendron* communities, can be reduced 2 to 2.5-fold. In one extreme case it declined by 20–30 times and in another by 40, while their yearly increment of annual *Haloxylon ammodendron* shoot growth declined by 4.5 and 15 times. The natural water content of these soils was recorded at a witting level of 1.3 – 1.5 %. In multi-year droughts the annual growth increment of saxaul dropped from 70 to 10 kg/ha (Kazantseva, 1983; Gunin, 1990 a, b). Therefore *Haloxylon ammodendron* communities become sparse or completely disappear, and are replaced by the more xeromorphic shrub *Iljinia regelii* on flat plateaus and in dry riverbeds.

Indeed, that process formed the Transaltai Gobi of extra-arid deserts as distinguished by E.I. Rachkovskaya (Rachkovskaya, 1977, 1993; Evstifeev, 1980). That type of desert is characterized by a complete absence of higher plants except in dry river beds *with Haloxylon ammodendron, Iljinia regelii, Ephedra sinica* and *Zygophyllum xanthoxylon*, whose root systems can penetrate 3 m deep or more (Baitulin, 1993).

All of these processes result in more desertification. According to P.D.Gunin (1991), intensification of landscape aridization in Central Asian deserts and the emergence of a large uniform extremely arid desert region appears to have occurred within the recent historic period. The disappearance of *Haloxylon* forests on flat plateaus and the formation of major sand sediments also were promoted by human activities. Fairly often in the most stringent extra-arid desert conditions, free from any current human settlements, Neolithic camps can be found. In addition, on dry stream flats the development of arboreous species communities is supported by frequent findings of poorly decomposed *Haloxylon* roots and a fairly high density (1–2 colonies per ha) of abandoned *Rhombomys opimus* Licht burrows, which earlier was a prominent species in the extra-arid desert zone. Thus, desertification appears to have intensified within the current historic period, indeed extra-arid deserts have developed within the last 100–200 years.

In addition to a direct effect on plant water consumption, precipitation decreases trigger a number of side effects and processes whose extent is intimately associated with plant cover changes. These include wind-driven processes such as transfer and redeposition of sand materials, and increases in the number and

Figure 4.8. Distortion of mounded sands by ground road Sukhbaatar – Altanbulag (photo by A.V.Prishchepa).

duration of dust-sand storms that spread from desert zones into steppes.

The loosening of sand surfaces also is promoted by agricultural practices including livestock overgrazing and on weak rangelands with light textured soils, by frequent livestock driving and by heavy truck traffic pressure (Figures 4.8 & 4.9). Amongst such areas are ones with little resistant to transportation effects including loam-sands and loamy plains with a 30–50% vegetation cover or sandy plains with a 30% cover (Evdokimova et al., 1992).

Natural and induced wind-driven movement of sand deposits at one site and their transfer and redeposition at other sites can cover huge desert territories in Mongolia. Satellite photography can monitor and track these movements effectively (Vostokova et al., 1994, 1995). These photos show sandy features formed near shrubs and dry plants; and in addition photos track the movement trends of dust-sand air currents. Constant and very strong winds can transfer sand long distances. When any natural or artificial obstacle is present, sands are deposited in a wind «shadow» in the form of a cape (Figures 4.10 & 4.11), or develop as typical mounds. In either case these loose materials bury any vegetation remains on a site. These sandy soils are not fixed and are very unstable even under favorable moist conditions. For example, from experience trying to protect built-up area in Davst from such migratory sands, man-made shield obstacles proved very ineffective. Such huge migratory sands, 10 m in height, have been observed in the Gobi deserts, and northward in steppe deserts and steppes. Specifically such sand ranges are found between Orog-Nur and Tatsyn-Tsagan-Nur Lake, and on the southern ridge of Tatsyn-Tsagan-Nur where they seem to creep over the foothills.

Figure 4.9. Slope microterraces along a grazing route (photo by A.V.Prishchepa).

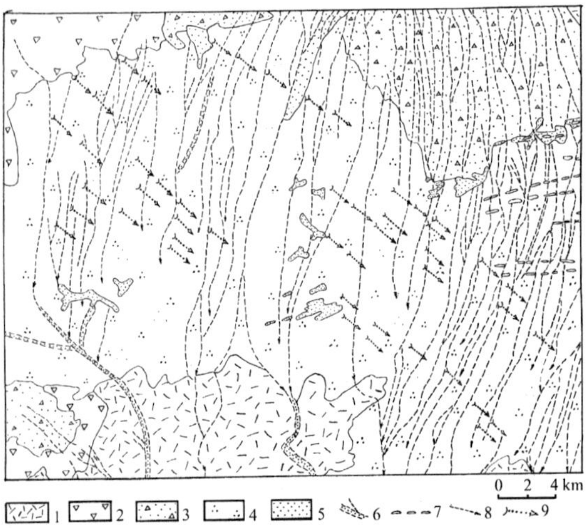

Figure 4.10. Wind flows in the Transaltai Gobi (according to Vostokova et al., 1995)

1 – medium height mountains with communities of *Anabasis brevifolia*, *Sympegma regelii* and participation of *Stipa glareosa*; low mountains with sparse groupings of *Sympegma regelii, Anabasis brevifolia, Ephedra przewalskii*; 3 – inclined sand-detrital plains with groupings of *Zygophyllum xanthoxylon* at sairas; 4 – flat loam – loamy sand plain with groupings of *Ephedra przewalskii* at sairas; sand massifs with communities of *Krascheninnikowia ceratoides*, *Artemisia xerophytica*; 6 – large sandy sairas and small erosion furrows with groupings of *Haloxylon ammodendron*, *Tamarix* spp., *Amygdalus mongolica* and others; 7 – chains of blown-in mounds near bushes of *Tamarix* spp., *Haloxylon ammodendron* along the watered fault; 8 – directions of water flows; 9 – directions of wind sand transportation.

Figure 4.11. Wind transport evidence: accumulated as sand «mantles» (photo by A.V. Prishchepa).

In addition to wind erosion processes, desertification also causes the silting over of soil surfaces. This process is readily observed with a single intensive rain in the mountains, where a heavy mud flow forms in dry river beds, that soon is deposited on the plains. Such a flow forms a silt-covered surface over buried plants (Figure 4.12). However intensive rains are rare even in the mountains so surface silting is not widely distributed. Under desert conditions the dried silt also can be scattered as a source of dust particles in soils typical of Gobi (Gunin, 1990a, b). In extra-arid deserts, silt deposits play a very detrimental role because they promote an unfavorable water supply condition. Silt deposits are poorly water permeable (with a filtration range of 0.02 to 0.07 mm/min) hence only a small portion of the scanty precipitation can penetrate even to 30 cm because most rain is intercepted by the crust horizon (Gunin, 1990a, b). Under these conditions over 75% of the rainfall is lost by soil evaporation.

Still another aspect of desertification occurs with declining rainfall, namely, an increase in surface salinization. Water-soluble salts are an inseparable aspect of arid zone soils (Kovda, 1968, 1984; Pankova, 1986, 1992). Salinization can be a limiting factor for soil fertility, affecting vegetation and its productivity. Also at the same time, salinization strongly reduces the water supply for plant communities, since large amounts of salts in a soil profile bind much of the input moisture into a dead resource for plants.

Generally, these processes occur in arid and semiarid regions, forming complex ecological conditions, which result in complexes of plant communities. For example, in the northern Transaltai Gobi in an intermountain

Figure 4.12. Silty crusts replacing a dried up temporary watercourse in Gobi (photo by A.V.Prishchepa).

depression, Ingenii Khovraiin Kholoi, the gentle foothill plains are dissected by numerous dry riverbeds. Some of them have several terraces, indicating that they developed in previous wet epochs (Kalicki & Prokop, 1995). Large areas are taken up by semi-fixed sand dunes. Sand dunes accumulate near the final rocks of mountains with migratory wind currents. The depression is an unstructured saline soil at the site of a former lake and ancient rivers. The lake shoreline is identified by sand mounds formed around shrubs developed around *Tamarix hispida*, which intercepted the wind-transferred sands (Figure 4.13).

In addition, wind and water flows do not always have the same direction of movement, which then further complicates vegetation patterns. East of the Ingenii Khovraiin Kholoi depression, between two parallel ridges, satellite photos convincingly show different directions for water erosion channels and riverbeds versus wind currents.

In wet conditions hydrogen salts accumulate whereas in dryer conditions salt accumulation also is strongly affected by salt and gypsum-containing rocks. When rocks are acted upon by environmental physico-chemical factors they become the main sources of salts within root-growing soil horizons. This is characteristic of Transaltai Gobi, the Lake Valley, Eastern Gobi, etc., where Cretaceous-Paleogenic red clays are salinized along with gypsum- and carbonate containing rocks (Murzaev, 1952; Sinitsyn, 1962; Marinov, 1967). In Transaltai Gobi alone, the total water-soluble salts in soils formed with red clays, reaches 5.0 t/ha. The basic redistribution of salts is promoted by water and wind erosion. Rainwaters are enriched with salts lifted into

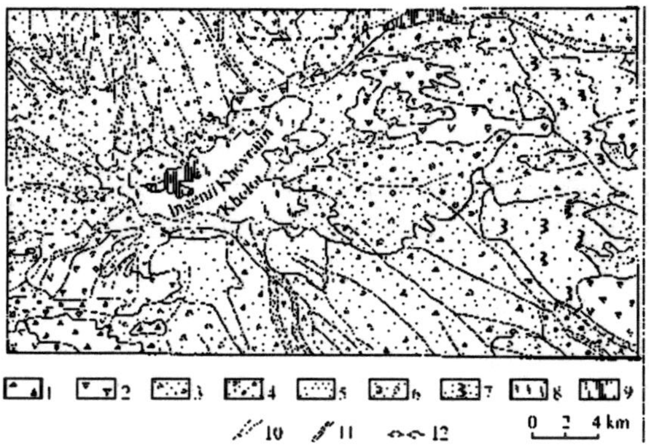

Figure 4.13. Combinations of plant communities at a dried up part of lake Ingenii Khovraiin Kholoi, formed under the influence of wind and salinization processes

1–3 – mountains and piedmont plains: 1 – medium height mountains with petrophyte desert communities with prevalence of *Sympegma regelii, Anabasis brevifolia, Ephedra przewalskii*; 2 – low mountains with single *Iljinia regelii* plants; 3 – inclined piedmont plain with loamy sand – detrital and stony – detrital cover and with groupings of *Haloxylon ammodendron, Ephedra przewalskii* mainly at sairas; 4 – loamy sand – pebble old alluvial plain with communities of *Haloxylon ammodendron* and participation of *Salsola passerina, Reaumuria songarica*; 5 – sands with gentle mounds fixed by *Stipa gobica, Caragana leucophloea, C. pygmaea*; 6 – sand «mantles» covering old surfaces, without vegetation; 7 – moving barkhan sands, almost without vegetation; 8 – salinized plains of ancient alluvial deltas, sandy at some places, with sparse communities of *Iljinia regelii, Sympegma regelii*; 9 – solonchaks with groupings of *Kalidium foliatum, K. gracile*; 10 – small sairas with single bushes; 11 – sandy beds of ancient sairas with communities of *Haloxylon ammodendron* + *Krascheninnikowia ceratoides, Stipa gobica*; 12 – large mounds with *Tamarix hispida.*

the air by wind currents. In Transaltai Gobi rain water minerals exceeded 0.12 g/l (Gunin, 1990a, b). The mineralization of strong water currents temporarily passing along ravines is even greater. As water moving from an upper ravine reaches to its mountain exit, the water mineral content can increase by 2.5 to 3 times. In the ravine, Ekhiin-Gol, a temporary stream in its lower reaches attained 0.54 g/l (Gunin, 1990a, b).

An even more rapid salt increase was found in the waters of streams that arise as fresh ground water. With springs along the tectonic formations of Obot-Khural (Alashan Gobi), water mineralization was 0.3–0.4 g/l and in terms of composition it was hydrocarbonate-calcium. Downstream the water minerals ranged from 0.42 g/l to 1.26 g/l, and in terms of water composition, the lower reaches stream was magnium-sodium, and sulfate-hydrocarbonate (Table 4.10). Simultaneously the vegetation changed: from a reed-liquorice community at the spring to a *Nitraria* community in lower reaches (Vosotkova & Kazantseva, 1995; Vostokova et al., 1993).

Salts transferred with water or wind currents in deserts accumulate in the crust or in the undercrust horizons, where salt layers develop. The depth of these layers normally does not exceed 30 cm, i.e., salts occur at a depth of maximally moistened soil. Hence rainfall, which is not physically evaporated, is bound by these salts. This restricts the effectiveness of rainfall to practically zero for plant growth (Zolotokrylin & Gunin, 1986).

Each of these processes cause regressive successions of plants communities and strongly handicap any progressive regeneration of plant community successions even with the onset of increased precipitation. And such degradations of plant cover are aggravated even more by irrational nature management practices. Combined, these promote an expansion of desertification in Mongolia!

4.3 Progressive plant community regeneration

During favorable rainfall years, disturbed and degrading plant communities can be partially regenerated in virtually all plant ecosystems of Mongolia.

Much attention has been given to forest regeneration after intense fires and clear fellings (Korotkov & Dorzhsuren, 1988; Dugarzhav, 1990). Successful forest regeneration is a strong function of: maturity of the main tree stand generation; undergrowth at

Table 4.10. Ground water mineralization changes downstream along the Barun-Sukhain river (Alashan Gobi) and the ecological sequence of *Achnatherum splendens* (deris) communities (Vostokova & Kazantseva, 1995).

Habitat and deris community	Total mineral content in water (g/l)	Type of salinization	Water pH
Upper reaches:			
reed-licorice	0.42	Calcium-sodium chloride-hydrocarbonic	8.13
Middle reaches:			
Flood-plain, reed-rush	0.65	Calcium-sodium Sulphate-hydrocarbonic	8.6
terrace, licorice-nitraria	–	–	
Low reaches:			
1 terrace, *Puccinellia tenuiflora – Leymus*	1.26	Magnesium-sodium sulphate-hydrocarbonic	9.86
II terrace, *Nitraria*	–	–	

Note: – means absence of data

the felling of mature trees; the number and quality of mature seed-bearing trees; and the site remoteness from dense forest areas. It is very important that forest tree seed yields occur in favorable years in terms of rainfall and sustainable conditions for seed germination and seedling growth. Seed yields of Siberian pine, larch, and pine undergo an annual variation with the biannual climatic cycle, e.g., with larch the years of 1983 and 1985 were productive; whereas 1988 was unproductive (Milyutin et al., 1988; Dugarzhav, 1990). In very favorable years, larch penetrates into steppe communities (Korotkov & Dorzhsuren, 1988). But a substantial limiting factor in Mongolia for natural forest regeneration is grazing animals feeding on seeds or injuring young shoots. And, in addition, tree crops often are injured or completely destroyed by insect pests.

General forest regeneration successions patterns for Mongolia are presented in Table 4.11.

With steppes, regeneration successions have received the greatest attention on fallow lands (Miklyaeva, 1996). According to I.M. Miklyaeva, the plant indicators of fallow lands are: *Artemisia scoparia, A. macrocephala,*

Table 4.11. Generalized scheme of forest renewal successions.

Vegetation		Renewal successions		
Initial community	After burn or clear cutting	⟶		
Larch forests	*Chamaenerion angustifolium* + regrowth of *Larix sibirica*	Birch - larch forests	Larch - birch forests	Larch forests
	Chamaenerion angustifolium + tall herb communities	Birch forests	Birch-larch forests	Larch forests
Fir-cedar forests	*Chamaenerion angustifolium* + single *Abies sibirica*	Fir	Firs + single cedars (*Pinus sibirica*)	Fir-cedar forests
Cedar-larch forests	*Chamaenerion angustifolium* + single larches	Larch-birch	Larch-birch with cedars	Larch-cedar forests
Pine forests	Meadow herb	Birch, Birch-pine forests	Pine-birch	Pine forests
	communities	Steppes		

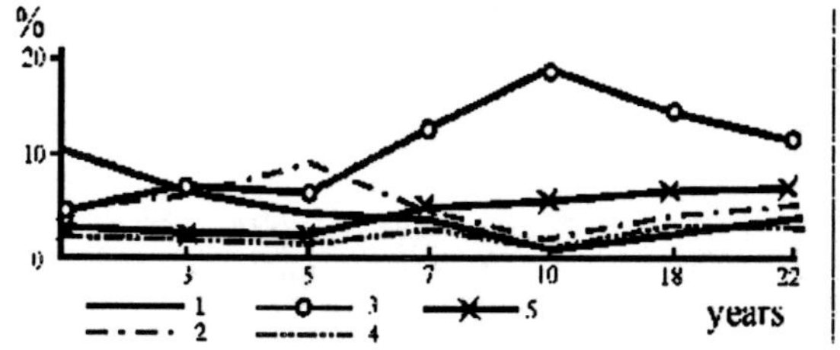

Figure 4.14. Changes of vegetation (number of species, %) on idle lands of different years (Miklyaeva, 1996; Gunin et al., 1996).

1–3 – biomorphs: 1 – annuals; 2 – biennials; 3 – perennials; 4 – weed species; 5 – mean number of species.

Fallopia convolvulus, Aconogonon divaricatum, Linaria acutiloba, Setaria viridis, Potentilla bifurca, and *Axyris amaranthoides*. These species play a strong role in the open communities of fallow lands and hardly occur in established communities of slightly disturbed steppes. The first regeneration stage involves 1 to 5 year fallows of tall weeds. The proportion of annual and biannual species is 44% and weed species are over 55%. Weeds expend the accumulated soil nutrients (Figure 4.14) and the weed species composition is similar to tilled croplands, with species unpalatable to livestock dominating, i.e., *Artemisia scoparia* and *A. macrocephala*. Here the community structure is simple and not established, the floristic richness is low and economic conditions are poor. The state of fallows at the loose sod stage, involving 7-year-old fallows, is moderate. Although the community structure has not been established, the floristic composition increases to near 40 species. The proportion of annual and biannual species declines slightly along with weed species. Species characteristic of dry steppes, e.g., *Stipa krylovii, Poa attenuata* subsp. *botryoides, Leymus chinensis* and others are present, hence the forage value of grass stands increases.

The condition of dense-sod stage fallows (18–22-year-old) is quite satisfactory. The community structure is fairly complicated, the floristic species richness is greater, the proportion of annuals and biannuals is negligible, and *Stipa sibirica* and *S. krylovii* are dominant. The contribution of weed species is still high, although it has declined to 43 %. In terms of their functioning regime, these fallows are similar to natural steppe plant ecosystems and the effects of burrowing mammals are similar to those on virgin plots.

Since their existence, the fallow lands in Eastern Mongolia have been exposed to unregulated grazing and burrowing animals. Virtually, they are exposed to the same extent as natural ranges. To find the natural-species indicators of grazing effects and their representation in community vegetation regeneration series (Table 4.12), a floristic analysis was performed with communities exposed to moderate or to weak range pressure.

On wheat crops and in their fallow land regeneration community series, species that do not tolerate grazing are poorly represented, except *Pulsatilla turczaninovii*. Also poorly represented are species for which moderate grazing pressure is optimal, except *Artemisia palustris*. The most constant species

Table 4.12. Pasture load tolerances of Eastern Mongolian steppe species (according to Gunin et al., 1996).

Plant Species		
Persistent to moderate load	Preferring moderate load	Optimally developing under moderate load
Cymbaria daurica	*Heteropappus biennis*	*Veronica laeta*
Bupleurum scorzonerifolium	*Serratula centauroides*	*Artemisia palustris*
Medicago ruthenica	*Dontostemon integrifolius*	*Orostachys malacophylla*
Astragalus tenuis	*Salsola monoptera*	*Festuca lenensis*
Potentilla tanacetifolia	*Lappula intermedia*	*Potentilla acaulis*
	Chenopodium album	

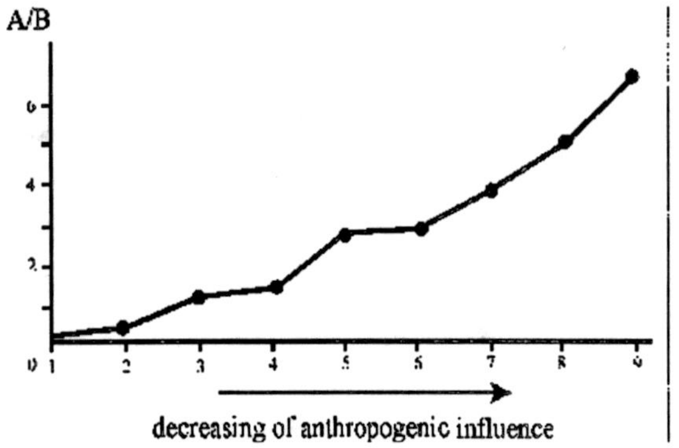

Figure 4.15. The degree of plant community maturity (according to ratio between perennials and annuals and biennials – A/B) during decreases in anthropogenic influences (Gunin et al., 1996)

Types of ecosystems: 1 – arable lands; abanded lands: 2 – one year (I stage of restoration), 3 – five years (I stage of restoration), 4 – seven years (II stage of restoration), 5 – eighteen years (III stage of restoration); pastores: 6 – IV stage of degradition, 7 – III stage of degradition, 8 – II stage of degradition, 9 – I stage of degradition. A/B : A – perrenial plants; B – annual and biennial plants.

irrespective of grazing pressure (except very heavy pressure) are species that prefer a moderate grazing pressure (Table 4.12) and have a constant occurrence in dense sod stage fallow regeneration communities. The extent of their occurrence is close to communities at stages III and IV of pasture degradation with the most common being *Heteropappus biennis, Medicago ruthenica* and *Potentilla tanacetifolia.* Typical dry steppe plants follow different patterns. Their presence in the tall-weed stage of regenerating communities is only minor. But beginning in the loose-sod stage, they noticeably increase. There are reasons to believe that they mostly account for the regeneration of fallow lands. And it should be noted that the role of dry steppe species increases under the effects of fires; that was shown in a 10-year-old fallow land fire study.

Thus the natural regeneration of fallow lands proceeds slowly. Full regeneration takes over 20 years, although spring fires considerably accelerate plant regeneration. Purposeful setting of dry grass stands on fire is a common method used for soil enrichment with ash elements. Normally after fires herbaceous plants develop more rapidly and vigorously.

Other research has shown that the resistance of plant communities to human-controlled effects is correlated to the degree of community establishment, which can be estimated from the ratio of perennial, annual, and biannual plants. By comparing communities at various degrees of establishment, they can be arranged in a series from croplands to steppes, which are the least affected by grazing (Figure 4.15).

Assessing the degree of plant community establishment allows us to follow the course of natural fallow lands regeneration and range degradation. On croplands the ratio of biological plant groups is 0.3, and one-year-old fallows have a similar ratio. In 7-year-old fallows that are in a loose sod regeneration stage the degree of plant community establishments increases roughly fourfold. With 18-year-old fallows, a maximum ratio between dense-sod perennials: annual and biannual plants occur near 2.5. The plant community species composition at this regeneration stage virtually matches that of steppes, which are not exposed to cropland effects. Evenso the plant cover of these fallow lands is not completely established, as indicated by the ratio of plant groups, which is only half that in-disturbed steppes. Also, the proportion of annual and biannual species is comparable to that on degraded ranges at stage IV. Traditionally, when ranges are used moderately by livestock, perennial plant species dominate over annuals. When ranges are undergrazed, grass stands become sparse due to aging and the ranges acquire annual and biannual herbs.

The ratio of plant groups can be regarded as a quantitative manifestation of plant cover by species characteristic of primary communities, for example, under range or cropland use, or at various stages of pasture degradation, or the regeneration of fallows. As illustrations,

Table 4.13. Ratio of different biological plant groups of cereal seedlings and in idle lands (Miklyaeva, 1996).

Biological groups of plants and species diversity of communities	Cerol Seeding	Idle lands of different age, years						
		1	3	5	7	10	18	22
Annuals	52	34	23	21	17	2	10	18
Biennials	24	31	46	25	21	7	18	24
Perennials	24	35	31	54	62	91	72	58
Mean number of species in description	15	12	11	18	24	27	32	33
weeds among them	11	7	6	8	12	2	14	14
Number of species in an area 20×40 m	35	29	26	28	42	52	67	70

18-year fallows are impoverished 2 times more than annual fallows and 12 times compared with undergrazed ranges. At the same time an undergrazed range has a 1.7-times greater species impoverishment than slightly degraded ranges.

Judging from our analysis of community formations and the species paucity, undergrazed ranges can be regarded at an initial stage of natural recovery, as observed in the absence of grazing, for instance, in reserve conditions (Gunin, 1990a, b).

Thus, the entire fallows recovery series falls into two main groups in term of structural differences, floristic composition, the ratio of obligate plant groups, and weeds. The first group, with a simple structure, includes one to seven year-old fallows and the second group, with more complex structures 10 to 18 to 22 year-olds (Table 4.13).

Similar successions of plant communities also occur even now where abandoned dirt roads are overgrown. During recent years, asphalt roads have been constructed along some heavily used routes. Naturally, traffic along parallel dirt roads is either completely discontinued or sharply reduced. As a result as early as 1–2 years after paving a highway, ground roads start overgrowing. For instance, the road from Ulan-Bator to Arvaikhere was paved in 1991–1993, by the year 1996 nearby roads were overgrown with plants that previously occupied only road edges (Hilbig, 1988). These groups were dominated by *Artemisia scoparia, A. macrocephala* and *A. frigida*, accompanied by numerous weeds (*Descurainia sophia, Fallopia convolvulus, Lappula myosotis, Polygonum arenastrum, Kochia densiflora*, etc.). Thanks to the bright green color of *Artemisia* spp., overgrown roads could be easily traced visually. This recovery corresponds well to the tall-weed stage of fallows vegetation.

Recovering vegetation successions have been observed in the Transaltai Gobi desert. By studying both natural and artificial regeneration of desert vegetation, a well-defined relationship was found between regeneration processes and the level of moisture. In fact, an anomalously high precipitation level in 1993 (Table 4.14) triggered a succession regeneration.

The uniqueness of 1993 was highlighted by the emergence of *Haloxylon ammodendron* and *Iljinia regelii* plants growing in inter-ravine watersheds (flat areas between steam beads) in extra-arid deserts. Previously, regeneration of higher plants in these habitats was not recorded. Moreover, it was exactly the absence of vegetation on built-up areas in the middle of stream flats that gave the main argument for distinguishing extra-arid deserts into a special subzonal area (Rachkovskaya, 1977, 1993).

However, despite the almost ubiquitous presence of *Haloxylon ammodendron*, only its seedlings recovered exceptionally. The shoot numbers ranged from 0.02 individuals/m^2 on flat areas to 1.12 individuals/m^2 in dry steam beds. Overall, in depressions and ravines, the shoot numbers were 1.3 to 3.4 times higher than on flats between streams. In contrast, in central Transaltai Gobi, west-north-west of Ekhiin-Gol station in the middle of a flat plain, 388 shoots were recorded per 20 m^2, i.e.,

Table 4.14. Transaltai Gobi precipitation measured at Ekhiin-Gol meteorological station, mm.

Year	Month												Total
	I	II	III	IV	V	VI	VII	VIII	IX	X	XI	XII	
1991	0.0	0.0	0.0	0.0	0.0	0.0	6.0	16.2	0.4	0.0	0.0	2.0	24.6
1992	0.0	0.0	7.8	0.0	0.0	18.4	68.8	3.3	0.0	0.0	0.0	1.2	99.5
1993	0.0	0.0	53.0	0.0	10.0	19.0	46.3	34.0	0.0	0.0	2.3	0.0	164.6
1994	0.0	0.0	1.0	2.4	0.0	11.3	25.3	16.3	0.8	3.8	0.0	0.6	61.5
1995	0.0	0.0	0.9	0.0	2.5	2.2	3.8	87.8	29.3	0.0	0.0	0.0	126.5
1996	0.0	0.0	0.0	7.9	0.0	25.2	43.5	4.8	0.0	0.0	0.0	0.0	81.4
1997	0.0	0.0	0.0	0.4	0.8	20.2	11.2	10.7	9.4	–	–	–	52.7

Note: – means absence of data

Table 4.15. Numbers and survival of dominant plants in typical extreme arid desert plant communities of Transaltai Gobi extreme arid deserts (Slemnev et al., 1997a).

Species	Year of observations	Watershed	Depression	Sair
Haloxylon ammodendron	1993	33.3	43.3	112.0
	1994	18.0	25.3	61.0
		54.0	58.4	54.5
	1995	12.0	14.7	31.0
		36.0	33.9	27.7
Iljinia regelii	1993	3.2	6.9	–
	1994	2.9	5.7	–
		90.6	82.6	
	1995	1.8	8.8	–
		56.2	127.5	

Note: Numerator – number of young plants (sp.100 m^2), denominator – survival rate (percentage to number in 1993)

19.4 individuals/m^2 in 1993, with a wide range in growth and development (Figure 4.16). The first year these young *Haloxylom ammodendron* plants developed a deep root system, 4 to 5 times greater than aboveground shoots (Figure 4.17). Thus, young plants can reach constantly moist soil horizons under favorable moisture conditions. If this situation does not occur, seedling mortality is inevitable!

1994 studies across Mongolia show that the entire area of *Haloxylon ammodendron* regenerated very slowly, except the Transaltai Gobi with its very active regeneration. There repeated observations show a very high viability of shoots – up to 60–70% (Table 4.15). Measurements of soil water content show that even on flats, soils were moistened to a depth of 1.7–1.8 m and in some cases the moisture of horizons reached 5%. Such a high moisture content on flats was not recorded during the previous 15 years by the Ekhiin-Gol Desert Station.

Simultaneously across the Transaltai Gobi steppe-like deserts, a high number of new shoots occurred on dominant species, i.e. *Anabasis brevifolia* and *Stipa gobica*. Indeed, on a sample site of 10 m^2 located in a large plain, 380 shoots of *Anabasis brevifolia* and 830 of *Stipa gobica* were counted. All shoots were roughly at the same developmental stage. Repeated observations in 1994 measured 30–40% viability. Thus, during wet cycles when moisture conditions that are unique for the Gobi occur, a new population generation can develop and dominant on flats previously vegetation free. According to our estimates

Figure 4.16. Seedlings of *Haloxylon ammodendron* in 1993 (photo by A.V.Prishchepa).

Figure 4.17. Young sprouts of *Haloxylon ammodendron* in 1996 (photo by A.V.Prishchepa).

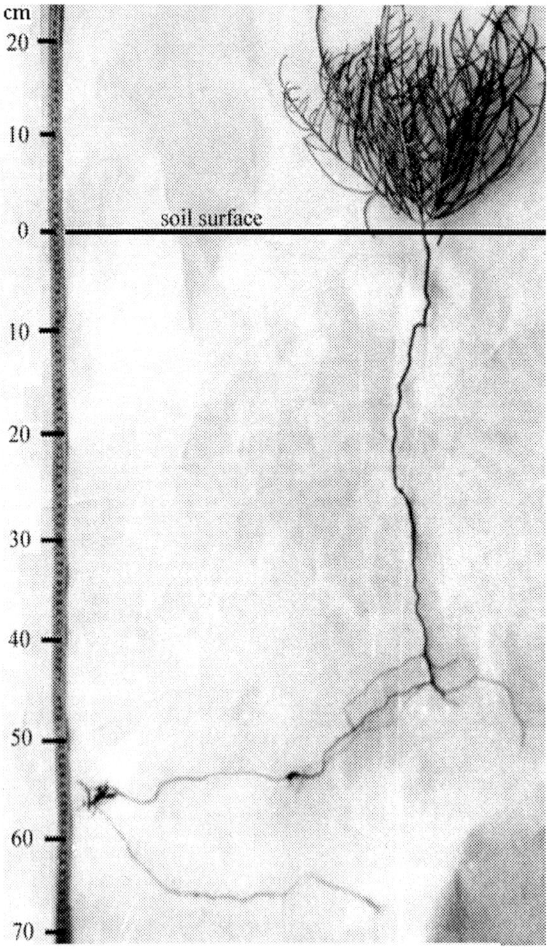

Figure 4.18. Root system of one-year old *Haloxylon ammodendron* (photo by A.V.Prishchepa).

from older shoot numbers and growth, such conditions may have existed there some 38–40 years ago. This almost corresponds to a 44 year climatic cycle.

Supporting observations also were made in experimental plots where special grooves were ploughed out with mounds facing west or east (Slemnev et al., 1994, 1997; Figure 4.18). These sand and moisture-accumulating grooves were dug out on flats between streambeds of the Tsagan-Bogdo piedmont plain (near the Ekhiin-Gol Station). Furrows, 40–50 cm deep and 50 m long, were ploughed every 3 m. They destroyed the crust horizon and a broken-stone shield. After 10 years of existence the furrows were partly filled with sand, but no vegetation had recovered. Only after the anomalously high 1993 precipitation, shoots of *Haloxylon ammodendron* as well as *Iljinia regelii* emerged on massive scales. Shoot viability during subsequent years with *Haloxylon ammodendron* in furrow positions ranged from 25.7 to 56.3% relative to the 1993 shoot numbers. Shoot viability with *Iljinia regelii* in 1995 varied from 16.4 to 88.6% relative to the shoot numbers in 1993. (Table 4.16).

Prior to that in extra-arid deserts *Haloxylon ammodendron* was observed only in dry streambeds. Therefore its regeneration with viable shoots both in experimental plots and naturally was regarded as a successful regeneration (Slemnev et al., 1997).

Hence, under favorable climate conditions, especially water, regeneration or even new successions have been recorded across Mongolia in many zones. But, on balance, a number of unfavorable processes result in sharp deteriorations of habitat ecological conditions. When these occur during plant community succession development a strong deacceleration occurs. Similar deaccelerations

Table 4.16. Survival of dominant plants around furrows in extreme arid deserts of Transaltai Gobi. 1994–1995, % of the number in 1993) (Slemnev et al., 1997a).

Species	Year	Mould to the west			Mould to the east		
		depression	mound	water shed	depression	mound	water shed
Haloxylon ammodendron	1994	67.7	94.8	66.7	56.3	55.7	111.1
	1995	44.6	56.3	32.2	38.7	25.7	44.4
Iljinia regelii	1994	155.6	39.6	90.1	85.6	74.0	105.0
	1995	51.4	16.4	62.2	53.6	44.6	88.6

Figure 4.19. Seedlings of *Haloxylon ammodendron* (above) and *Iljinia regelii* (below) in ploughed sand-accumulating furrows. Eikhiin-Gol, 1996 (photo by A.V.Prishchepa).

also are promoted by intensification of irrational nature management practices. On the other hand, construction of paved roads greatly reduced the extent of traffic pressure on ranges and dirt roads, promoting healthy vegetation regeneration.

4.4 Mapping vegetation dynamics

Observations over the last 25 to 50 years show the cyclic nature of natural vegetation successions. All studies during periods of stable decreases in annual precipitation show exceptionally degrading successions. Recovery successions, including the regeneration of forest vegetation, occurred very rarely or were absent altogether. Unfavorable natural conditions aggravated by intensive human economic activities, together have resulted in an expansion of areas occupied by disturbed plant communities.

Climate cycles with increased precipitation levels promote the emergence of progressive generations of plant communities successions. Presumably with time, the plant cover would attain its optimal condition. But often this is prevented by at least two factors. First, during periods when degrading successions emerge, frequently other intensive processes develop. Covering large areas, these processes worsen ecological conditions so profoundly that the regeneration of primary vegetation often became either virtually impossible or it takes a protracted time period. Thus unfavorable climate periods can cause a new set of degrading successions. Secondly, incessant human-associated effects, e.g., livestock grazing, particularly during unfavorable years, results in the fixation of degrading successions and handicaps any natural regeneration of disturbed vegetation.

However, even local environmental protection measures, such as a reduction in transport pressures on rangelands via the construction of asphalt roadways, immediately and positively affects plant cover!

Hence, thoughtful nature management necessitates programs supporting the conservation of unique plants and plant communities, plus supporting the use and regeneration of plant resources. For these to occur, specific maps are necessary that show and assess ecosystems distributions plus give the degree of plant community disturbances.

One such set of maps is *Ecosystems of Mongolia*, scale 1:1000000 (1995). Specific features of those maps are firstly the use of indicator vegetations as an integral feature of natural environments modern state and, secondly, the presentation of graphic work on the basis of satellite photos (Gunin et al., 1992; Gunin & Vostokova, 1993; Sokolov et al., 1990).

The sensitive responses of plants and plant communities to changes in ecological conditions forms a good basis for their use as indicators of modern environmental states. The use of plants as an index to environment changes is a trend of indication geobotany (Viktorov et al., 1962). Vegetation serves as an index to the integrated condition of an ecosystem involving all species plus natural and human-associated effects. To plot these maps, the use of plant cover indicators was used because plants are one of the strong physiological elements of a landscape.

Based on information about the state of vegetation, supplemented by changes in relief and in soils, generalized criteria were developed to assess the disturbance level in a whole ecosystem (Gunin & Vostokova, 1993). These assessments were plotted and given both in legends and as a supplementary index for each ecosystem. These indices reflect the current stages of degradative successions in primary plant communities. For each vegetation type, similar criteria were developed to assess the stages of degradative vegetation successions.

Our experience in applying these criteria has shown their applicability in characterizing the status of ecosystems as a whole. For example, tables to assess the vegetation disturbances in altitudinal zones of Mongolian Altai were collected (Table 4.17), along with those for mountain-forest, steppe, and desert communities (Tables 4.18, 4.19 & 4.20).

Table 4.17. Assessment of plant cover disturbances on altitude belts.

Altitude belt	Height range, meters above mean sea level	Vegetation	Values of main characteristics of vegetation: Specific richness	Coverage, %	Participation in grass cover, %: grasses	sedges	Forbs: total number	Forbs: including wormwoods	Character of anthropogenic impact	Assessment of the state of ecosystem (ranking points)
Alpine belt (mountain tundra)	2800-3800	*Kobresia*, *kobresia* - sedge, and sedge communities	15-25	80-100		50-60	20-35		Summer grazing, permanent pastures	0-1
			<15	up to 10					brigade centers.	3-4
	2850-3000	Steppized grass - cobresia communities	15-17	90-95	5-10 up to 30	70-80	10	5	Summer grazing permanent pastures,	0-1
			<15	up to 10	up to 5	<70	>10		brigade centers.	3-4
Sub-alpine belt	2750-3400	Cryophyte cushion plant formations:	11-15	65-90					Summer grazing.	0-1
		a) steppized grass	up to 20		30-50		15-30			
		b) with little participation of grasses		40	15-20		>30			2-3
	2600-3200	Cryophyte grass-forb steppes		40	8-10		>30		Concentrations of	3-4
				20	2-5	up to 3	>30		yurtas and herds.	4
Mountain steppe belt	2500-2750	Bunch grass - forb moderately dry steppes	13-18	40-50	20-30		10-20	3-5	Winter grazing	0-1
			50-60	30-40				15		
			20-40	>5-10				up to 40	Overgrazing	3-4
	2100-2500	Meadow forb-grass and sedge steppes	up to 50	40-30	20-40	25	25-30		Winter grazing	0-1
	1700-2400	Fairway grass - low feathergrass - forb dry steppes	8-12	60-70	15-30 up to 50		10-25	5-15	Winter grazing	0-1
				up to 60	<15		>25	up to 15		2-3
				<10	<3		>25	>15	permanent pastures, watering points, transport routs.	3-4
	1700-2300	Desertified steppes	10-15	50-60	25-35		10-15	5-10	Winter grazing	0-1
					2-3		up to 40	>10		2-4

Table 4.18. Assessment of human-related vegetation disturbances based on mountain-forest communities of Mongolia.

Disturbance	Ecosystems	Characteristics	Criteria (quantitative assessments)	Degree of disturbance
Cuttings	Mountain – forest	Elimination of tree stands	% at the test area:	
			30	weak
			30–60	medium
			>60	strong
Fires	Mountain – forest	Disturbance of subordinate layers of phytocenoses (soil-covering vegetation, boscage) during low-layer fire.	Partial burning	weak
		Destruction of wood stands as a result of upper fire.	% at the test area:	
			30	weak
			30–60	medium
			60	strong
		Destruction of forest phytocenosis during low fire.	% at the test area:	
			30	weak
			30–60	medium
			60	strong
	Forest-steppe	Destruction of dry grass litter.	Burning of litter %:	
			fragmentary 50	weak
			continuous 50	medium
			complete burning	strong

Table 4.19. Expert assessment of anthropogenic disturbance in steppe communities.

Influencing factor	Zonal type of ecosystem	Processes	Characteristics	Criteria (quantitative characteristics)	Degree of disturbance
Grazing	Steppe	Suppression of plant survival.	Disturbance of life cycle.	Ratio between vegetative and generative plants, %	
			Changes in habitus of plants.	Height and diameter of crown, cm. Decrease in coverage (% at the experimental site).	Weak (1)
		Changes of relationships and of the role of dominating species in a community.	Substitution of dominants by co-dominants and appearance of weeds.	Ratio between coverages.	Mean (2)
		Changes in qualitative composition of a community.	Increase in coverage of weed and non-edible species.	Coverage by weed and non-edible species (% at experimental site: 50 % of community coverage).	Strong (3)

Table 4.19. Continued.

Influencing factor	Zonal type of ecosystem	Processes	Characteristics	Criteria (quantitative characteristics)	Degree of disturbance
			Development of species with compensation ability (creeping-rooted, vegetatively regenerated).	Increase in quantity of bushes and dwarf bushes (% at the experimental site).	
		Change of indigenous communities.	Substitution of indigenous dominants and co-dominants by weed and non-edible species.	Abundance, percent coverage of weed and non-edible species (% at experimental site).	Very strong (4)

Table 4.20. Signs of human-associated disturbances Mongolian desert communities of depending on the factor and impact degree.

Factor	Degree of influence	State of vegetation of separate components and ecosystems
Grazing of domestic animals	No or very weak	Vegetation possesses complete set of dominants and accompanying species, good survival ability.
	Weak	Phytocenoses have constant composition of species dominants, survival ability is satisfactory. The floristic composition is disturbed. On more than 10% of the area there are considerable disturbances of vegetation and decreased survival.
	Medium	The dominant composition of communities remains the same, survival ability of plants is decreased, bushes and sods are deformed. The composition of accompanying species is disturbed considerably. Up to 15 % of the area with considerable changes in dominants. Up to 15 % of the territory subject to general trampling. Sandy soils show wind processes.
	Strong	Dominant composition of communities damaged, locally destroyed completely. Alteration of communities takes place, often with weed species domination. Accumulation of sandy sediments near bushes
	Very strong	Complete trampling of natural vegetation.
Wood collection	Medium	Up to 10 – 15 % of vegetation (*Haloxylon* and large bushes) extracted.
	Strong	More than 15 % of bushes and trees eliminated.
Agriculture (irrigated)	Very strong	Spontaneous plant communities destroyed, agroecosystems formed.
Settlements of urban and rural types and their vicinities	Very strong (irreversible changes)	Natural vegetation destroyed completely and substituted by technogenic-ruderal vegetation (within settlements).
	Strong	Dominant composition of plant communites disturbed (20–25 km from towns).

Table 4.20. Continued.

Factor	Degree of influence	State of vegetation of separate components and ecosystems
Mining industry	Very strong (irreversible changes)	All the components are disturbed (relief, lithogenic base, soil-vegetation cover) as a result of quarries and slag-heaps with vegetation (mainly ruderal species).
	Strong	The dominant natural vegetation,, composition of natural vegetation and composition of accompanying species disturbed. Often communities are substituted by weeds (up to 1 km around mining sites).
Transport Roads	Weak	Once or twice car tracks disturb survival ability of plants linearly.
	Medium	Roads of seasonal use, destroying vegetation linearly. Linear disturbances can reach 10 – 15 m in width with ruderal vegetation between them.
	Strong	Multi-rut roads of constant use. Vegetation cover substituted or partly substituted by ruderal groupings (between ruts). The width of linear disturbances near 100–200 m. At slopes ruts cause erosion furrows and narrow gulches form.
	Very strong	Large roads with multiple ruts, the width of influence up to 0.5 – 1.5 km. Sometimes with weed groupings between ruts. Furrows and gulches of considerable size. Wind movements deflation of sandy grounds. The width of degraded sites increases due to presence of parallel multi-rut roads.

Developing a current map is a creative integrative compilation of a priori data obtained by conventional terrestrial geobotany studies plus the entire complex of remote-sensing information available today. Of great importance to us were remote-sensing pictures available at 1:1000000 and magnified to 1:500000 that were used to form maps and plant photos in transformed pictures. For some plots, photos and color synthesized pictures were made at a 1:200000 scale. Plant cover dynamics on some plots was traced with asynchronous photos. Deciphering of remote-sensing information was performed using previously developed methods (Vostokova et al., 1988a, b). Note that vegetation photography on the basis of satellite pictures has a number of distinctions from conventional geobotany mapping. The distinctive peculiarities of satellite mapping are presented in Table 4.21.

Therefore map investigations of the main trends in plant cover dynamics provide an insight into present-day ecosystem conditions. However note that the same degree of disturbance in different plant communities and ecosystems can be caused by different disturbances and that future development of soils and plant cover can take different courses at different rates. Hence it is necessary to reflect on those factors that cause each change. For example some satellite data indicates detrimental processes that were triggered by degradative vegetation successions.

Evenso, it should be noted that maps were developed on the basis of vegetation dynamics information. This occurred because the period of declining precipitation and the current increase in human-associated effects on vegetation extended for almost the exact decade during which the major satellite data were obtained. Therefore to develop

Table 4.21. Main vegetation mapping features at medium and low scales, based on space images.

Parameter Analyzed	For vegetation mapping based on space images
Mapped objects	Taxonomic unit classifications of vegetation with determination of micro- and macro-vegetation complexes.
Principles of identification	Landscape typological reflecting landscape interrelations, for instance, of vegetation with ecological conditions (relief, surface sediments).
Features of map contents	Contemporary vegetation cover plus structure.
Sequence of determinations	From large combinations to micro, i.e. from general to particular.
Method of generalization	Based on general optical features of objects on land surfaces as seen in photoimages.
Contour Changes	To shown natural vegetation boundaries.
Main method of mapping	Laboratory geobotanical and landscape image interpretations, plus field investigations as a check.

comprehensive maps of vegetation dynamics, it will be necessary to take into account regenerative successions also.

This multi-theme map development permits their use to inventory ecosystems as a basic reference to conduct subsequent vegetation dynamics monitoring. For example a comparative analysis of current vegetation changes Vs those at the time of developing the maps «Ecosystems of Mongolia». Or to investigation present-day vegetation and its dynamics, larger-scale geobotany maps of human-associated disturbances or natural human-induced vegetation successions were developed for some plots of the Joint Russian-Mongolian Biological Expedition such plots were in the area around where (Figure 3.1).

4.5 CONCLUSIONS

Our considerations and analysis of natural plant community dynamics gives grounds to believe that Mongolia's vegetation dynamics are strongly effected by climate (or solar) periodicity. Before the onset of the next multi-age climate cycle, natural plant communities should have reached some equilibrium. Actually, of course with a biological system equilibrium is not reached, for instance here there is an increase in degradative plant successions and the continuous presence of detrimental wind and water erosion processes. These processes combine strongly to influence the behaviour of these biosystems and prevent the regeneration of primary vegetation.

In addition, climate cycles are associated with huge human-associated pressures on the natural environment, primarily on vegetation, in the grazing-based society of Mongolia. Indeed human-associated factors cause changes in natural environments frequently equaled to the effects of multi-age climatic oscillations and they can be compared to geo-tectonic processes (Kotlov, 1978; Glazovsky et al., 1991; Medous et al., 1994).

The powerful detrimental roles of human-driven activities have been noted before. For instance, several decades ago S.V. Viktorov lamented the detrimental consequence of anthropogenic effects on desert plant cover (Viktorov, 1973). But it took a long time for the scientific and public community at large to become aware of the hazards of plant community degradation across the Earth; though today this is almost universally recognized. Throughout the world, numerous projects and programs associated with global environment changes are in-progress, including vegetation dynamics and soil cover. Regrettably, an intensification of human economic activities due to the need to provide an expanded population with food, dwellings, and other social needs, often contradicts the need for environmental conservation work such as plant cover on the extensive grazing lands of Mongolia.

Certainly Mongolia is no exception in these worldwide problems. But happily, the dynamics of human-associated vegetation degradations are not yet disastrous for much of Mongolia. Nevertheless firmly implemented systems of sound nature conservation and management that are acceptable to the Mongolian people are needed to preserve and to enhance the economic development of this unique and beautiful country.

CHAPTER 5

STRATEGIES FOR NATURE MANAGEMENT AND VEGETATION CONSERVATION

5.1 Introduction. Methods for vegetation conservation

Our analysis of modern plant cover dynamics has shown bi-directional successions in various plant communities with some ecosystems recovering and some being degraded currently in Mongolia. Across Mongolia the largest areas are occupied by plant communities that have been exposed to human-associated effects to varying extents and which are at different degradation stages. This vegetation degradation is a result of immediate and long-term human-associated modifications, the widespread effects of natural climatic cycles, water erosion processes, and strong wind movement of soils. Human effects of great importance now are: an expanded and intensive economic exploitation of natural resources; population growth caused by new enterprises to develop mineral resources; the growth of cities; and the widespread construction of new built-up areas. These are illustrated in the maps Ecosystems of Mongolia (1995) partially shown in Figure 5.1, which demonstrate some ecosystem disturbances with respect to vegetation degradation. Figure 5.1 shows some affects on plant cover with reference to grazing livestock breeding, to tree fellings, and to fires in forest steppes.

Some vegetation regeneration processes were started in the early 1990s, but these cover only minor areas. Current developments in regenerating vegetation are intimately associated with economic activity patterns, which have become increasingly intensive. For example, traditional branches of agriculture are

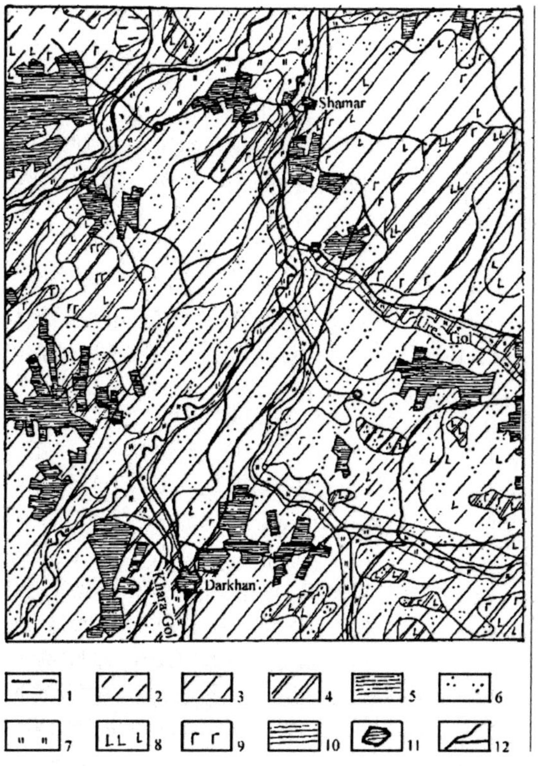

Figure 5.1. Expert assessment of vegetation. The scheme was created based on the map, Ecosystems of Mongolia (space image interpretation by E.A.Vostokova)

1–5 assessment of degree of plant cover distortion: 1 – weakly distorted; 2 – moderately distorted; 3 – heavily distorted; 4 – extremely heavy distorted; 5 – natural plant communities substituted completely by agrocenoses and ruderal groupings. 6–12 – main factors of vegetation disturbance: 6 – grazing; 7 – grazing and hay-cutting; 8 – forest cutting; 9 – forest fires; 10 – ploughing; 11 – settlements; 12 – transport.

developing therefore steppes are being used increasingly for agriculture, felled out forests areas are increasing, and mining complexes which can fully destroy vegetation along with transportation pressures are greatly increased. All of these activities deplete plant resources, degrade plant cover, and they can also chemically pollute the environment. These human-associated ecological changes strongly affect the current status of vegetation and its recovery.

Therefore there is an acute need to implement measures for: i) the encouragement of plant regeneration successions, ii) conserving, increasing, and maintaining vegetation resources, and iii) reducing degradation successions. These are especially important in arid areas where an expansion of desertification processes is occurring now, as in the most vulnerable dry and steppe-like deserts. Unfortunately these exact areas are where the main grazing ranges are concentrated. If it is not done quickly, vegetation recovery will be very expensive or impossible. As a result, some unique plant communities of Mongolia likely will perish. Note Mongolia is the only country of Eurasia to retain huge areas of steppes vegetation. In other parts of Eurasia steppes either have become extinct or occupy small patches or are partially replaced by agricultural communities (Lavrenko et al., 1991).

Hence, one of the central tasks in conserving plant cover and increasing its resource potential is to maintain and to promote the successful regeneration of plant community successions.

Vegetation is a natural resource of greatest importance to this grazing-based society and its botanical genetic diversity is a valuable resource for Mongolia. After N.M. Novikova (1997), here we regard botanical diversity as both floristic and plant community diversity with a complicated integration at the levels of species, communities, and ecosystems.

For a long time, Mongolia was considered a floristically poor country, but over the last 10 to 15 years, a greater plant floristic richness has been established (Grubov, 1984; Gubanov & Kamelin, 1988b, 1991, 1992; Gubanov, 1996). Now 2823 vascular plants of 662 genera and 128 families have been recorded in Mongolia (Gubanov, 1996). Botanists have found that numerous species, which were considered rare, actually are distributed over a much wider range. Long-term floristic studies in Mongolia guided by R.V.Kamelin under the Joint Russian-Mongolian Complex Biological Expedition yielded more insight into higher plants in Mongolia than in contiguous countries. In Central Siberia 2400 species and subspecies of vascular plants are listed (Malyshev & Peshkova, 1984) and Inner Mongolia (China) lists 2176 species (Flora Intramongolica, 1977 - 1985).

Mongolia's floral richness is explainable by its location at the junction of two subcontinents – Northern and Central Asia and the florist richness of its mountain systems - Mongolian Altai, Khangai, and Khentii. Considering the largest flowering plant families, the flora of Mongolia combines features of boreal and ancient Mediterranean floras, i.e., a high status of Rosaceae, Cyperaceae, and Ranunclulaceae as boreal families and Leguminosae, Cruciferae, and Chenopodiaceae as Mediterranean. The continental nature of its flora is supported by the paucity of ferns, Lycopsida, and only 13 conifer species.

Mountain features play important roles in the richness of Mongolia's flora. Floral composition is poorest in plains even those up to 700 m above the sea level. Indeed plains of higher elevations (up to 1000 - 1400 m above sea level), or of rocky hill topography or low mountains are almost twice as rich. Floristic richness is concentrated in the middle mountains around Lake Hubsugul, along the Selenga River, and in the Great Khyangan piedmont. These middle mountain zones are isolated from one another, but they are closely connected floristically. High mountain zone flora is fairly rich if lichens and bryophytes also are considered. The high-mountain flora of Khentii is sharply distinguished from the western mountains.

Despite its continental location, in most of Mongolia, mesophytes and xeromesophytes dominate and there are many, 41, evergreens species. A great number of semishrub plants of xeromorphic appearance are in regions of Mongolia characterized by xeromorphic plant structures. At the same time a few bulbous herbs plus annual and biannuals are common.

Mongolia's flora is characterized by a fairly high percentage of endemics and subendemics, near 10% of the entire flora (Grubov, 1984). The last version of the Flora of Mongolia lists 140 endemic species, and even more subendemics (Gubanov, 1996). Subendemics are species naturally endemic partly in Mongolia and partly in adjacent countries. For example subendemics in Southern-Siberian – Mongolia are *Vicia tsydenii, Sajanella monstrosa, Hedysarum sangilense*, and *Thymus pavlovii*. Important subendemics in Dagurian-Manchuian-Mongolia are *Stenosolenium saxatile, Diarthron linifolium, Clausia trichosepala* and other species of the complex associated with *Armeniaca sibirica* communities, which is a Dagurian, Manchurian-Northern-Chinese endemic.

On the whole, Mongolian subendemics are highly diversified. Amongst the species in Central-Asia and Gobi proper (Mongolian-North-Chinese) one finds species such as in eastern Gobi-Ordos (*Potaninia mongolica*), Kashgar-Jungarian (*Zygophyllum kaschgaricum*), Tuva-Mongolian (*Stevenia sergievskajae, Hedysarum sangilense*), and Buryat-North-Mongolian (*Vicia tsydenii*). In the flora the number of Dagurian, Dagurian-Mongolian-North-Chinese, and Dagurian-Manchurian subendemics is considerably greater than other subendemics of boreal Mongolia. Indeed, according to tentative estimates, near 80 species.

The distribution of some rare endemic plants is given in Figure 5.2. Note that the Red Data Book of Mongolia includes far from all the endemic and subendemic species, and some rare species (Table 5.1). At the same time, these protected species include some widely distributed community-formers. Indeed amongst Red Data Book species are *Haloxylon ammodendron, Tamarix* spp., *Populus diversifolia, P. pilosa*, and *Halimodendron halodendron*. They were listed in the Red Data Book as endangered species due to their intensive use for firewood. Another group of species also listed in the Red Data Book are medicinal and ornamental plants. Folk medicine plants are widely used and extensively destroyed, e.g., *Rhodiola quadrifida, R. rosea, Peganum harmala, Valeriana alternifolia, Sophora flavescens, Paeonia anomala*, and *Rhaponticum carthamoides*. A fairly wide use of wild food plants is common (*Allium altaicum, A. galanthum, Elaeagnus moorcroftii*, and *Hippophaë rhamnoides*). Near cities flowers of ornamental-type wild plants often are collected, e.g., *Hemerocallis lilio-asphodelus, Lilium dauricum, Nymphaea candida, Paeonia lactiflora, P. anomala, Adonis mongolica*, and *Gentiana algida*. Flower gatherers are great predators on these and other species which, fortunately, are still widely distributed.

In Mongolia plant community diversity and endemic types and subtypes of vegetation remain fairly high. Indeed unique communities exist such as ultra-continental larch forest-steppes (Bannikova, 1980), meadow tansy steppes (Lavrenko et al., 1991), and cold-resistant relatively dry low-grass *Kobresia* highland meadows (Yunatov, 1950; Karamysheva, 1988). Also one could include some deserts, for example, rock debris, stony and sandy-rubble deserts which are characterized by an absence of ephemerals. Characteristic communities occur in steppe-like deserts and grasslands (Yunatov, 1950, 1974; Rachkovskaya, 1989, 1993). These communities occupy semi desert zone successions and in terms of composition and structures they are unprecedented in other arid and subarid regions (Lavrenko et al., 1991). They comprise unique vegetation on a large portion of Mongolia.

Mongolia is derived from mountains, hence, latitudinal zones are universally complicated by altitude zones. Mongolia's altitude zones principally are of two different types: boreal and arid (Gunin & Vostokova, 1995a).

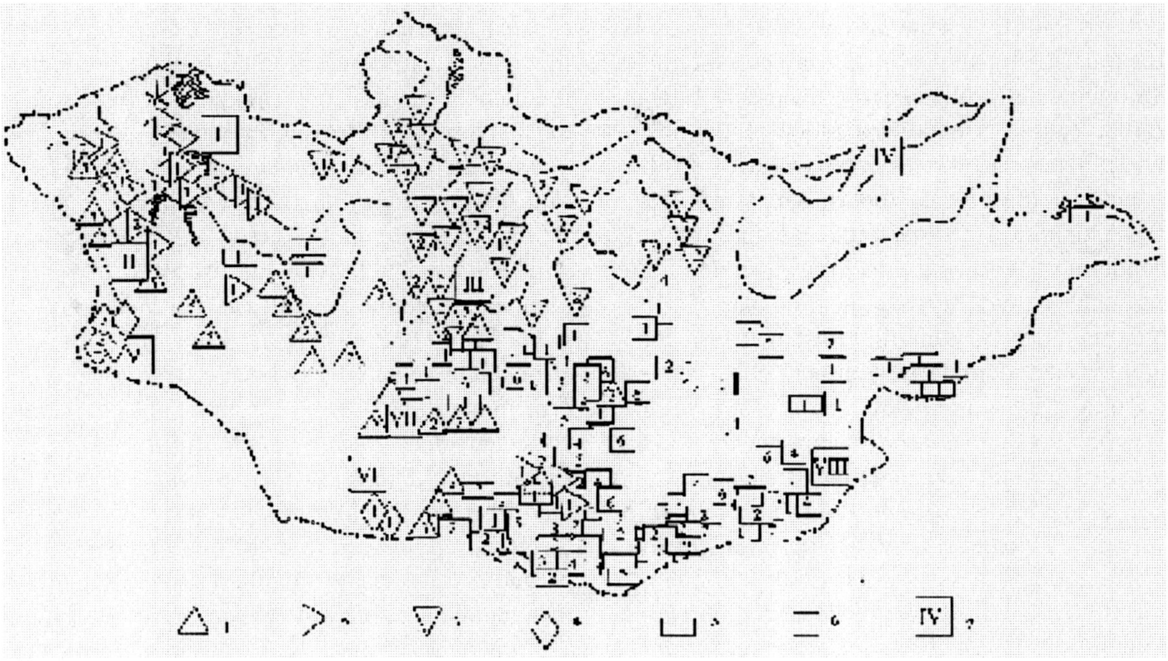

Figure 5.2. Distribution of certain endemic Mongolian plants (according to the National Atlas of Mongolia, 1990; added by Gubanov, 1996)

1 – Gobi Altai and Mongolian Altai species (*Oxytropis fragilifolia*, 2. *O. bungei*, 3. *Potentilla ikonnikovii*, 5. *Stellaria pulvinata*). 2 – West Mongolian (1. *Asterothamnus heteropappoides*, 2. *Prionotrichon kamelinii*, 3. *Potentilla inopinata*). 3 – Khangai and Central Mongolian 1. *Oxytropis diantha*, 2. *Adonis mongolica*, 3. *Potentilla chenteica*, 4. *Astragalus changaicus*). 4 – Dzungarian and West Gobian (1. *Astragalus gobicus*, 2. *Ajania grubovii*, 3. *Zygophyllum neglectum*, 4. *Nanophyton mongolicum*). 5 – Gobian (1. *Psammochloa villosa*, 2. *Amygdalus mongolica*). 6 – Central Gobi and North-Eastern Gobi (1. *Iris bungei*, 2. *Sibbaldianthe sericea*, 3. *Ammopiptanthus mongolicus*, 4. *Caragana brachypoda*, 5. *Pugionium dolabratum*, 6. *Euphorbia kozlovii*, 7. *Astragalus grubovii*). 7 – Strictly local endemic plants: I. of the Uvsunur depression (*Bromopsis ubsunurica, Astragalus gubanovii*). II. of the Mongolian Altai (*Potentilla laevipes, Astragalus alexandri, Delphinium gubanovii, Astragalus granitovii, Hedysarum kamelinii, Swertia banzragczii*). III. of the Khangai highlands (*Luzula changaica, Papaver rubro-aurantiacum* subsp. *changaicum, Alchemilla changaica, A. gubanovii, A. pavlovii, Potentilla leucophylla, P. hilbigii, P. mongolica, Astragalus klementzii, Adenophora changaica, Delphinium changaicum*). IV. of the Erden-Daba ridge (*Astragalus pseudochorinensis, A. viridiflavus, Dracocephalum junatovii*). V. of the Baitag-Bogdo ridge (*Aconitum gubanovii, Papaver baitagense, P.pseudotenellum, Smelowskia mongolica, Rosa baitagensis, Astragalus baitagensis*). VI. of the Transaltai Gobi (*Microstigma junatovii, Chesneya grubovii, Saussurea grubovii*). VII. of the Gobian Altai (*Papaver rubro-aurantiacum* subsp. *saichanense, Arabis mongolica, Galitzkya macrocarpa, Astragalus koslovii, Asperula saxicola, Valeriana saichanensis*). VIII. of the Eastern Gobi (*Brachanthemum mongolicum, Polygonum intramongolicum, Caragana gobica, Limonium gobicum*).

Table 5.1. Plants present in Mongolia's «Red Book» (according to Gubanov, 1996).

Family	Species	Distribution
Pinaceae	*Abies sibirica*	Khentii, Hubsugul region
	Larix dahurica	Eastern Mongolia, Eren-Daba range
Cupressaceae	*Juniperus davurica**	Khentii
Ephedraceae	*Ephedra glauca*	Transaltai and Eastern Gobi
Poaceae	*Stipa pennata*	Khangai, Daguria
Araceae	*Acorus calamus**	Khyangan region, Daguria, Khangai
Alliaceae	*Allium altaicum*	Almost everywhere, excluding Eastern Gobi and Eastern Mongolia
	*A. condensatum**	Khyangan region, Eastern Mongolia
	A. galanthum	Mongolian Altai, Gzungarian Gobi (Baitag-Bogdo range)
	A. macrostemon	Eastern Mongolia
	*A. obliquum**	Mongolian Altai
Asphodelaceae	*Anemarrhena asphodeloides**	Eastern Mongolia, Khyangan region
Hemerocallidaceae	*Hemerocallis lilio-asphodelus**	Eastern Mongolia, Khentii
Liliaceae s.str.	*Lilium dauricum*	Khentii, Khyangan region
	*Tulipa uniflora**	Great Lakes Depression, Mongolian Altai, Khangai
Iridaceae	*Iris dichotoma*	North-Eastern part of Mongolia
Orchidaceae	*Calypso bulbosa**	Daguria
	*Corallorhiza trifida**	Hubsugul region, Khentii, Daguria
	Cypripedium calceolus	Hubsugul region, Khentii, Daguria
	C. macranthum	Khentii, Khyangan region
	*Dactylorhiza fuchsii**	Hubsugul region, Daguria
	Epipogium aphyllum	Khentii, Khangai, Daguria
	*Goodyera repens**	Hubsugul region, Mongolian Altai, Khangai, Khentii
	Gymnadenia conopsea	Hubsugul region, Khentii, Khangai, Daguria, Khyangan region
	*Neottia camtschatea**	Hubsugul region, Khentii, Khangai
	Neottianthe cucullata	Khentii, Khangai, Daguria
	*Orchis militaris**	Khangai, Daguria
	Platanthera bifolia	Khangai, Khentii, Daguria
Salicaceae	*Populus diversifolia*	Gobian Altai, Dzungarian and Transaltai Gobi
	P. pilosa	Mongolian and Gobian Altai, Khangai, Great Lakes Depression, Daguria
Chenopodiaceae	*Haloxylon ammodendron*	Gobi
	Iljinia regelii	Gobi
Caryophyllaceae	*Gymnocarpos przewalskii*	Gobi
Nymphaeaceae	*Nuphar pumila*	Hubsugul region, Khangai, Great Lakes Depression
	Nymphaea candida	Khangai, Great Lakes Depression
Paeoniaceae	*Paeonia anomala*	Hubsugul region, Daguria, Khentii, Khangai
	P. lactiflora	Khyangan region
Ranunculaceae	*Adonis mongolica***	Hubsugul region, Khangai, Daguria
	A. sibirica	Hubsugul region, Khangai, Khentii, Middle Khalkha, Mongolian Altai
Brassicaceae	*Pugionium pterocarpum***	Great Lakes Depression
Crassulaceae	*Rhodiola quadrifida*	Hubsugul region, Khentii, Khangai, Mongolian and Gobian Altai
	R. rosea	Hubsugul region, Khentii, Khangai, Mongolian and Gobian Altai
Rosaceae	*Amygdalus mongolica***	Eastern and Alashan Gobi, Gobian Altai
	*Potaninia mongolica***	Gobian Altai, Lake Valley, Eastern and Alashan Gobi
	Sorbaria sorbifolia	Khyangan region, Daguria

Table 5.1. Continued..

Family	Species	Distribution
Fabaceae	*Ammopiptanthus mongolicus***	Eastern and Alashan Gobi
	*Gueldenstaedtia monophylla***	Mongolian and Gobian Altai, Great Lakes Depression, Eastern and Alashan Gobi
	Halimodendron halodendron	Great Lakes Depression, Dzungarian and Transaltai Gobi
	Sophora flavescens	Eastern Mongolia, Daguria
Peganaceae	*Peganum harmala*	Gobi
Rutaceae	*Dictamnus dasycarpus**	Khyangan region
Celastraceae	*Euonymus maackii*	Khyangan region
Rhamnaceae	*Rhamnus ussuriensis**	Khyangan region, Daguria
Hypericaceae	*Hypericum attenuatum*	Khangai, Khentii, Eastern Mongolia
Tamaricaceae	*Tamarix elongata**	Lake Valley
	T. gracilis	Gobian Altai, Transaltai and Alashan Gobi
	T. hispida	Gobian Altai
	T. karelinii	Great Lakes Depression, Dzungarian, Transaltai and Alashan Gobi
	T. leptostachys	Great Lakes Depression, Gobian Altai, Dzungaria, Transaltai and Alashan Gobi
	T. ramosissima	Great Lakes Depression, Dzungarian, Transaltai and Alashan Gobi, Gobian Altai
Elaeagnaceae	*Elaeagnus moorcroftii*	Transaltai and Alashan Gobi
	Hippophaë rhamnoides	Khangai, Daguria, Mongolian and Gobian Altai, Great Lakes Depression, Lake Valley, Dzungarian Gobi
Cynomoriaceae	*Cynomorium songaricum*	Great Lakes Depression, Lake Valley, Gobi, Gobian Altai
Ericaceae	*Oxycoccus microcarpus**	Khentii
	Rhododendron aureum	Hubsugul region, Khentii
	*Vaccinium myrtillus**	Khentii
Primulaceae	*Androsace longifolia*	Eastern Mongolia
Gentianaceae	*Gentiana algida*	Hubsugul region, Mongolian and Gobian Altai, Khentii, Khangai
Apocynaceae	*Poacynum pictum*	Hubsugul region, Khentii, Khangai, Mongolian and Gobian Altai
Verbenaceae	*Caryopteris mongholica***	Everywhere
Scrophulariaceae	*Lancea tibetica**	Khangai
Bignoniaceae	*Incarvillea potaninii***	Transaltai Gobi
Caprifoliaceae	*Viburnum mongolicum*	Eastern Mongolia
	V. sargentii	Khyangan region
Valerianaceae	*Valeriana alternifolia*	Hubsugul region, Khentii, Khangai, Daguria, Khyangan Region
Asteraceae	*Brachanthemum gobicum***	Eastern and Alashan Gobi
	*B. mongolicum***	Dzungarian Gobi
	*Dendranthema sinuatum**	Mongolian Altai
	*Pyrethrum changaicum***	Mongolian Altai, Khangai
	Rhaponticum carthamoides	Mongolian Altai, Dzungarian Gobi (Baitag-Bogdo range)
	*Saussurea ceterachifolia***	Mongolian Altai, Khangai
	S. involucrata	Hubsugul region, Khentii, Khangai
	Synurus deltoides	Khyangan region

* – Rare species for Mongolia

** – Endemics for Mongolia

Figure 5.3. Exposed forest-steppe in taiga on the South-Western Khentii (above) and Central Khentii (below) (photos by E.N.Matyushkin).

Boreal zones are characterized by a replacement of steppe communities with forest-steppe and forest. At higher elevations there is a transition to xerophyte herbaceous communities dominated by *Kobresia* spp., *Carex* spp. plus lichen and shrub tundras. But nowhere do forests form continuous zones. Even in the Hubsugul Lake area of Mountains along the Selenga river and Khentii under bald mountains on south-facing slopes, meadow-steppes frequently wedge into larch forests or Siberian pine and into larch-Siberian pine open woodlands. Southward, forests soon become patchy and only remains on north-facing slopes (Figure 5.3). A few patches of *Larix sibirica* forests penetrate far to the south, beyond the exposed forest-steppe zone. Farther south, in the Gobi Altai and Gobi Tyan-Shan Mountains, typical arid zone vegetation dominates (Figure 5.4).

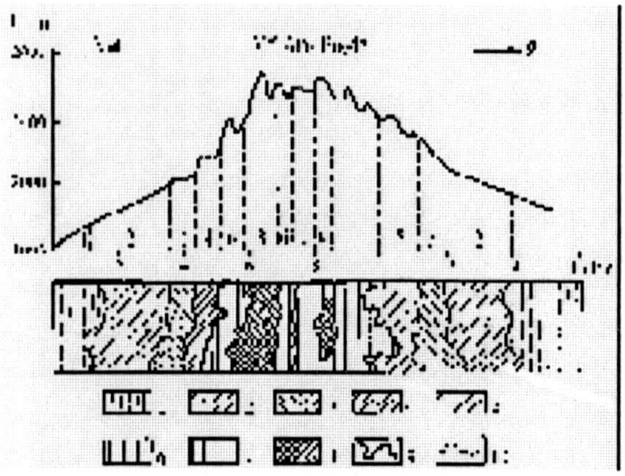

Figure 5.4. Arid altitudinal belt structures (Volkova, Rachkovskaia & Fedorova, 1986)

1 – extreme arid deserts with complicated bush communities in sairas; 2 – communities of *Sympegma regelii*; 3 – communities of *Stipa glareosa* + *Anabasis brevifolia* + *Krascheninnikowia ceratoides*; community *Artemisia frigida* + *Krascheninnikowia ceratoides* + *Stipa gobica*; 5 – community *Stipa gobica* + *Artemisia sublessingiana*; 6 – community *Agropyron cristatum* + *Artemisia frigida* + *Stipa orientalis*; 7 – steppes *Stipa gobica* + *Artemisia frigida* + petrophillic forbs; 8 – steppes *Artemisia frigida* + *Agropyron cristatum* + petrophillic forbs; 9 – belt boundaries; 10 – sub-belt boundaries.

Some unique plant community combinations and complexes exist on altitude zones with large species diversity. In fact, the forest-steppe zone, so typical for Mongolia, is characterized by its highest botanical diversity. The territories with the highest indices of floristic and community diversity are found in northeastern Mongolia on the Eren-Daba ranges, in Eastern Khangai, and around Cis-Khyangan.

In deserts, the greatest botanical diversity is in Dzungarian Gobi and in some regions of Gobi Tyan-Shan. The latter represent natural oases in the foothills and in intermountain depressions. These oases give the desert a unique pattern, and without a doubt, enrich its flora. In fact, about half of all the species of higher plants species known in Mongolia occur in the Barun-Khurai depression and its mountain (Dzungarian Gobi); but they occupy only 2.5% of Mongolia (Kamelin & Gubanov, 1993). The middle range, Baitag-Bogdo in the extreme southwest, is very endemic-rich.

Another arid region characterized by high botanical diversity is the Lake Ubsu-Nur depression and, to the south, the adjacent Khankhukhiin Nuruu range north-facing slope is distinguished by the most complicated pattern of altitude zones (Karamysheva & Banzragch, 1976). In the small area of the Ubsunur depression, desert steppes and true deserts combine rock-debris, semi-fixed sand dunes, saline clays, and hydromorphic communities from the Tesiin-Gol river delta. Of course the lake itself is the largest salt lake of Mongolia situated 759 m above the sea level with an area of 3350 km^2; and its water mineralization attains 18.8 g/l. This is the final runoff basin for the Tesiin-Gol, Nairyin-Gol, and Dzakhain-Us rivers plus a number of small rivers, which often dry up. The isolation of this depression, the diversity of its landscape, its floristic and plant community richness – all of these features led scientists to study it as a natural biosphere model (Bugrovsky, 1986; 1995; Bugrovsky et al., 1990; Vostokova, 1986; Buyan-Orshikh, 1992).

Table 5.2. Main types of plant conservation and renewal efforts.

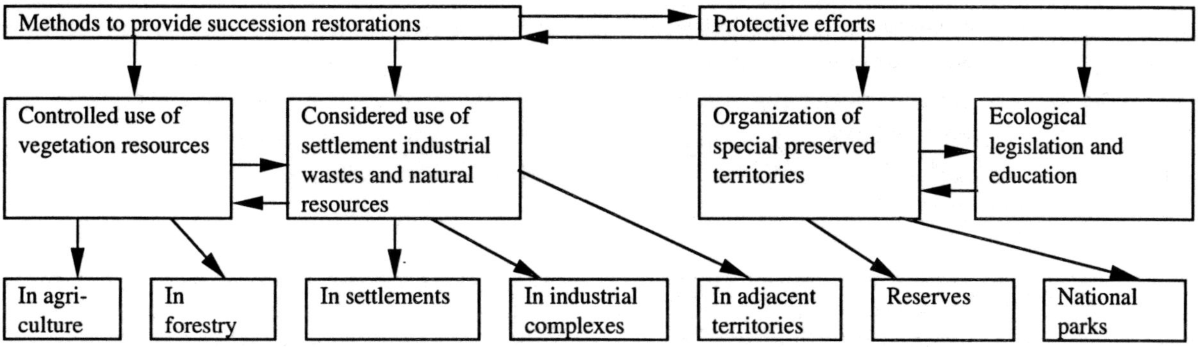

But programs of study and conservation are needed to support regeneration type plant successions plus saving and protecting botanical diversity. Nature management measures must extend across the entire country. Consequently, integrated programs of nature management and nature conservation are required to preserve the unique resources within the botanical diversity of Mongolia. These Mongolian programs should include (Table 5.2):

– an integrated complex of measures aimed at the restoration of vegetation successions and the considered use of all natural biological resources including plants, animals, microbes, soils, and water.

– a system for the conservation of floristic and plant community diversity in different regions of Mongolia.

5.2 Restoration and conservation of botanical successions

Current nature management programs in Mongolia involve maintaining some biological resources and reducing some negative degradation processes (Efremov, 1976; Reimers & Shtilmark, 1978). These efforts are part of important objectives to conserve biological diversity and to create conditions, which guarantee survival of rare and endangered plant species.

In Mongolia the three main trends for the restoration and conservation of its unique natural complexes are as follows:

1. The use of different types of management programs to provide maximal conservation of natural resources, their sustainable functioning, and an enhanced development of regeneration successions.

2. The recultivation of disturbed areas and land improvement by establishing natural complexes for recreational uses; an organized clean up of built-up areas; the organized utilization of industrial and domestic wastes; and the construction of a network of resort institutions, tourist stations, and camps.

3. Establishing a nationwide network of strictly protected areas (reserves, preserves, etc.) for biodiversity conservation, plus saving endangered species, unique plant communities, and ecosystems.

In much of Mongolia the cold-arid soil pattern is aggravated, temporally and spatially, by reoccurring extremes of radiation heat and cyclic water conditions, which lead to unstable plant communities. These extreme environments activate aridization and cryodization with a synergistic effect on disturbed lands leading to the formation of natural- and human-induced communities with new negative properties such as unpredictable behavior and unbalanced matter-energy flow. Such conditions emphasize the need to develop more precise, yet culturally acceptable ecological principles of nature management; specifically in the skillful use of plant resources plus developing and maintaining landscape recovery successions.

Of course each type of vegetation and human disturbance requires specific nature management measures (Table 5.3). In addition, the local environmental features must be taken into account. This is very important in Mongolia, where the established plant cover has developed under very harsh climate conditions; otherwise a more rapid destruction of soil and plant cover would have occurred.

Livestock breeding is the leading economic component of the agrarian Mongolian society. It accounts for 70 to 75% of the total agricultural production. Traditionally, livestock breeding included sheep, horse, camel, cattle, and goat breeding (Romanov et al., 1988); but it also includes large wild ungulate herds such as gazelles. However a recent qualitative and quantitative transformation in favor of domestic goat breeding has sharply and negatively affected vegetation status with the development of unfavorable plant successions that have caused numerous points of steppe desertification. Specifically the destruction of shrub and herbaceous cover on sand mounds by goats causes extensive wind erosion of sand deposits.

In this connection a complete system of range rotation needs to be developed not only for each farm, but also for all natural and economic regions across Mongolia. A system of range management should take into account the modern vegetation state of ranges, their degree of degradation, plus the seasonality of range usage, while on the other hand, it should promote optimal numbers of farm livestock. Thus an estimate of naturally-admissible range pressures are needed along with regulating the domestic animals species ratio to bring their

Table 5.3. Main recommendations to ensure renewal of vegetation successions.

Economic Area	Vegetation communities		
	Forests	Forest–steppes and steppes	Deserts
Forestry	Restrict clearings	Only sanitary tree cuttings	Prohibit breaking and uprooting of *Tamarix* spp., *Haloxylon ammodendron*
	Restrict cutting areas to 5 ha		
	Preservation of seed resources		
	Artificial forest planting (mainly *Larix sibirica, Pinus sylvestris*).		Seeding of *Haloxylon ammodendron*
	Fire protection furrows and forest–cut strips		
Agriculture		Introduce soil–protection technologies	
		Decrease in rainfed agriculture	
		Arable lands melioration	
Stock–raising	Decrease of grazing at forest cuttings	Organize pasture rotations	
		Pasture melioration (grass seeding, irrigation)	
		Temporary restrictions on grazing in heavily degraded pastures	
		Create forage reserves	
Transportation	Build solid cover roads		

Table 5.4. Plant community disturbance aimaks (1995, %).

Aimak	Ecosystem disturbance degree:				
	Very weak	Weak	Moderately	Strong	Very strong
Bayan-Ulgii	1	14	50	30	Locally
Khovd	–	10	55	35	Locally
Uvs	5	25	40	30	Locally
Gobi–Altai	3	20	63	14	Locally
Zavkhan	–	15	54	31	Locally
Khuvsgul	10	20	40	30	Locally
Arkhangai	3	22	55	20	Locally
Bayankhongor	2.5	20	60	17.5	Locally
Bulgan	–	20	75	5	Locally
Uvurkhangai	–	10	40	45	5
Umnugobi	–	62	28	10	Locally
Selenge	5	30	20	40	5
Tov	3	11	41	40	5
Dundgobi	–	30	60	7	3
Khentei	5	32	43	20	Locally
Dornogobi	–	37	50	13	Locally
Dornod	–	20	69	11	Locally
Sukhbaatar	–	10	80	10	Locally

numbers in line with the range vegetation status, i.e., the animal carrying capacity of each range. Today in Mongolia degraded ranges only comprise 5 to 6% of the land, but some already require a temporary grazing ban and in some aimaks (regions), such as the Tov, Uvurkhangai, Selenge, 10 to 15% of the territory has grazing bans now (Table 5.4).

One of the main indices of range degradation in plant cover is the prevalence of range weeds. These regressive plant community successions are characterized by a replacement of quality forage plants by unpalatable, poisonous, or little-consumed plants that form unstable groups (Table 5.5). The worst affected vegetation is around livestock collecting sites near wells or water sources. Today, the unique natural oasis plant communities of the semidesert and desert zones are in a disastrous situation. These excessive human-animal-induced pressures have virtually completely killed arboreous plants such as *Ulmus pumila, Populus diversifolia* and the hydromorphic plants accompanying them. Where *Phragmites australis* and *Achnatherum splendens* used to grow, one only finds groups of little-palatable *Artemisia annua, Euphorbia humifusa, Lappula myosotis, Peganum nigellastrum* and other plants (Table 5.5). A full or partial ban on domestic livestock grazing in such areas for a certain period of time would undoubtedly favor the regeneration of arboreous-shrub vegetation communities.

Unfortunately a complete absence of improved ranges and a lack of scientific approaches to these issues highlights the need for new knowledge and solutions. Experimental studies should be directed at the regeneration of degraded ranges and at the establishment of cultivated rangeland ecosystems which take into account long-term climate changes and experiences in other countries (Nechaeva, 1985). In artificial range improvement it is wise to use the indigenous flora gene potential that is well adapted to local ecological conditions. Some promising results have been obtained with experiments in extra-

Table 5.5. Principal unedible pasture plants (Gunin & Vostokova, 1989).

Species	Main regions of distribution
Acroptilon repens	River flood-plains, oases
Arenaria capillaris	Khentii, Khangai, Hubsugul region
Artemisia annua	Khentii, Khangai, Hubsugul region, Khalkha, Eastern Mongolia, Lake Valley, Great Lakes Depression, Southern Gobi, river flood-plains
A. changaica	Khentii, Khangai, Hubsugul region
A. laciniata	Khentii, Khangai, Hubsugul region
Bistorta alopecuroides	Khentii, Khangai, Hubsugul region
Cardaria pubescens	Lake Valley, Great Lakes Depression, Gobi
Hypecoum lactiflorum	River flood plains
Cynanchum sibiricum	Southern Gobi
Euphorbia humifusa	Lake Valley, Great Lakes Depression, Gobi
Hedysarum dahuricum	Khentii, Khangai, Hubsugul region
Lactuca serriola	Southern Gobi
L. tatarica	Khentii, Khangai, Hubsugul region
Lappula myosotis	Khentii, Khangai, Hubsugul region, Khalkha, Eastern Mongolia, Lake Valley, Great Lakes Depression, Gobi
Lepidium densiflorum	Lake Valley, Great Lakes Depression, Gobi
L. obtusum	Lake Valley, Great Lakes Depression, Gobi, river flood plains
Oxytropis glabra	Lake Valley, Great Lakes Depression, Northern Gobi, river flood plains
Pedicularis flava	Khentii, Khangai, Hubsugul region
P. longiflora	Everywhere
Peganum harmala	Lake Valley, Great Lakes Depression, Gobi
P. nigellastrum	Khentii, Khangai, Hubsugul region, Khalkha, Eastern Mongolia, Lake Valley, Great Lakes Depression, Northern Gobi
Plantago depressa	Khentii, Khangai, Hubsugul region, Eastern Mongolia, Lake Valley, Great Lakes Depression, Southern Gobi, river flood plains
Polygonatum odoratum	Khentii, Khangai, Hubsugul region
Saussurea salicifolia	Khentii, Khangai, Hubsugul region
Stellera chamaejasme	Khentii, Khangai, Hubsugul region, Khalkha, Eastern Mongolia
Tribulus terrestris	Lake Valley, Great Lakes Depression, Gobi
Urtica cannabina	Everywhere, except Southern Gobi
Veronica pinnata	Khentii, Khangai, Hubsugul region

arid deserts with plant regeneration in sand and water-accumulating ruts (Gunin et al., 1993; Slemnev et al., 1997).

One of the unfortunate trends of modern nature management in Mongolia is an increase in non-irrigated and irrigated farming. Even though the entire cereal and fodder crops only account for about 1% of the land, these plots use the best rangelands and have the most favorable natural conditions of moderately wetted steppes. This system of non-irrigated farming is not adapted to the natural conditions of Mongolia. Hence it has led to a rapid loss of soil fertility and to a loss of farmlands altogether (Table 5.6). Moreover the long-term process of natural regeneration on these lands results in the formation of weed communities for over two decades. Therefore, it is necessary: i) to introduce into this farming new soil-saving technologies; ii) to reduce the ploughing of virgin steppes; and iii) to develop scientifically tested methods to recover and to maintain soil fertility.

On plain steppe and semidesert areas, it is possible to conduct traditional forms of grazing livestock breeding and to essentially eliminate

Table 5.6. Degradation processes at arable lands of Mongolia.

Processes and characteristics	Degree of degradation processes		
	weak	medium	strong
Eolic processes leading to deformation of surface of ploughed lands (eolic mantle blown-out depressions):			
– % of arable land disturbance	<25	25–50	>50
– thickness of eolic accumulation or deepness of blown-out depressions cm	<5	5–20	>20
Water erosion processed resulting in deformation of ploughed land surface:			
– sheet erosion: decrease in area of ploughed are humus horizon, % of initial state	<25	25–50	>50
– sheet erosion: formation of sinkholes at pastures, % of pasture surface subject to water erosion	<10	10–25	>25
depth of sinkholes cm.	<5	5–20	>20
number of sinkholes deeper than 20 cm per 100 m	<3	3–7	>7
Processes of soil degrading:			
– loss of humus in ploughed layer (0 – 20 cm), % of background values	<15	15–30	>30
– appearance of carbonate srots due to ploughing up of carbonate horizon, % of arable land surface covered by the spots	<15	15–30	>30
– increase in sand content in ploughed layer, loss of fine material (<0.01), % of background values	<25	25–30	>50
Specific soil degradation processes connected with anthropogenic and natural factors:			
– salinity (according to sodium in water extract, me per 100 g of soil)	<2	2–6	>6
– alkalinity (according to the content of absorbed sodium, % of the sum of absorbed bases)	<5	5–15	>15
– stoniness of the ploughed layer (assessed as percentage of arable area, subject to the process, %)	<25	25–50	>50
Lands not subject to active degrading processes, either experience improvement due to anthropogenic influence (increase in humus content, improvement of water-physical properties of soils, etc.)	Not classified		

any widespread ploughing of lands for cereal crops. Small vegetable farms should be developed to provide fresh vegetables to residential areas and recreation complexes plus growing some reserve fodder for livestock breeding (Romanov et al., 1988).

Widespread historic traditional uses of plant resources in Mongolia are encouraged but these need to be well integrated with present-day situations. For example, a historic range improvement custom is to burn last-year's dry grass. Each spring fires cover many miles in the steppes. But in forest-steppe zones, fires can readily burn and destroy forests. Hence, special measures need to be taken to protect forest lands.

The mountain pattern in parts of Mongolia prevents extensive farming, plus a low population density creates conditions in the future for more extensive use of these lands. Thus grazing livestock breeding domination should be combined with some highly intensive forms of farming. Amongst these are the development of vegetable crops and hot house farms and the establishment of valuable fur animals. In addition, it seems wise to expand the breeding of yaks, yak-cows, red deer, and to improve cattle breeding. Certainly all of these will affect plant cover and emphasizes the need to promote progressive plant successions.

In mountain forests one measure directly aimed at forest regeneration is to reduce clear felling sites to 5 ha (Korotkov & Dorzhsuren, 1988). It would be very advantageous to restrict fellings to selective cuts combined with good forest regeneration measures. With these methods the volume of timber needed in Mongolia could be sustainably obtained. Today, only half of the felled out forest reaches consumers– the rest is lost in poorly designed clear-fellings and in poor distribution.

To artificially regenerate forests, a forest nursery network was established in Mongolia. However, to be successful, the number of such nurseries should be increased. Work on the introduction of a number of trees should be conducted, in particular, fruit crops because there are virtually no fruit orchards in Mongolia or cultivated berry shrubs. Nevertheless, fruit trees and shrubs can grow well in Mongolia. For instance, on the south-facing slope of Mongolian Altai in the Bulgan-Gol River valley (Dzungaria) apple trees remain with good yields that were planted by Chinese settlers. Apparently, properly selected fruit trees could be widely cultivated in Mongolia. A number of wild berry shrubs occur fairly widely, for instance, *Ribes nigrum*, *R. rubrum*, *Berberis sibirica*, *Armeniaca sibirica*, *Amygdalus mongolica* and *Prunus pedunculata*.

Mountain-forest territories have many natural resources including aesthetic and health resources (Figures 5.5 & 5.6). The presence of medicinal mineral springs, drug containing plants and picturesque scenery makes these regions invaluable. Using these natural resources an organized tourism industry with a network of recreation sites, which are so few in Mongolia today, can be set up. True enough that would require special infrastructures and funding. But such investments can be rapidly compensated with good organization and foreign tourist publicity.

Concentrating people in resorts, town and cities accentuates the problems of plant cover conservation not only for greenery and grazing but also for residential areas. The establishment of recreational zones and solutions for industrial and domestic waste problems can create favorable conditions for plant cover regeneration around residential areas. Otherwise the most beautiful nature sites, or forests, or rare endemic plants and communities will be destroyed. Without implementing broad general measures for controlling environmental pollution, environmental degradation will continue and areas occupied by weedy vegetation will expand.

Only a few sanitaria and resorts even exist today as in the picturesque mountains of Khentii, near the outcrops of hot springs in Khangai. But even these already need recultivation and plant improvement efforts! For instance, a sanatorium in the Khudzhirt region needs some extra green plantations and a running water supply now. With Mongolia's growing population and to attract foreign tourists, simply to set recreational sites aside is not sufficient. Because vegetation adjacent to these sites can be extensively disturbed, indeed essentially destroyed; thus such lands will be occupied by weedy or unpalatable or unattractive plants.

To maintain plant cover, including regenerative successions, organized recreational zones must be created near residential areas. These zones should be well provided with public services and amenities so that an accumulation of visitors would not cause excess environmental pollution nor heavily disturb the plant cover. For residential areas and recreation zones, it is wise and feasible to use locally adapted ornamental plants, both shrubs and herbs (Figures 5.7 & 5.8). Table 5.7 gives an abridged list of well-adapted ornamental plants. Many of these are endangered and listed in the Red Data Book, so nursery-grown plants must be used. Even today, at the Botanical Gardens, Mongolian

Figure 5.5. Larix sibirica on a steep bank of the Orkhon river (photo by A.V.Prishchepa).

Academy of Sciences, Ulan-Bator, work is being done on the introduction of wild plants and a green belt is being built around the city (Zhavzan, 1996).

Figure 5.6. Cedar forests of Southern Khangai (above) and Central Khankhukhiin (below) (photos by E.N.Matyushkin).

Figure 5.7. Achnatherum splendens + Populus diversifolia near Obot-Khural springs (photo by A.V.Prishchepa).

Figure 5.8. Flourishing onions on the Transaltai Gobi (photo by A.V.Prishchepa).

Table 5.7. Some decorative grasses and bushes.

Species	Natural habitats	Distribution
Juniperus sabina	Cliffs, loose slopes, upper mountain-steppe belt	Mongolian Daguria, Middle Khalkha, Khyangan region, Great Lakes Depression, Gobian Altai
Hemerocallis lilio-asphodelus	Flood-plain, steppe and forest meadows	Eastern Mongolia
H. minor	Forest edges, meadows, bush thickets	Hubsugul region, Khyangan region, Khentii, Khangai
Allium leucocephalum	Steppe stony slopes	Mongolian Daguria, Middle Khalkha, Khyangan region, Great Lakes Depression, Gobian Altai
A. microdictyon	Larch stands, birch stands, forest glades	Khentii, Mongolian Daguria
Lilium buschianum	Dry steppe meadows, sandy steppes	Khyangan region
L. dauricum	Pine stands, glades, bush thickets	Khentii, North-Eastern Mongolia, Khyangan region
L. martagon	Larch stands and Pine stands	Hubsugul region, Khentii, Khangai, Mongolian Daguria, Khyangan region
L. potaninii	Scarce bushes at stony slopes	Eastern Mongolia – Dariganga
Asparagus trichophyllus	Sea-shore solonchaks, deris communities	Great Lakes Depression, Eastern and Alashan Gobi
Polygonatum odoratum	Larch and pine forests	Hubsugul region, Khentii, Khangai, Mongolian Daguria, Khyangan region
Iris bungei	Sandy desertified steppes, detrital slopes	Middle Khalkha, Lake Valley, Eastern Gobi
I. dichotoma	Sands, shore sands	Khentii, Mongolian Daguria, Eastern Mongolia, Khyangan region
I. dichotoma	Sands, shore sands	Khentii, Mongolian Daguria, Eastern Mongolia, Khyangan region
I. tigridia	Detrital and stony slopes, sandy-pebble steppes	Hubsugul region, Khentii, Khangai, Mongolian Daguria, Middle Khalkha
Betula fruticosa	River valleys	Khentii, North-Eastern Mongolia
Acanthophyllum pungens	Desertic stony slopes, sandy slopes	Mongolian Altai, Dzungarian Gobi
Dianthus superbus	Larch and cedar-larch forests	Hubsugul region, Khentii, Khangai, North-Eastern Mongolia, Khyangan region
Dianthus superbus	Larch and cedar-larch forests	Hubsugul region, Khentii, Khangai, North-Eastern Mongolia, Khyangan region
D. versicolor	Detrital and stony slopes, cliffs	Hubsugul region, Khentii, Khangai, Mongolian Daguria, Mongolian Altai
Gypsophila paniculata	Sands, sand steppes	Mongolian Altai, Great Lakes depression
Paeonia anomala	Larch and mixed forests	Hubsugul region, Khangai, North-Eastern Mongolia
P. lactiflora	Piedmont meadows	Hubsugul region, Eastern Mongolia
Aconitum anthoroideum	Nigh mountain meadows, banks of streams	Mongolian Altai

Table 5.7. Continued

Species	Natural habitats	Distribution
A. baicalense	Larch and birch forests	Hubsugul region, Khentii Khangai, North-Eastern Mongolia
A. barbatum	Birch and larch forests, forest edges, meadows	Hubsugul region, Khentii, Khangai, Mongolian Daguria, Mongolian and Gobian Altai
A. chasmanthum	Slopes with meadows and steppes, river banks, bush thickets	Khangai and Gobian Altai
Adonis mongolica	Mountain forests	Khangai, North-Eastern Mongolia
Anemone crinita	Larch forests, forest meadows and edges, stream banks and bush thickets	Hubsugul region, Khentii, Khangai, Mongolian Daguria and Mongolian Altai
A. sylvestris	Larch and birch forests and their edges, bank meadows and bushes	Hubsugul region, Khentii, Khangai, Khyangan region, Mongolian Daguria and Mongolian Altai
Aquilegia glandulosa	River bank meadows	Mongolian Altai, Mongolian Daguria
A. sibirica	High mountain meadows and open woodlands, stony slopes	Hubsugul region, Khentii, Khangai, Mongolian Altai and Mongolian Daguria
A. viridiflora	Dry stony mountain slopes	Khentii, Mongolian Daguria, Eastern Mongolia, Mongolian and Gobian Altai
Atragene sibirica	Larch and mixed forests	Hubsugul region, Khentii, Khangai, Mongolian Daguria, Mongolian and Gobian Altai
Clematis intricata	Mounded and salinized sands	Middle Khalkha, Lake Valley, Eastern Gobi, Gobian Altai
C. tangutica	Stony and detrital slopes, slopes and bottoms of sairas	Khangai, Mongolian Daguria, Mongolian and Gobian Altai, Middle Khalkha
Delphinium grandiflorum	Meadow and forb steppes, forest edges	Hubsugul region, Khentii, Khangai, Mongolian and Gobian Altai
Pulsatilla ambigua	River bank meadows and bushes, pebble and sand deposits, stony slopes	Hubsugul region, Khentii, Khangai, Mongolian Altai
P. bungeana	Cliffs, stony and detrital slopes	Khentii, Khangai, Mongolian Daguria, Mongolian and Gobian Altai, Middle Khalkha
P. turczaninovii	Sandy and stony slopes with steppes	Hubsugul region, Khentii, Khangai, Mongolian Daguria, Eastern Mongolia
Thalictrum baikalense	Birch and elm forests	Khentii, Khyangan region
Trollius asiaticus	Forests and wet forest edges	Hubsugul region, Khentii, Khangai, Mongolian Altai
Berberis sibirica	Cliffs, stony slopes	Hubsugul region, Khentii, Khangai, Mongolian and Gobian Altai
Papaver nudicaule	Sandy banks, steppes and slopes with steppes	Hubsugul region, Khentii, Khangai, Mongolian Daguria, Middle Khalkha, Eastern Mongolia
Rosa davurica	Larch forests, bush thickets	Khentii, Khyangan region, Eastern Mongolia
R. laxa	Stony canyons, along streams	Mongolian and Gobian Altai, Dzungarian Gobi

Table 5.7. Continued

Species	Natural habitats	Distribution
R. pimpinellifolia	Stony canyons	Khangai, Mongolian Altai, Dzungarian Gobi
Tamarix elongata	Mounded sands	Lake Valley
T. gracilis	Mounded sands, solonchaks	Gobian Altai, Borzon-Gobi
T. hispida	Solonchaks, sands	Gobian Altai
T. ramosissima	Mounded sands, solonchaks	Great Lakes depression, Gobian Altai, Gobi
Viola altaica	High-mountain meadows, larch open woodlands	Mongolian Altai, Khangai
Rhododendron aureum	Balds, stony placers, cedar forests	Hubsugul region, Khentii
Goniolimon speciosum	Steppe and desert stony and detrital slopes	Hubsugul region, Khentii, Khangai, Great Lakes depression, Gobian and Mongolian Altai, Mongolian Daguria, Middle Khalkha
Limonium aureum	Solonchaks, dry shore sands	North-Eastern Mongolia, Middle Khalkha, Lake Valley, Gobi
L. bicolor	Sandy and detrital steppes	North-Eastern Mongolia, Middle Khalkha, Lake Valley, Eastern Gobi
L. flexuosum	Steppe stony and detrital slopes	Great Lakes depression, Mongolian Altai
Gentiana dahurica	Steppe slopes	Khyangan region, Eastern Mongolia
G. decumbens	Steppe and meadow slopes, alkaline shore meadows and pebble deposits	Hubsugul region, Khentii, Khangai, Mongolian Daguria, Middle Khalkha, Great Lakes depression, Mongolian and Gobian Altai
G. macrophylla	Larch and mixed forests, shore meadows, bush thickets	Hubsugul region, Khentii, Khangai, Mongolian Daguria, Khyangan region, Mongolian Altai
Thymus gobicus	Sandy and sandy-detrital steppes	Hubsugul region, Khangai, Mongolian Daguria, Mongolian and Gobian Altai, Middle Khalkha, Grate Lakes depression, Lake Valley
Veronica daurica	Steppe and meadow-steppe slopes	Khentii, Mongolian Daguria, Khyangan region
V. incana	Steppe and meadow-steppe slopes, dry edges of larch and pine forests	Hubsugul region, Khentii, Khangai, Mongolian Daguria, Khyangan region, Mongolian and Gobian Altai, Middle Khalkha, Eastern Mongolia, Great Lakes depression
Adenophora stenanthina	Steppe and meadow slopes	Hubsugul region, Khentii, Khangai, Mongolian Daguria, Khyangan region, Middle Khalkha
Campanula altaica	Larch forests and their edges, shore and forest meadows	Mongolian Altai
Aster alpinus	Larch forests, mountain steppes	Hubsugul region, Khentii, Khangai, Mongolian Daguria, Mongolian and Gobian Altai, Eastern Mongolia – Dariganga
Asterothamnus centrali-asiaticus	Desert detrital and stony slopes	Middle Khalkha, Mongolian Altai, Lake Valley, Eastern Gobi, Gobian Altai, Transaltai and Alashan Gobi

Table 5.7. Continued

Species	Natural habitats	Distribution
Dendranthema zawadskii	Larch forests and shore meadows	Hubsugul region, Khentii, Khangai, Mongolian Daguria, Khyangan region, Eastern Mongolia
Gnaphalium baicalense	Sandy-pebble banks, swampy meadows	Khentii, Khangai, Mongolian Daguria, Dzungarian Gobi
Rhaponticum uniflorum	Steppe slopes	Hubsugul region, Khentii, Khangai, Mongolian Daguria, Khyangan region, Middle Khalkha

The most promising wild flora for cultivation includes: *Dendranthema zawadskii, Echinops latifolius, Dianthus versicolor, Goniolimon speciosum, Iris dichotoma, I. tigridia, Thymus gobicus*, and *Veronica incana* (Zhavzan, 1996). In addition to the Ulanbator Botanical Gardens, native plant introductions can be carried out at breeding stations and nurseries.

An important place for disturbed plant cover regeneration is the recultivation of mining spoils and roadways (Figure 5.9). Mining and even geological prospecting cause human-badlands to develop. Such territories

Figure 5.9. Abandoned quarry near Bayan-Teg (photo by A.V.Prishchepa)

are of practical importance for a grazing-based society and ecologically hazardous since, in the course of mining, rocks often are brought to the surface which contain compounds deleterious for plants. And as a result of renewed water and wind erosions these toxic materials can travel great distances. Such conditions occur now where there is extensive mining, e.g., for gold in the Zolter region (Western Khentii), for brown coal in the Baganur region, and for polymetals in the Erdenet region.

To recover the ecological potential of such regions intense measures are demanded. These are expensive operations first using heavy machinery and then skillfully integrating proper vegetation plantings and water controls with minimal wind erosions. On disturbed mining plots, one must take into account its geomorphological conditions, the altitude zone of the area, the nature of surrounding plant communities, and the level of vegetation disturbances. All such ecosystem factors complicate plant improvement and recultivation of such greatly disturbed lands.

At this same time, it seems impossible to arrest the increasing use of mineral resources in these very rich regions. Therefore, strict control is needed to implement nature-saving technologies with minimal emissions of deleterious substances into the surrounding environment. Recultivation measures should be a prerequisite condition for intensive industrial utilization of natural resources as is now being done in other countries (Motorina & Ovchinnikov, 1975).

Every form of industrial, tourism, or economy development requires the establishment of road networks, plus the development of aviation and associated infrastructures. Particularly in Mongolia with its merger paved road network, the construction of paved roads would preservation countless hectares of rangelands. Now the unregulated passage of vehicles across virgin lands by multi-rut dirt roads leads to an extensive degradation of rangeland vegetation. But from some little experience on abandoned roads, the native vegetation regenerates extensively. So far, only the first stages of such regeneration successions have been observed, for instance, the plants along the newly asphalted road from Ulan-Bator to Arvaikhere.

Of great importance for sound nature management for vegetation conservation and recovery is a strict enforcement of water conservation zones. Today Mongolia has several water conservation zones situated in high mountains, where main rivers originate (Figure 5.10). In addition, it is possible to add water conservation zones in the Mongolian Altai involving the upper reaches of the Khovd River, its sources in the mountains along the Selenga River, and also to expand the water conservation zone in the Khangai Upland. In addition, these should include the upper reaches of southern slope rivers in the Khangai Range. Also note these would strongly promote forest regeneration, since forest felling is banned in water conservation zones.

One of the most acute problems hindering regeneration successions in hydromorphic vegetation floodplain habitats is river water purity, primarily with waters downstream from larger industrial-residential centers. Heavy pollution occurs in areas of the Selenga, Orkhon, Tuul and Kerulen rivers (Figure 5.10). For instance, water pollution increases in the Kerulen river downstream from Underkhan almost twice, and to an even greater extent downstream from Choibalsan (Table 5.8).

Waters of the Tuul River are heavily polluted with industrial and domestic effluents from the city of Ulan-Bator and are purified only 100-km downstream (Gunin & Vostokova, 1993). Large irrigated lands also are heavy polluters of river water for instance, Kharkhorin in the Orkhon river valley. Direct discharge of polluted waters into the river causes strong disturbances of ecological equilibrium in the river and badly affects the floodplain vegetation.

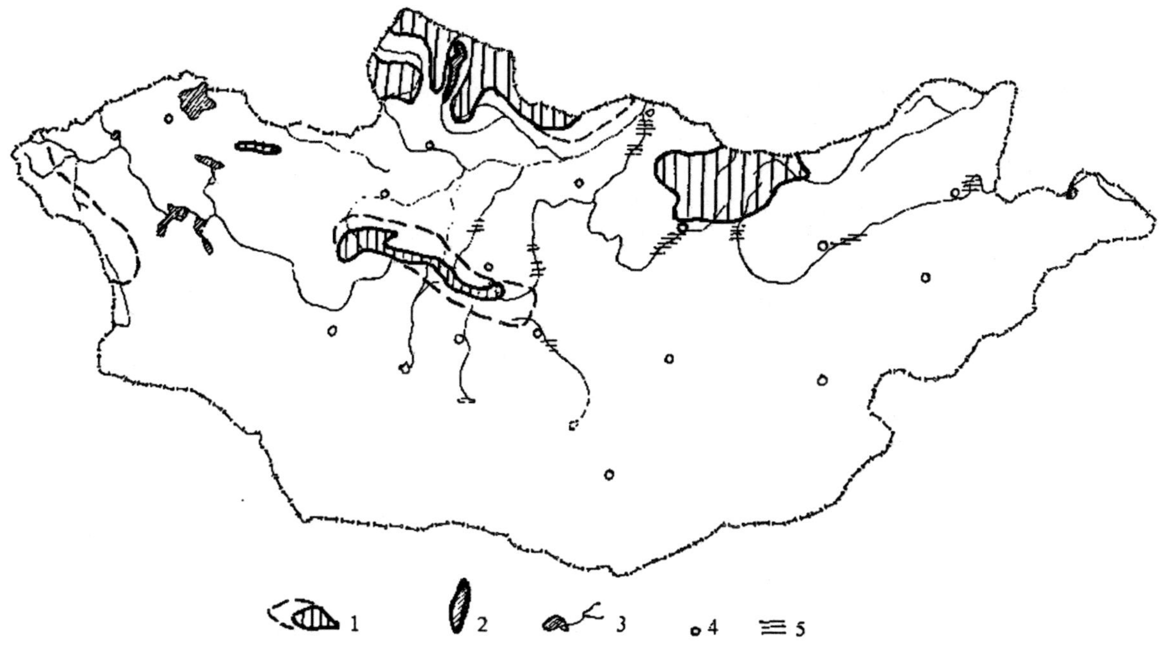

Figure 5.10. Water protection zones

1 – existing and recommended (dotted line) water protection zones; 2 – the zone of strict regime for Hubsugul lake; 3 – lakes and rivers; 4 – towns; 5 – polluted parts of rivers.

An important creation is a water-protection zone for the Selenga River basin, the main artery charging Lake Baikal (Figure 5.11). Two thirds of the Selenga River basin is situated within Mongolia and only one third in Buryatiya and the southwest Chita Region (Russia). Along the river valley and its numerous tributaries, particularly in the middle and lower reaches, are big industrial complexes and fairly large croplands, which combine to cause pollution and degradation of river waters and shore vegetation. Pure surface water is necessary for direct consumption and also to ensure more natural ecological conditions for bank and aquatic vegetation. Some special measures are needed now, specifically, industrial enterprises must add reliable water treatment plants.

Table 5.8. Characteristics of river waters near Kerulen in 1988 (Sevastyanov et al., 1990).

Gauge at Kerulen river	Total minerals, mg/l	Turbidity, g/m^3	Solid runoff, millions m^3	Suspended load discharge m^3/s
Bayandelger	126.3	70	749	52
Underkhan	206.4	108	808	87
Choibalsan	250.2	170	819	105

Some characteristics of water pollution in the Kerulen river (April, 1989, according to data supplied by UGKS of MPR)

Characteristic	Above Underkhan	Below Underkhan	Above Choibalsan	Below Choibalsan
O_2 mg/l	8.9	8.7	8.9	8.6
BPK_5 mg/l*	0.61	0.80	0.48	1.34
Suspended matter mg/l	43.4	322.4	138.8	185.2

* Biocamical oxygen consumption by water for 5 days

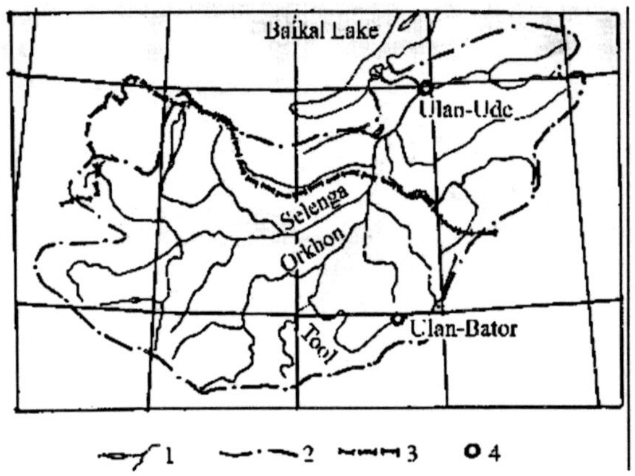

Figure 5.11. Basin of the Selenga river within Mongolia and Russia

1 – lakes and rivers; 2 – boundary of Selenga River basin; 3 – Russian-Mongolian boundary; 4 – capitals of Mongolia and Buryatia.

Today some consistent monitoring-type activity is needed for all types of natural resources usage. As quickly as possible an organized monitoring system should be strictly implemented. One recommendation is to organize satellite mapping monitoring including use of remote-sensing information to develop maps of the current status of plant cover and environment in all of Mongolia (Vostokova & Kelner, 1986; Vostokova, 1995). Also it is feasible, to conduct ground monitoring in areas of special importance. Amongst these are industrial and residential complexes plus their adjacent territories, cultivated lands, and biosphere reserves.

To prevent subsequent detrimental effects of economic activity on natural resources, their considered use should include an ecological risk assessment for each big project of industrial or domestic construction. This risk assessment should promote more ecologically and economically sound projects. To accomplish that, Mongolia must train a cadre involving biologists, geobotanists, ecologists, geographers, etc. for assessing and monitoring the use of natural resources.

Nature conservation is an important aspect of considered nature management. Hence work on restoring plant cover or conservation of botanical diversity should be integrated with well-developed programs on rangelands organization, grazing migrations, sites where large concentrations of livestock are stationed and other problems of large livestock complexes. Such programs also should include aspects of economic land use, for example, placement and organization of industrial complexes, their infrastructure, and residential areas; construction of paved roads; machine-based farming; purification and sanitation of polluted lands; establishment of organized recreational zones and sites; and legally protected lands.

Thus, the complete system of considered nature management is aimed at providing good normal living conditions for the entire population of Mongolia; and that goal must be skillfully integrated with the conservation and restoration of plant cover in this vast animal grazing-based society.

5.3 Systems for the conservation of botanical diversity

It is noteworthy that Mongolians historically have a love of preserving the landscape because its citizens set aside the Bogd Khan Uul Mountain range south of Ulan-Bator as perhaps one of the world oldest National Parks in 1778.

Mongolia now has several special protected territories of different sizes and with different restrictions on economic activities. Amongst them are more strictly protected biosphere reserves. Some territories are now planned from the Biodiversity Conservation Action for Mongolia in 1995 with a less strict protection regime, i.e., national parks, zakazniks (preserves) and nature monuments. However, until the late 1980s, there was no such strict system of protected territories in Mongolia (Neronov et al., 1988). Occasionally the so-called reserves contained very small areas, but were not provided with safe guards. Only the enactment in 1995, by the Great Khural of Mongolia, of the Law of Specially Protected

Nature Territories (Mongolian Environmental Laws, 1996) made a clear distinction for reserves, national parks, and other protected natural territories.

Reserves and national parks with special protected areas are of central importance for the protection of plant cover and botanical diversity. Table 5.9 presents a list of reserves and national parks Mongolia as of 1996. Their distribution is shown in Figure 5.12. For the conservation of vegetation, small and virtually unprotected preserves or nature monuments are of little importance because these areas customarily are used as ranges and for traffic along with nearby unprotected lands.

Reserves are the best-protected areas from human-associated effects. A unique type of reserve is that of cluster reserves, for instance the Uvs Nuur and the South Gobi Reserves. They include several separate protected areas while somewhat limited economic activity is allowed in the remainder of the territory. The Uvs Nuur cluster is an international reserve - it includes 4 protected areas in Mongolia (Figure 5.13) and 4 protected areas in Tuva (Russia).

Currently national parks are for nature conservation and ecological educational institutions. They include natural complexes and objects of particular aesthetic, ecological, or historical values through which they combine the objectives of nature conservation and tourism. These large parks are divided into functional zones. A particularly well-protected area includes a core, where visitation is banned or is restricted to defined marked routes. Such protected areas are of primary importance for the preservation of botanical diversity.

Before the enactment of the Mongolian Law of Strictly Protected areas in 1995, 21 plots with a total area of 8717.3 thousand ha were reserved. They ranged from the biggest, the Great Gobi Reserve with 5300 thousand ha to tiny areas in inactive volcanoes of only 0.9 thousand ha. A critical analysis of the state and location of all protected areas in Mongolia was

Figure 5.12. Mongolian protected areas.

1–3 – boundaries: 1 – regions, 2 – provinces, 3 – districts; 4 – indeces of zoning units; 5–6 – natural protected areas (by Table 5.9).

Table 5.9. Mongolian protected areas (by Mongolia's Wild Heritage, 1996)

	Aimaks located in	Code*	Area (thousand ha)	Established
Strictly Protected Areas (Natural reserves)				
Great Gobi A and B (Two sites)	Gobi-Altai, Bayankhongor, Khovd	1A, 1B	5311.7	1975
Khokh Serkh	Bayan-Ulgii, Khovd	2	65.9	1997
Bogd Khan Uul	Tov	3	41.6	1778
Khasagt Khairkhan	Gobi-Altai	4	27.4	1965
Khan Khentii	Tov, Khentii, Selenge	5	1227.1	1992
Nomrog	Dornod	6	311.2	1992
Dornod Mongol	Dornod, Sukhbaatar	7	570.4	1992
Mongol Dagurian	Dornod	8	103.0	1992
Otgon Tenger	Zavkhan	9	95.5	1992
Uvs Nuur Basin (Four sites)	Uvs	10	771.6	1993
Little Gobi	Umnogobi, Dornogobi	11	1839.1	1996
			10305.4	
National Concervation Parks				
Khovsgol Nuur	Khovsgol	12	838.1	1992
Khorgo Tsagaan Nuur Terkh	Arkhangai	13	77.3	1965
Gobi Gurvansaikhan Uul	Umnogobi	14	2171.7	1993
Gorkhi Terelj	Tov	15	286.4	1993
Altai Tavan Bogd	Bayan-Ulgii	16	636.2	1996
Khangai Nuruu	Arkhangai, Uvorkhangai	17	888.5	1996
Khara Us Nuur	Gobi-Altai	18	–	1997
			4904.8	
Nature Reserves				
Nagalkhan Uul	Tov	19	3.0	1957
Batkhan Uul	Tov, Uvorkhangai	20	21.8	1957
Lhachinvandad Uul	Sukhbaatar	21	58.8	1965
Bulgan Gol	Khovd	22	7.6	1965
Khustain Nuruu	Tov	23	49.9	1993
Ugtam Uull	Dornod	24	46.2	1993
Sharga-Mankhan (Two sites)	Khovd, Gobi-Altai	25	390.0	1993
Zagiin Us	Dornogobi, Dundgobi, Umnogobi	26	273.6	1996
Alag Khairkhan	Gobi-Altai	27	36.4	1996
Burkhan Buudai	Gobi-Altai	28	52.1	1996
Ergeliin Zoo	Dornogobi	29	60.9	1996
Ikh Nart	Dornogobi	30	43.7	1996
			1044.3	
Natural Monuments				
Bulgan Uul	Arkhangai	31	1.8	1965
Uran-Togoo Tulga Uul	Bulgan	32	1.6	1965
Eej Khairkhan	Gobi-Altai	33	22.5	1992
Khuisiin Naiman Nuur	Uvorkhangai	34	11.5	1992
Ganga Nuur	Sukhbaatar	35	32.9	1993
Suikhent Uul	Dornogobi	36	4.8	1996
			75.1	
Total of Protected Area			**16 329.9**	

* Identification code for Figure 5.12.

Note: – not available.

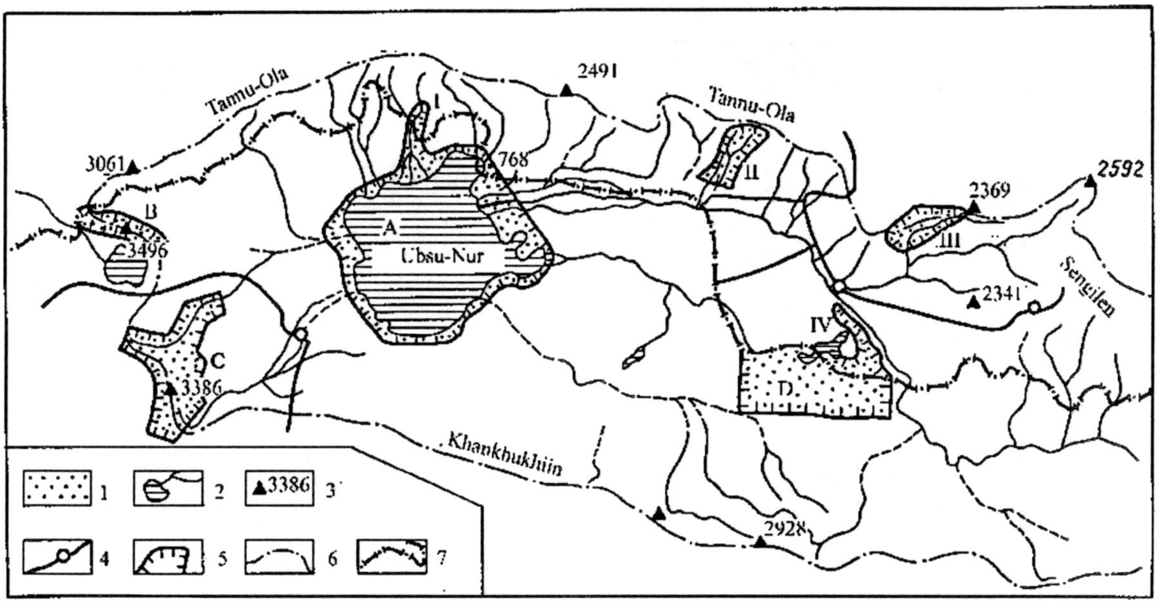

Figure 5.13. Protected sites in the Uvs Nuur cluster transboundary reserve of Mongolia (A–D) and Russia (I–IV).

1 – protected areas; in Mongolia: A – Uvs Nuur, B – Tsagan-Shibetu, C – Turgen, D – Altin-Els; in Russia: I – Torgalic river, II – Khiralig-Khem, III – Ular river, IV – Tsuger-Els. 2 – lakes and rivers; 3 – elevations; 4 – roads and settlements; 5 – boundaries of protected sites; 6 – Watershed of Uvs-Nur lake; 7 – Russian-Mongolian boundary.

conducted by P.D.Gunin (1993) to form a basis for developing a modern network of reserves and national parks. Reserve status was assigned to 9 of the previously existing reserves and a new transboundary cluster reserve was established in the Uvs Nuur Basin. In 1996 two previously proposed reserves were established – Mountain Mongolian and South Gobi. And by adding area to previously existing reserves, national parks were set up. A large area was added south of the Valley of Yol reserve to establish the Gobi Gurvansaikhan Uul National Park (Gunin et al., 1998).

But even after these additional and changes, the reserve coverage is highly irregular (Figure 5.12). The greatest number of reserves are located in high- and middle mountains and on the border with China. Thus the most protected now are plant communities of mountain tundra, partly, mountain-forest zones of high and middle mountains, and the desert zone. Only a small steppe and hydromorphic vegetation area has been reserved (Tables 5.10 & 5.11).

But some steps have been taken to conserve hydromorphic ecosystems including plant cover. Thus, the Mongol Dagurian Reserve is situated in northeastern Mongolia near the Dagurian Reserve, around the Torei lakes in Russia, and eastward of the Dalai Nuur Reserve in China. In 1996, a joint agreement was reached to establish a joint Russian-Mongolian-Chinese Reserve with a large buffer zone. The main objective of this international cluster reserve is to conserve wetlands with their adjacent steppes. Implementation of this agreement makes it possible to create conditions necessary to conserve the unique flora of eastern Asian sector Eurasian steppes (Figure 5.14). Unfortunately today steppes are not covered by a sufficient number of reserves. The absence of large steppe zone reserves creates the danger of

Table 5.10. Number of plant types in reserves.

Vegetation types according to position in altitude belt and zone	Vegetation									
	Mountains				Plains		Hydrogenic habitats			
	High mountains	Medium-height mountains	Low mountains	Hills	Uplands	Low	Mountain rivers	Plain rivers	Lakes, lakes and rivers	Sairas
Glacial-nival	1	–	–	–	–	–	–	–	–	–
Tundra (high mountain)	6	4	–	1	2	–	–	–	–	–
Cryophyte-meadow	5	2	–	1	–	–	–	–	–	–
Mountain-forest	1	11	10	2	4	–	–	–	–	–
Forest-steppe	–	2	1	1	1	1	–	–	–	–
Meadow-steppe	–	1	4	2	2	–	–	–	–	–
Steppe	2	3	6	3	4	2	–	–	–	–
Dry-steppe	–	3	–	–	2	1	–	–	–	–
Desertified-steppe	–	1	3	–	1	1	–	–	–	–
Semi-deserts	–	4	3	1	2	2	–	–	–	–
Desert	–	3	7	2	5	5	–	–	–	–
Extra-arid	–	–	3	–	5	4	–	–	–	–
Non-zonal:										
hydromorphic in mountains	–	–	1	2	–	1	5	–	–	–
hydromorphic at plains	–	–	–	–	–	–	–	2	5	4
aquatic	–	–	2	–	–	1	–	–	1	–
TOTAL	15		40	15	28	18	5	2	6	4
TOTAL:	104				63					

extinction for Mongolia steppe ecosystems and for Inner Asia steppe biomes.

Studies on the present-day status of plant cover status in reserves shows that human-associated disturbances in most of the protected territories are essentially minor or absent. But even in the long-established Great Gobi Reserve, the vegetation on its margins is at a moderate state of rangeland degradation because of unregulated grazing by domestic livestock. Near springs located on the edge of this Reserve the vegetation is heavily disturbed. In the western part of the Great Gobi Reserve (Cluster B situated in the Dzungarian Gobi) the high botanical diversity of this natural oasis is very strongly upset.

Table 5.11. Areas of protected plant communities in Mongolia.

Main reserve ecosystems	% of country area
Glacial-nival	0.24
Tundra and cryophite-meadow	4.65
Mountain forest	6.69
Forest-steppe	3.45
Mountain-steppe (predominantly)	11.07
Semi-deserts	1.98
Deserts	8.92
Extra-arid	50.88
Hydromorphic	2.94
Aquatic	3.18

The vegetation in reserves covering mountain-forest and forest-steppe zones often has been exposed to fires and illegally disturbed by tree fellings and livestock grazings. But the vegetation in high mountain reserves is virtually unchanged.

Most of the reserves suffer little effects of big industrial enterprises, residential areas, or military testing grounds. These factors can affect vegetation not only directly but indirectly also due to soil accumulations of heavy metal salts and radionuclides. But two of the leading reserves in Mongolia are effected and both are in the world biosphere reserve network: The Great Gobi is located near the atomic testing ground of China. The southern edge of the Bogd Khan Uul Biosphere Reserve is effected by industrial aspects of the

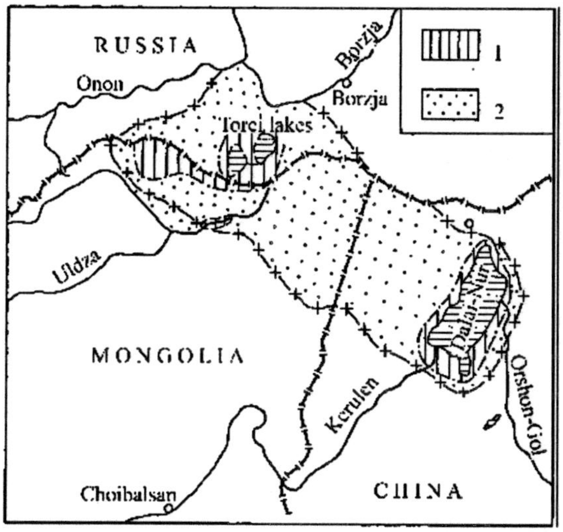

Figure 5.14. Infrastructure of the Russian – Mongolian – Chinese transboundary reserve:

1 – reserves; 2 – buffer zone.

thermoelectric power station for Ulan-Bator (Kasimov et al., 1992; Lebedeva et al., 1997). And the boundary location of a number of reserves–the Mongol Dagurian, Nomrog, Dornod Mongol–do have an increase in vegetation disturbance caused by traffic along the national border. But on the whole, the vegetation there is only slightly disturbed because domestic livestock grazing is virtually absent.

Vegetation in national parks is strongly exposed to human-associated disturbances. These territories are criss-crossed by ground roads with heavily disturbed vegetation along them. Numerous territories along newly organized parks have long been used as ranges, for instance, the Gobi Gurvansaikhan Uul National Park. Even in the world-famous Khovsgol Nuur National Park, which was a reserve until 1995, despite its remoteness and difficult access many ground roads are around the lake, caused by economic activities. Roads connect the northern village Khankh to Russia (across the border) and to the Khatgal village at the south of the lake, and with numerous wintering sites on the east side. The recreation-use load of the Bogd Khan Uul Reserve is associated with the nearby capital concentrating over 350 thousand residents, and with built-up villages, cottages, and recreation centers, plus communication and electric power lines. Recreational pressures also strongly affect plant cover on the former Valley of Yol reserve, which is today part of the national park Gobi Gurvansaikhan. This reserve occupies the Dzun Nuruu range in the Gurvansaikhan mountain system with a big active tourist center situated on its border. The negative role of recreational pressures on plant cover is associated not only with littering and trampling plants down along numerous paths, but the very destructive gathering of beautiful plants, many of which need strict protection. Wild food and drug plants are harvested to a much lesser extent.

Overall, we believe that the vegetation in reserves, having clear-cut boundaries with protected territories and special guards, are in good condition. Plant community disturbances are minor and most function at a close to natural state.

Today the location in Mongolia of reserves and national parks mostly in mountains is justified and sufficient for the conservation of mountain vegetation botanical (floristic) diversity. But reserves to conserve diverse plant communities, including the unique Mongolian steppe ecosystems, have failed. In this connection, there is an acute need to establish a big reserve in the plain steppes of Mongolia. The central nature of that problem is mainly associated with the fact that these territories have long been used for livestock grazing, and for occasional farming. Encouragingly, fairly recently in portions of Mongolia, a number of Bhuddist temples and monasteries exist and no grazing is permitted in their neighborhood (Dashnyam, personal communication). Today these lands are distinguished by a greater vegetation preservation. In planning a new steppe reserve, the condition and location of such lands should be strongly considered.

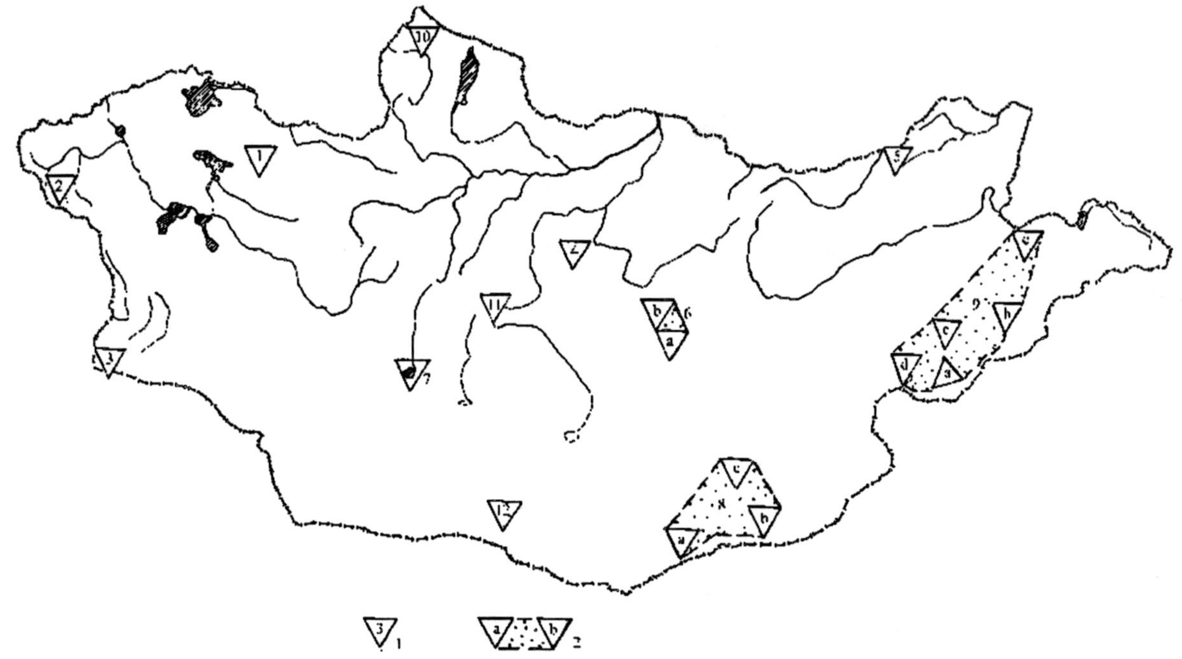

Figure 5.15. Recommended strictly preserved territories:

1 – preserved sites; 2 – buffer zones. 1–9 – recommended reserves: 1 – Khankhukhiin; 2 – Altai Mountains (Mongolian Altai); 3 – Baitagdalin (Dzungarian Gobi); 4 – Khunge-Khanul (Eastern Khangai, Orkhon and Tuul interfluve); 5 – Erdene-Daba (North-eastern Mongolia); 6 – Dry steppe (Middle Khalkha): a. Baga-Gadzuirin-Nur, b. – Undul; 7 – Bon-Tsagan-Nur (Lake Valley); 8 – Southern Gobi: a. Borzon Gobi, b. Galbyn Gobi, c. Ulan Gobi; 9 – Great Steppe reserve (Eastern Mongolia): a. Dariganga, b. Matad-Lagnur, c. Asgat-Tsagan-Toloi, d. Bayan-Tukhemeiin Gobi, e. Menengiin Tal; 10–12 – preserved territories to be added to national parks: 10 – Shishkhid Gol, 11 – Upper reaches of Orkhon river, «Cedar Island», 12 – Khuren-Khana ridge.

Also, it should be noted that on existing reserves, hydromorphic vegetation is not represented adequately (Table 5.10). Only 4 reserves contain lake water areas and sections of river valleys.

Hence, the existing network of protected areas does not fully meet the needs for botanical diversity, either floristic or ecosystem. Even the protected areas being designed would not fully cover all of the unique ecosystems and plant communities of Mongolia! Hence, to deflect the increasing human-related destruction of plant cover and to preserve its floristic-community richness, the existing network of protected lands must be expanded considerably. Against an intensive background of plant degradation processes and the destruction of rare and endemic species and communities, the need to organize new nature conservation territories becomes increasingly clear (Figure 5.15).

An establishment of legally protected lands implies not only isolation and delimitation of such lands, but also establishment of a guard system and studies of their diversity. Normally reserves have a broad task to protection the entire ecosystem complex, particularly, the biotic part, i.e., vegetation, wildlife and microbes. Most frequently only the animal population is protected, perhaps to a less extent some rare and endemic plant species, while the protection of particular plant communities or microbes are neglected completely. The conservation of botanical diversity calls for strong measures to retain both individual plants and plant communities.

To ensure the protection of botanical diversity on designated protected lands it is possible to set up a buffer zone near each land with a limited use of natural resources. Meriting special attention is the limited use of plant resources. Simultaneously, inside of strictly protected reserve areas, strict conservation areas should be clearly designated where only research activity will be allowed.

A landscape-ecology approach is possible because reserved areas are designed to conserve ecosystems. The approach is based on analysis of the natural distribution of ecosystems plus considering the patterns and levels of human-associated effects. A landscape-ecology approach provides the isolation of protected areas because it considers geological and geomorphological features, plus landscape structure, biological diversity (including botanical diversity), and the modern environment status as well as changes in the structure and functioning of ecosystems under the effects of economic activities. In this case one can widely use plant cover indicator properties both as indices of abiotic environment conditions, and as indicators of the levels of human-related effects.

In planning the new protected areas (Puzachenko et al., 1983; Gunin & Neronov, 1986) ecological principles were employed to develop recommendations for expanding the network of reserves in Mongolia. One principle states that the network density should conform to the landscape-ecological complexity and structure and to species diversity saturation. Therefore, the establishment of new protected areas is recommended in: Khalkha (Central Mongolia), Lake Valley, Southern Gobi, the Khankhukhiin Nuruu mountains and Mongolian Altai (Table 5.12). The first proposal priorities are to organize: a reserve in Khankhukhiin Nuruu; a Dry-Steppe Reserve in Middle Khalkha; a Greater Steppe Reserve in Eastern Mongolia; Eren-Daba in the

Table 5.12. Addition recommendations for special protected areas.

Status	Name	Location
Reserve	Mountain-Altai***	North-West of Mongolian Altai
	Khankhukhiin Nuruu*	Central and Eastern parts of Khankhukhiin Nuruu range
		North-Eastern part of Eren-Daba range
	Eren-Daba	Eastern Khangai, Tuul and Orkhon interfluve
	Forest-steppe	Middle Khalkha
		Eastern Mongolia
	Dry-steppe	
	Great-steppe**:	
	A. Dariganga	
	B. Matad-Lagnur	
	C. Asgat-Tsagan-Tolgoi	Eastern Gobi
	D.Bayan-Tukhemiin-Gobi	
	Southern Gobi***:	
	A. Borzongiin Gobi	
	B. Galbyn Gobi	Lake Valley
	C. Uran Gobi	Lake Valley
	Bon-Tsagan-Nur	
	Orog-Nur	

* Can be one of reserve sites of Uvs Nuur Reserve
** With large buffer zone
*** Included in reserve list in 1996

northeastern Khentii; a Forest-Steppe in Eastern Khangai; and a South Gobi reserve. Other preserves will be proposed in the future. Of the reserves recommended in 1996, decisions have been made to organize the Altai Mountain (Altai Tavan Bogd National Park) and The Southern-Gobi (Little Gobi Strictly Protected Area) reserves.

In Mongolian Altai the Altai Mountain Reserve is recommended to preserve unique forest «islands» of *Picea obovata* plus *Larix sibirica*. These forests are located on the western macroslope of Mongolian Altai in the Elt river valley (Black Irtysh basin) at an altitude of 2000 to 3500 m. Herbaceous plants are represented by colorful forb meadows which are characterized by a rich floristic composition and the presence of rare and endemic species, including *Bergenia crassifolia, Schultzia crinita, Aconitum septentrionale, Ranunculus altaicus, Woodsia ilvensis, Geranium albiflorum, Swertia obtusa, Silene altaica* and *Viola macroceras*. The reserve is planned for about 300 thousand ha.

In western Mongolia, a high botanical diversity and a specific geobotany is characteristic of the Khankhukhiin Nuruu Range (Figure 5.16). The recommendation is to establish either an independent reserve or one additional area in the Uvs Nuur Basin Cluster Reserve. The Khankhukhiin Nuruu Range stretches almost latitudinally from the south to separate the Lake Uvs Nuur depression from the remaining Big Lake Depression. The range is normally considered as the Khangai Upland westernmost edge and represents an interesting biogeographical feature (Karamysheva & Banzragch, 1976; Gunin, 1993; Matyushkin, 1995). The Gobi floristic effect is distributed from the south to the range foothills. The Khankhukhiin Nuruu Range itself, in terms of structural zones with a north-facing macroslope, is quite Siberian (Figure 5.16). High-mountain flora is fairly well represented above the forest border. This Range is unequaled in Mongolia in terms of plant community contrasts and the spectrum of diversity in its altitude zones. A reserve of 28-300 thousand ha should be isolated into central and eastern Range parts.

Great floristic and plant community diversity is characteristic of the Eren-Daba, low-mountain range situated east of Khentii in the Onon and Uldza river valleys. This vegetation is representative of a peculiar combination of pine, larch, and birch forests and larch-polar floodplain forests, i.e., river valley forests. The herbaceous cover floristic composition also is very peculiar. In the foothills and on gentle flats, steppes of the

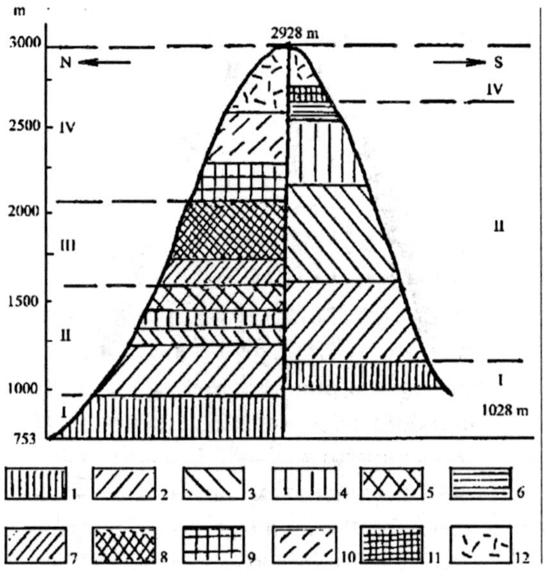

Figure 5.16. Belt vegetation cover structure on the Khankhukhiin Nuruu ridge (Karamysheva & Banzragch, 1976):

N – Northern slope; S – Southern slope. I–IV – altitude belts: I – desert, II – steppe, III – forest, IV – high mountain. 1 – 11 – vegetation: 1 – grass – semi-dwarf bush deserts; 2 – desertified semi-dwarf bush – bunch-grass steppes; 3 – dry bunch-grass steppes; 4 – forb – bunch-grass steppes; 5 – meadow steppes; 6 – high mountain steppes; 7 – grass-bush and grass steppized larch stands; 8 – cedar – larch forests; 9 – sub-bald cedar – larch open woodlands; 10 – high mountain tundras in combination with barrens and meadows; 11 – high mountain barrens and meadows; 12 – stony placers without continuous plant cover, with fragments of high mountain barrens and meadows.

Daurian and Eastern Mongolian type occur which contain *Filifolium sibiricum, Lespedeza davurica*, etc. (Kamelin et al., 1992). All the stony slopes and bases are occupied by a rich complex of steppe shrubs and low xerophilous trees including *Armeniaca sibirica, Ulmus macrocarpa, Rhamnus parvifolia, Ribes pulchellum, Spiraea aquilegifolia* and *S. pubescens*. This reservation is designed to conserve and to promote regeneration of vegetation successions that have been heavily disturbed by forest fellings, fires and grazing. According to R.V. Kamelin et al., 1992, a natural national park should be established there with a strictly protected core and a Chinghiz Khan memorial.

Conservation of the botanical diversity in steppes and forests of Mongolia is very important. Therefore a new reserve is proposed including forest-steppe complexes of the middle Khugne Khan Uul Mountains in the Orkhon and Tuul river valleys (Eastern Khangai). The larch-Siberian pine forests on the northern slopes of Eastern Khangai differ in composition and structure from those in the Bogd Khan Uul Reserve. The hummocky areas and valley plains are occupied by steppe communities with a dominance of *Stipa krylovii, Cleistogenes squarrosa, Koeleria cristata* subsp. *mongolica, Caragana microphylla, C. stenophylla* and also communities with *Festuca lenensis, Poa attenuata, Koeleria cristata* subsp. *mongolica, Oxytropis filiformis*, and *O. nitens*. Such plant groups are quite representative of these isolated mountain-forest and -steppe communities.

In the Middle Khalkha, to conserve dry steppes, typical communities should be reserved of *Festuca lenensis + Stipa krylovii, S. grandis* and *Cleistogenes squarrosa + Stipa* spp. A central position here is occupied by the low granite Mountain Baga-Gazaryn Nuruu. At granite outcrops in cliff fractures, *Ulmus pumila* and *Amygdalus mongolica* are characteristic. Groves of *Ulmus pumila* also are located in the mountain foothills at outcrops of subterranean spring waters that form small brooks surrounded by communities of *Achnatherum splendens*. On these steppe plots, wild ungulates can compensate for undergrazing by domestic livestock and they will prevent the natural degradation of steppe vegetation. Granite rocks in the mount Undzhul steppe also form a very interesting area in the reserve (Figure 5.17).

The Great Steppe Reserve should be established to preserve Mongolia's unique steppe ecosystems. The reserve should consist of several clusters, of over 1 million ha,

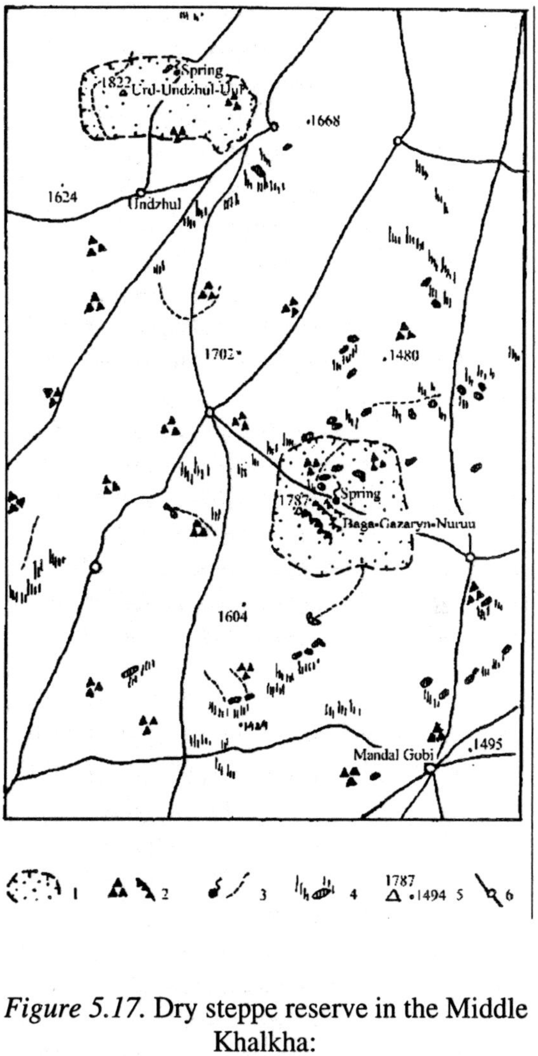

Figure 5.17. Dry steppe reserve in the Middle Khalkha:

1 – recommended sites and their boundaries; 2 – granite rocks and cliffs; 3 – springs and intermittent water streams; 4 – solonchaks and intermittent salt lakes; 5 – elevations; 6 – roads and settlements.

representative of eastern Asia steppes. In general the protected reserves fall into 4 clusters: Matad-Lagnuur, Dariganga, Asgat-Tsagan-Tolgoi and Bayan-Tukhemiin-Gobi (Figure 5.14).

Gazelle protection also is necessary here in addition to the steppe vegetation, which is still only weakly, disturbed. Here large accumulations of *Procapra gutturosa* are found during calving. These steppes are the so-called «nurseries» for protected mammal species (Lushchekina et al., 1983; Sokolov et al., 1982). In addition to protected areas, a large buffer zone of about 2-3 million ha should be set aside for the disbursement of animals and as corridors for their mass disbursements on winter ranges. All types of natural resources exploitation should be excluded that damage the gazelle population. Without a doubt, the primary steppe vegetation, which provides the main gazelle forage resources, should be retained. Dry steppes are dominated by *Stipa krylovii, Cleistogenes squarrosa, Leymus chinensis, Koeleria cristata* subsp. *mongolica* with *Sibbaldianthe adpressa, Haplophyllum davuricum,* and *Astragalus galactites*. Often one finds *Caragana microphylla, C. stenophylla*; and on stony rocky areas *Stipa krylovii, Arenaria capillaris, Arctogeron gramineum* and other petrophytes. For steppe vegetation conservation, the construction of large residential areas and long dirt road ruts or mining complexes should not be permitted. Ploughing of large land areas also should not be permitted. Recognize that the removal of human-related pressures will not lead to a natural degradation of steppe vegetation from undergrazing because gazelle herds fully compensate for the absence of domestic livestock.

The organization of a Great Steppe Reserve on the plain steppes of Mongolia with its rich set of plant communities and their representative characteristic species will be an event of world importance. Both in Europe and North America, the vegetation of plain steppes has long been replaced by agriculture. Organization of this reserve should be done quickly while the steppes are in a relatively satisfactory condition. This reserve will merit qualification as a World Natural Heritage object.

It should comprise several clusters in Eastern Gobi: Borzongiin Gobi, Galbyn Gobi and Ulan Gobi (Figure 5.18). This region is distinguished by the presence of unique Gobi arboreous flora. Many dominants and indicators of plant communities are listed in the Red Data Book, including *Potaninia mongolica, Gymnocarpos przewalskii, Ammopiptanthus mongolicus* and *Amygdalus mongolica*. Natural conditions create diverse plant community combinations on both dry and moist soils with the dry soil plant communities on rock-debris and sandy deserts naturally dominating (Figure 5.19). In addition, there are many picturesque and exotic sites – cliffy granite outcrops breaking through sand mounds, dead volcanoes, sandstone rock precipices, tall *Haloxylon* stands on sands and moist soil plant communities at outcrops of subterranean waters on slopes, and unstructured saline soils with typical halophyte vegetation along with clay soils without vegetation. The reserve will retain some plant communities of true Alashan and Eastern Gobi desert characteristics. These deserts are dominated by *Salsola passerina, S. laricifolia, Psammochloa villosa* and *Ammopiptanthus mongolicus* which are absent in the Great Gobi Reserve.

Borzongiin Gobi concentrates a great botanical species and community diversity for a desert. Much of the plant cover composition is made of *Haloxylon ammodendron* communities, frequently along with *Nitraria sphaerocarpa* and *Ammopiptanthus mongolicus*. *Haloxylon ammodendron* is noted for its arboreous form of growth as high-grade forage. Jointly with *Haloxylon ammodendron* large areas in sandy deserts are taken up by *Ammopiptanthus mongolicus*, which can frequently form independent communities.

Communities of *Achnatherum splendens* + *Phragmites australis* are found at springs that

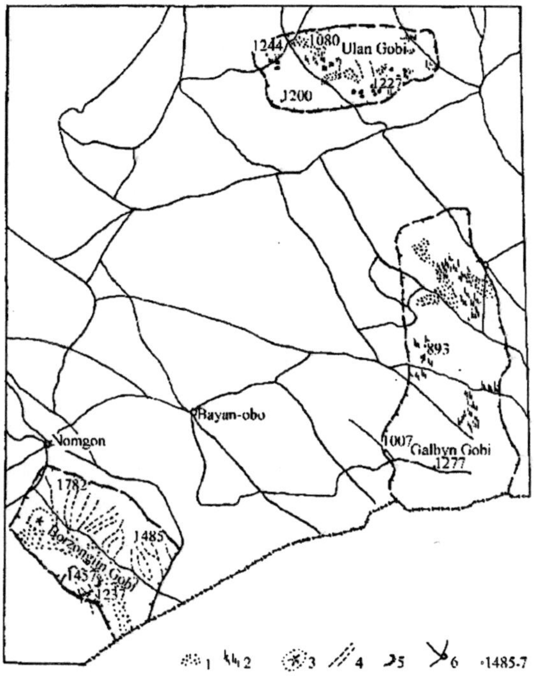

Figure 5.18. Infrastructure of the Little Gobi reserve (Borzongiin Gobi, Galbyn Gobi, Ulan Gobi)

1 – sand massifs; 2 – solonchaks; 3 – takyrs; 4 – sairs (dry valleys); 5 – terrace scarps; 6 – settlements and roads; 7 – altitudes of mountains and plains/

are replaced by thickets of *Nitraria sphaerocarpa* and groupings of *Kalidium foliatum* on saline soils. Stream valleys are characterized by tree chains of *Ulmus pumila* (Figure 5.20), occasionally accompanied by *Populus diversifolia, Tamarix ramosissima, Amygdalus mongolia* or *Achnatherum splendens*. The total reserve area is about 1.5 million ha.

To conserve desert grass stands that exist widely in Mongolia (desertified steppes and steppe-like deserts) it is feasible to attach an area around the former monastery, Oldrakh-Khinu, in the Southern Gobi Reserve. This protects communities of *Stipa gobica* and *S. glareosa* + *Allium polyrrhizum* that characterize northern Gobi deserts (Rachkovskaya, 1989, 1993; Kalinina, 1974). Also it is possible to attach areas around Saikhan Dulan and Altan Shira containing communities of *Ajania trifida* and desert grass stands of *Caragana leucophloea, C. pygmaea* and *C. korshinskii* growing on sandy loams.

To conserve moist soil botanical diversity, it is recommended that a reserve be established in the Lake Valley by including it with Lake Bon-Tsagan-Nur and the Baidrag-Gol delta plus nearby Lake Adgiin-Tsagan-Nur which will periodically dry up. This will protect the moist soil vegetation of the delta. But on the whole, this reserve also should be directed toward the conservation of waterfowls, other birds, and fishes.

In addition to these Reserves, to help conserve rare, endemic, or ornamental plants, we encourage the setting aside of reserved areas in national parks, in both existing and proposed parks.

One such site is a low-mountain range, Khuren-Khana, situated south of the Gobi Gurvansaikhan national park that can readily be attached. There, in addition to interesting landscapes, *Caragana altaica, C. alaschanica, C. gobica* and *C. tibetica* are endemic along with several ornamental species proposed for protection.

Another area is the valley of Shishkhid-Gol adjacent to the Khovsgol Nuur National Park northern border. This would include a large mountain forest along with its steppe features plus an above-forest bald mountain zone. An inclusion of the mount Munku Sardyk system would completely conserve this range of altitude zones. The reserved area could be up to 350 thousand ha. The future establishment of a large Mongolian-Russian high-mountain reserve is quite feasible since the Tunkin National Park of Russia joins the Khovsgol Nuur National Park.

Within the proposed Eastern-Khangai National Park a reserved area should be set aside in the upper Orkhon river basin where the forest zone is dominated by *Pinus sibirica.* This site is 450 - 500 km away from the main *Pinus sibirica* range and is situated at the southern forest border in Mongolia. Indeed this is the southernmost population of *Pinis sibirica* in Eurasia. In addition, the floristic composition of the adjoining forest-steppe is very interesting where one finds *Festuca venusta, Saxifraga hieracifolia, Astragalus changaicus,* and *Vaccinium vitis-idaea*, and other boreal species.

Furthermore, the status of a strictly protected area should be given to the southernmost habitat of *Larix sibirica* which is detached from the major range of *Larix sibirica* by 545 - 550 km in the southwest of Mongolia. Being in the vicinity of the Great Gobi Reserve western cluster, it could become

Figure 5.19. Ammopiptanthus mongolicus in Borzongiin Gobi sands (photo by A.V.Prishchepa).

Figure 5.20. Ulmus pumila, Amygdalus mongolica in Borzongiin Gobi sairas (photo by A.V.Prishchepa).

part of that Reserve. Being surrounded by Dzungarian Gobi deserts, these small forest populations appear to be relicts. They are associated with Baitag-Bogdo mountain north-facing slopes. Today, these forests are heavily felled, and goatherds have destroyed the undergrowth. Without special nature conservation measures they may be completely destroyed. Reserve establishment should include a complete ban on fellings and plantation establishments in favorable *Larix sibirica* habitats (Podtyazhkin, 1992). On the north-facing slopes of mount Baitag-Bogdo the plant communities include a number of species which are not typical of these latitudes (*Cystopteris fragilis, Rheum undulatum, Stellaria dichotoma, Melandrium viscosum, Juniperus sabina, Rosa pimpinellifolia, Valeriana dubia, Saussurea involucrata, Bistorta vivipara, Gentianopsis barbata* and *Trollius asiaticus*). In addition, there are many endemics including some listed in the Red Data Book such as *Allium altaicum, A. galanthum, Saussurea involucrata, Aconitum gubanovii, Papaver baitagense, P. pseudotenellum, Smelowskia mongolica, Rosa baitagensis* and *Astragalus baitagensis*. Also found here are some original steppes with *Festuca valesiaca*.

One of the best conservation tactics for botanical diversity in Mongolia likely is the establishment of «transboundary reserves». These are located on international boundaries of large natural areas (Sokolov et al., 1987). Indeed there is great promise in establishing international reserves whose clusters are situated in adjacent countries but virtually forming a single natural-territorial complex. Examples are the Cluster International Uvs Nuur Reserve, a three-sided Russian-Mongolian-Chinese Reserve that unites the Mongol Dagurian Reserve in Mongolia, the Daurian Reserve in Russia, and the Dalai Nuur in China. Or similar international Reserves could be established in the Alashan Gobi or to the east of the Greater Khyangan foothills.

Such expansions of strictly protected areas are in full accordance with the main trends of considered nature management.

5.4 CONCLUSIONS

In the livestock grazing based society of Mongolia the very best strategies for solving the problems of nature management, vegetation conservation, and economic development are not available. Vegetation conservation and maintenance, the recovery of disturbed plants and plant communities, and the conservation of botanical diversity are integral problems that need solutions and yet often conflict with other needs of society. For instance, frequently the complex measures needed to regenerate a disturbed plant cover disagrees with the current social requirements, e.g., providing dwellings, food, grazing, roads, etc. Or using a non-renewable natural resources, e.g., brown coal, polymetals, precious metals, etc., is considered to be more important for society than the preservation of a unique endemic species habitat or than the conservation of vegetation as a natural foraging resource.

Still, Mongolia has done much in terms of nature conservation, particularly after the Great Khural law enacted in 1995. Evenso it is critical to enhance the status and implementation of regulations, boundaries, and goals for legally protected areas. And additional work is needed: i) to define the network of reserves and national parks; ii) to enact guidelines regarding the cultivation of disturbed lands; and iii) to establish and evaluate ecologically and economically justified systems of ecosystem management. Collectively these would promote conservation and recovery of plants, including floristic and community richness, and promote plant regeneration successions while reducing the rate of vegetation degradation processes across all of Mongolia. These conclusions and recommendations for conserving and maintaining the landscape and vegetation are of primary importance and are strongly supportive actions that will serve well to sustain the grazing-based economic culture of the entire Mongolian society.

SUMMARY CONCLUSIONS AND RECOMMENDATIONS

The livestock-grazing based society of Mongolia is directly dependent upon the primary production of its vegetation. Vegetation on the vast fenceless steppes of Mongolia has sustained this society for many centuries. Evenso the vegetation of Mongolia is in a very strained ecological condition, hence work on plant cover dynamics is of central importance to help sustain and develop sound considered nature management programs.

Current studies on the dynamics of vegetation across Mongolia show both natural and human-driven modifications of plant successions are prevalent. Plant community research shows many regressing degrading plant community successions. But other work shows that the regeneration of quasi-original primary plant communities with a high structural organization and a high biological productivity can be accomplished.

Trends in plant cover successions and the factors determining them permit predictions about the status of plant communities under different scenarios of natural or human-related pressures. Unfortunately, almost all of Mongolia now shows wide spread regressive vegetation changes, even with favorable precipitation conditions the last few years. The domination of regressive vegetation successions is primarily associated with increasing human-driven effects. Recognizably many human-driven effects markedly reduce the natural capacity of plant cover for self-regeneration.

In studying the degradation of plant communities and the lower plant productivity caused by both natural or human-associated changes we noted some relationships between community dynamics and certain ecological traits of some plant groups. In Mongolia with its low water supply, even in «wet» years, plants that depend mostly upon rainfall (ombrophytes) are amongst the first to deteriorate either by elimination or subordination in plant communities. With their weak water-searching root system, they are the most sensitive to lower precipitation and to animal-type compaction of surface soil horizons. This is true in both herbaceous and in arboreal ecosystems. In contrast, in forest communities, marked human-associated successions frequently are induced by timber harvesting, because removal of a water-consuming arboreous tree layer on a gentle terrain will cause temporary overmoistening; that then causes an intensive growth of a herbaceous layer. Most of Mongolia's forest ecosystems are located on slopes of various steepness. Hence, felling-type disturbances normally cause swamps to form or burns or extensive soil erosion. Commonly erosion ruts and sinks are formed such that collectively cause both water conditions and nutrition conditions for plants deteriorate badly. Therefore, on such former forest areas, vegetation now is either absent or plant groups grow whose needs for water and mineral nutrition are low.

In steppe plant communities, elimination of ombrophytes also leads to the growth of more resistant weed or unpalatable semi-weed species with stronger root systems that can reach at least the capillary water table edge.

From these studies on plant community dynamics related to both natural and human-associated factors we present the following recommendations that will aid the recovery of plant cover in community successions, help conserve botanical diversity, and continue to sustain the grazing-based economy of Mongolia's people.

First, to maintain the recovery of plant community successions a carefully considered nature management system is needed. Here of primary importance are correct measures to organize range rotations and to obtain the ecologically correct ratio of different livestock species. For instance, to reduce the number of goats and to simultaneously increase the number of cattle, including yaks and sarlyks. Also special attention should be paid to animal breeding to obtain greater biological production from a smaller number of animals.

Second, in mountain-forest zones, either selective fellings should be done or, when clear felling is necessary, the felling area should not exceed 5 ha. After any felling the undergrowth should be left and mature seed-bearing trees should be retained regularly. At the same time it is quite feasible to greatly expand the use of all mountain forest species for tourist, recreational, and therapeutic purposes.

Third, in deserts, where the restoration of disturbed Haloxylon forests is very difficult, the complete harvesting of tall Haloxylon forests needs to be banned. Their need for fuel can be compensated for by other energy sources such as coal.

Fourth, considered nature management must include recovery of plots heavily disturbed by agricultural technology. One should not rely only on natural regeneration of this agriculturally disturbed vegetation. A system of recultivation measures should be used on these lands to restore the plant covers productivity.

Fifth, all considered nature management program for Mongolia must include measures providing for the conservation of florist and plant community diversity. So far, the reserve and park network covering protected areas does not fully conserve the biodiversity of Mongolia. The enactment by the Great Khural of the Law of Strictly Protected Areas, has done much for nature conservation! Evenso additional refinement of the network of reserves and national parks is especially crucial to conserve botanical diversity, and to regenerate and conserve highly productive plant ecosystems of Mongolia.

Finally, to maintain a current assessment of vegetation dynamics and to predict future changes under different human-driven effects one must quickly and integratively access the changes in environments along with the ecological-economic trends of the changing Mongolian society. Such integrated efforts can be greatly assisted by using maps rapidly obtained by remote sensing. These provides an objective insights into the current landscape status, also newly developing dynamics, and can help predict changes in vegetation dynamics in various ecosystems for all of Mongolia.

REFERENCES

Aleksandrova, V.D., 1964. Studies on changes in plant cover. In: Field Geobotany. Vol. III, p. 300–450. Nauka, Moscow, Leningrad (in Russian).

Atlas of Hubsugul Lake. Mongolian People's Republic. 1989. GUGK, Moscow, 118 p. (in Russian).

Baitulin, I.O. (ed.), 1993. Phytoecological studies in Southern Gobi. Gylym, Almaaty, 164 p. (in Russian).

Bannikova, I.A., 1980. Peculiarities of the nature of the mountain forest-steppe of Mongolia (Eastern Hangai). In: Bannikova, I.A., Vipper, P.B., Ulziihutag, N. & Tsegmid, G. (eds) Biological resources of the MPR, their use and protection, p. 19–22. Ulan Bator (in Russian).

Bannikova, I.A., 1983. Vegetation cover. The structure of altitude belt pattern. In: Lavrenko, E.M. & Bannikova, I.A. (eds) Mountain forest-steppe of the Eastern Hangai (MPR). Natural conditions (somon Tevshruleh), p. 89–130. Nauka, Moscow (in Russian).

Bannikova, I.A. & Khudiakov, O.I., 1976. The soil-vegetation sub-belts of South-Eastern Hangai. In: Lavrenko, E.M. & Rachkovskaya, E.I. (eds) The structure and dynamics of the main ecosystems in MPR, p. 72–98. Nauka, Leningrad (in Russian).

Banzragch, D., Volkova, E.A. & Rachkovskaya, E.I., 1978. The vegetation of the medium high mountains Atas-Bogdo-ul in the Transaltai Gobi. In: Bannikova, I.A. & Medvedev, L.N. (eds) Geography and dynamics of vegetation and wildlife of the MPR, p. 30–34. Nauka, Moscow (in Russian).

Batzhargal, Z. (ed.), 1996. Nature and environment in Mongolia. Ministry of nature and the environment, Ulaanbaatar, 131 p. (in Mongolian).

Beideman, I.N., 1983. Reference book on water required by plants in different natural zones of the USSR. Nauka, Novisibirsk, 257 p. (in Russian).

Beresneva, I.A., 1983. The climate. In: Lavrenko, E.M. & Bannikova, I.A. (eds) Mountain forest-steppe of the Eastern Hangai (MPR). Natural conditions (somon Tevshruleh), p. 32–39. Nauka, Moscow (in Russian).

Beresneva, I.A., 1992. Peculiarities of studies and methods of assessment of climatic resources of complex ecosystems of Mongolia. In: Dorofeyuk, N.I., Ulziihutag, N. & Gunin, P.D. (eds) Ecology and nature management in Mongolia. p. 25–32. Scientific Center of biological studies of the Ac. of Sc. of the USSR, Pushchino (in Russian).

Berger, A. & Loutre, M.F., 1991. Insolation values for the last 10 million years. *Quaternary Science Reviews* 10: 297–317.

Berglund, B.E. & Ralska-Jasiewiczowa, M., 1986. Pollen analysis and pollen diagrams. In: Berglund, B.E. (ed.) Handbook of Holocene palaeoecology and palaeohydrology, p. 455–484. John Wiley & Sons, Chichester.

Biodiversity conservation action plan for Mongolia, 1996. The Ministry for Nature and the Environment, Ulaanbaator, Mongolia, 127 p.

Blazkova, D., 1985. *Ulmus pumila* L. und ihre Vergesellschaftung in der Nord-Mongolei. *Feddes Repertorium* 5–6: 433–444.

Bold, G., 1989. The history of pastoral use of Mongolian steppes. PhD Thesis (Biology). Moscow, 18 p. (in Russian).

Bugrovsky, V.V., 1986. Preface. In: Bugrovsky, V.V. (ed.) Informational problems of biosphere studies. «Experiment Uvs-Nuur», p. 3–5. Scientific Center of biological studies of the USSR Academy of Sciences, Pushchino (in Russian).

Bugrovsky, V.V. (ed.), 1987. The Ubsunur depression as a unique experimental site for biosphere studies. Scientific Center of biological studies of the USSR Academy of Sciences, Pushchino, 54 p. (in Russian).

Bugrovsky, V.V. (ed.), 1995a. «Experiment Uvs-Nuur». Vol. I. Intellect, Moscow, 336 p. (in Russian).

Bugrovsky, V.V. (ed.), 1995b. «Experiment Uvs-Nuur». Vol. II. Intellect, Moscow, 269 p. (in Russian).

Bugrovsky, V.V. (ed.), 1996. Global Monitoring and Uvs-Nuur basin. Reports of IV Intern. Symposium on results of Intern. Biosphere Monitoring Scientific Program «Experiment Ubs-Nuur». Intellect, Moscow, 286 p. (in Russian).

Bugrovsky, V.V., Zaletaev, V.S., Kerzhentsev, A.S. & Khakimov, F.I., 1990. Biosphere cluster reserve. In: Bugrovsky, V.V. (ed.) Informational problems of biosphere studies. Ubsu-Nur depression as a natural biosphere model, p. 445–455. Scientific Center of biological studies of the USSR Academy of Sciences, Pushchino (in Russian).

Bunge, A.A., 1835. New genera and species of Chinese and Mongolian plants. *Scientific Notes of Kazan. University* 4: 154–180 (in Russian).

Buyan-Orshih, H., 1977. Vegetation and forage resources of the sand massifs of the Great Lakes depression

(MPR). PhD Thesis (Biology). Ulan Bator, 30 p. (in Mongolian).

Buyan-Orshih, H., 1985. The vegetation of the sand massifs Borig-Del-Els in the Western Mongolia. *Problems of the Desert Development* 2: 60–64 (in Russian).

Buyan-Orshih, H., 1988. Vegetation of the Great Lake Depression. In: Gubanov, I.A., Dorofeyuk, N.I. & Neronov, V.M. (eds), Natural conditions, vegetation cover and wildlife in Mongolia, p. 159–169. Scientific Center of biological studies of the USSR Academy of Sciences, Pushchino (in Russian).

Buyan-Orshih, H., 1992. Vegetation of Great Lake Depression and South-Eastern part of Mongolian Altai (geography, classification and forage resources). Dr. Sc. (biology) Thesis, Novosibirsk, 31 p. (in Russian).

Cherednikova, Yu.S., Kuzmin, V.A., Zoeo, D., Chuluunbaatar, D. & Tegshzhargal, D., 1991. Anthropogenic disturbances of the Hentei forest ecosystems in Mongolia. *Geography and Natural Resources* 3: 154–161 (in Russian).

Chernova, G.M. & Dirksen, V.G., 1995. Peculiarities of formation of sub-recent spore-pollen spectra in the Khoit-Turgen-Gol valley (Mongolia). *Vestnik St.-Peterburgskogo Universiteta, Ser. Geologiya, Geographiya* 2 (14): 69–81 (in Russian).

Chognii, O., 1988. Changes in species composition during pasture degrading. In: Byazrov, L.G. & Mirkin, B. M. (eds) Phytocenological basis of improvement of natural forage lands of the Mongolian People's Republic, p. 54–57. Nauka, Moscow (in Russian).

Dashniam, B., 1976. The steppe flora of the Eastern Mongolia, and some problems of the botanical-geographical regionalization. *Proc. of the Institute of Botany of the Academy of Sciences of MPR* 1: 142–155 (in Russian).

Davitaya, F.F. (ed.), 1960. Climatic Atlas of the USSR, 1st volume. Gidrometeoizdat, Moscow, 181 p. (in Russian).

Devyatkin, E.V., 1981. The Cenozoic of Inner Asia (Stratigraphy, geochronology and correlation). Nauka, Moscow, 200 p. (in Russian).

Devyatkin, E.V., 1993. Meridianal distribution of Pleistocene ecosystems in Asia: basic problems. *Stratigraphy and Geological Correlation* 4: 439–444.

Devyatkin, E.V. & Shilova, G.N., 1970. Palynological characteristic of some key sections of Middle-Upper Pliocene and Lower Quaternary deposits of Western Mongolia. In: Devyatkin, E.V. (ed.) Mesozoic and Cenozoic geology of Western Mongolia, p. 128–142. Nauka, Moscow (in Russian).

Dinesman, L.G. & Bold, G., 1992. The history of grazing and pasture digression development in Mongolian steppes. In: Historical ecology of wild and domestic animals, p. 172–216. Nauka, Moscow (in Russian).

Dinesman, L.G., Kiseleva, N.K. & Knyazev, A.V., 1989. The history of steppe ecosystems of the Mongolian People's Republic. Nauka, Moscow, 215 p. (in Russian).

Dmitriev, P.P., 1985. On the relations of some bushes of Mongolian steppes to mammal settlements. *Journal of General Biology* 10: 661–669 (in Russian).

Dmitriev, P.P., 1992. Formation and development of steppe ecosystems of Mongolia. In: Dorofeyuk, N.I., Gunin, P.D. & Ulziihutag, N. (eds) Ecology and nature management in Mongolia, p. 81–93. Scientific Center of biological studies of the Russian Academy of Sciences, Pushchino (in Russian).

Dmitriev, P.P., Gugalinskaya, L.A. & Guricheva, N.P., 1982. The origin of the aeolian hills with (*Achnatherum splendens* (Trin.) Nevski) in dry steppes of Mongolia and the marmot (*Marmota sibirica* Radde). *Journal of General Biology* 5: 712–718 (in Russian).

Dmitriev, P.P. & Guricheva, N.P., 1978. Small mammals in the grazing biogeocenoses of the Eastern Hangai. In: Bannikova, I.A. & Medvedev, L.N. (eds) Geography and dynamics of vegetation and wildlife of the MPR, p. 124–131. Nauka, Moscow (in Russian).

Dmitriev, P.P. & Guricheva, N.P., 1983a. Diversity of vegetation types at the butanes of tarbagans in the Eastern Hangai as the indicator of stability of their locations. In: Protection, rational use, and ecology of marmots, p. 42–44. Nauka, Moscow (in Russian).

Dmitriev, P.P., Guricheva, N.P., 1983b. The main forms of spot pattern of mountain steppes vegetation cover in the Eastern Hangai (MPR) in the mammals habitats. *Reports of the Academy of Science of the USSR.* Vol. 271, 1: 250–254 (in Russian).

Dmitriev, P.P. & Khudiakov, O.I., 1989. Zoological factor of soil cover heterogeneity in dry steppes of Mongolia. *Reports of the Academy of Science of the USSR.* Vol. 304, 3: 757–762 (in Russian).

Dmitriev, P.P., Khudiakov, O.I. & Galsan, P., 1989. Micromosaic structure of solonchak depressions of East-Mongolian steppes as a result of fossorial mammal activity. *Ecology* 5: 7–12 (in Russian).

Dmitriev, P.P., Khudiakov, O.I. & Zhargalsaihan, L., 1990. Successional ranges of the dark-chestnut soils and steppe vegetation of Eastern Mongolia associated with burrowing mammal's activity. *Bull. of the Moscow Society of Nature Investigators. Div. of Biology.* 1: 3–15 (in Russian).

Dmitriev, P.P. & Shauer, I., 1987. Fossorial activity of mammals as the weathering factor of rocks and rubbish mantle formation. *Izvestia of the Academy of Science of the USSR. Div. of Geography.* 1: 92–100 (in Russian).

Dmitriev, P.P., Shvetsov, Yu.G. & Dulamtseren, S., 1992. Mammals of Hangai Highland. Fauna, ecology, significance in biocenoses. Nauka, Moscow, 199 p. (in Russian).

Dorofeyuk, N.I., 1988. Paleogeography of Holocene of the Mongolian People's Republic according to results of diatomic analysis of lake bottom sediments. In: Gubanov, I.A., Dorofeyuk, N.I. & Neronov, V.M. (eds) Natural conditions, vegetation cover and wildlife of Mongolia, p. 61–83. Scientific Center of biological studies of the USSR Academy of Sciences, Pushchino (in Russian).

Dorofeyuk, N.I., 1994. The history of development of water ecosystems of Mongolia according to the data of diatoms analysis of lake sediments. PhD Tesis (Biology), Moscow, 25 p.

Dorofeyuk, N.I. & Tarasov, P.E., 1998. Vegetation and lake levels in Northern Mongolia in the last 12500 years as indicated by data of pollen and diatom analyses. *Stratigraphy and Geological Correlation* 6: 70–83.

Dorzhsuren, Ch., 1979. Influence of major use cuttings on the subordinate layers and forest renewal on the larch stands of the Mongolian Central Hangai. *Forestry, Forest Cultures, and Soil Science* 8: 38–42 (in Russian).

Dugarzhav, Ch., 1990. Natural renewal of the sub-taiga larch stands of the Eastern Hentei. In: Dorofeyuk, N.I. (ed.) Ecology and nature management in Mongolia, p. 85–86. GUGK MPR, Ulan Bator (in Russian).

Efremov, Yu.K., 1976. Problems the theory of nature management. In: Actual directions of the contemporary geography, p. 71–88. Moscow division of the All-Union Geographical Society, Moscow (in Russian).

Evdokimova, A.K., Prishchepa, A.V. & Samoilova, G.S., 1992. Road digression in Mongolian natural complexes. In: Dorofeuyk, N.I., Ulziihutag, N. & Gunin, P.D. (eds) Ecology and nature management in Mongolia, p. 211–217. Scientific Center of biological studies of the Russian Academy of Sciences, Pushchino (in Russian).

Evstifeev, Yu.G., 1980. Extreme arid soils of Gobi. *Problems of Desert Development* 2: 20–30 (in Russian).

Evstifeev, Yu.G. & Pankova, E.I., 1984. Peculiarities of soil salinization in the desert zone of the Mongolian People's Republic. *Problems of Desert Development* 1: 44–52 (in Russian).

Evstifeev, Yu.G. & Rachkovskaya, E.I., 1976. On the problem of interrelations between soils and vegetation cover in the southern part of the MPR. In: Lavrenko, E.M. & Rachkovskaya, E.I. (eds) The structure and dynamics of the main ecosystems of the MPR, p. 125–144. Nauka, Leningrad (in Russian).

Evstifeev, Yu.G. & Rachkovskaya, E.I., 1991. Regularities of the spatial differentiation of desert ecosystems. *Problems of Desert Development* 3–4: 36–48 (in Russian).

Evstifeev, Yu.G., Amarzhargal, B., Buyan-Orshih, H., Gunin, P.D., Kazantseva, T.I., Rachkovskaya, E.I., Timofeev, D.A. & Yakunin, G.N., 1990. Methods for assessment and mapping of the contemporary state of the desert ecosystems of the MPR (Methodological recommendations). Ulan Bator, 52 p.

Flora Intramongolica 1977–1985. Huhhot (in Chinese).

Florensov, N.F. & Korzhuev, S.S. (eds), 1982. Geomorphology of Mongolian People's Republic. Nauka, Moscow, 260 p. (in Russian).

Gerasimov, I.P. & Nogina, N.A. (eds), 1984. Soil cover and soils of Mongolia. Nauka, Moscow, 275 p. (in Russian).

Giterman, R.E. Golubeva, L.V., Zaklinskaya, E.D., Koreneva, E.V., Matveeva, O.V. & Skiba, L.A., 1968. Main stages of the vegetation development of Northern Asia in Anthropogene. Nauka, Moscow, 269 p. (in Russian).

Glazovsky, N.F., Koronkevich, N.I., Kochurov, B.I., Krenke, A.N. & Sdasyuk, G.V., 1991. Critical ecological regions: geographical approaches and principles of studies. *Izvestiya vsesoyuznogo geographicheskogo obshchestva* 1: 9–17 (in Russian).

Golubeva, L.V., 1976. Vegetation of the North-Eastern Mongolia in Pleistocene and Holocene. In: Lavrenko, E.M. & Rachkovskaya E.I. (eds) The structure and dynamics of the main ecosystems in MPR, p. 59–71. Nauka, Leningrad (in Russian).

Golubeva, L.V., 1978. Vegetation of the Northern Mongolia in Pleistocene and Holocene (basins of rivers Selenga and Orhon). *Izvestia Academii nauk SSSR, Ser. geology*. 3: 68–81 (in Russian).

Gravis, G.F. & Lisun, A.M., 1974. Rhythmic stratigraphy of Quaternary Deposits in MPR according to palynologic data and history of permafrost evolution. In: Melnikov, P.I. (ed.) Geocryological conditions of the Mongolian People's Republic, p. 148–187. Nauka, Moscow (in Russian).

Grichuk, V.P., 1984. Late Pleistocene vegetation History. In: Velichko, A.A. (ed.) Late Quaternary environments of the Soviet Union, p. 155–173. Longman, London.

Grichuk, V.P. & Zaklinskaya, E.D., 1948. Analysis of fossil pollen and spores and its application to paleogeography. Geografgiz, Moscow, 224 p. (in Russian).

Grubov, V.I., 1955. Conspectus of flora of the Mongolian People's Republic. Izdatel'stvo AN SSSR, Moscow, Leningrad, 308 p. (in Russian).

Grubov, V.I., 1982. Determinator of the vascular plants of Mongolia (with the atlas). Nauka, Leningrad, 442 p. (in Russian).

Grubov, V.I. (ed.), 1963–94. Plants of the Central Asia according to materials of Komarov's Botanical Institute of the USSR Academy of Sciences. Iss. 1, 2, 3, 4, 5, 6, 7, 8, 9, 10, 11. Nauka, Leningrad, 168+136+120+246+209+84+138+126+148+119+123 p. (in Russian).

Grubov, V.I., 1984. Endemic Species of the Flora of the Mongolian People's Republic. In: Grubov, V.I. (ed.)

Novitates systematicae plantarum vascularium. Vol. 21, p. 202–220. Nauka, Leningrad (in Russian).

Grubov, V.I. & Mirkin, B.M. (eds), 1980. Vegetation of river flood plains of the Mongolian People's Republic (the experience of the syntaxonomic analysis with the help of the floristic criteria). Nauka, Leningrad, 284 p. (in Russian).

Grubov, V.I. & Yunatov, A.A., 1952. Main peculiarities of flora of the MPR in relation to its regionalization. *Botanicheskiy Zhurnal* 37: 45–64 (in Russian).

Gubanov, I.A., 1996. Conspectus of flora of Outer Mongolia (vascular plants). Valang, Moscow, 136 p. (in Russian).

Gubanov, I.A. & Hilbig, W., 1993. Bibliographia Phytosociologica: Mongolia. Pars II. *Excerpta Botanica Sec. B (Sociologica)* 30: 63–81.

Gubanov, I.A. & Kamelin, R.V., 1988a. The history of floristic investigations in the Mongolian People's Republic. In: Gubanov, I.A., Dorofeyuk, N.I. & Neronov, V.M. (eds) Natural conditions, vegetation cover and wildlife of Mongolia, p. 169–189. Scientific Center of biological studies of the USSR Academy of Sciences, Pushchino (in Russian).

Gubanov, I.A. & Kamelin, R.V., 1988b. Materials to the flora of the Mongolian People's Republic. New vascular plants, found in the MPR during recent years. In: Gubanov, I.A., Dorofeyuk, N.I. & Neronov, V.M. (eds) Natural conditions, vegetation cover and wildlife of Mongolia, p. 189–217. Scientific Center of biological studies of the USSR Academy of Sciences, Pushchino (in Russian).

Gubanov, I.A. & Kamelin, R.V., 1991. New Species of Vascular Plants Observed in Mongolian Flora. *Bull. of the Moscow Society of Nature Investigators. Div. of Biology.* 6: 112–116 (in Russian).

Gubanov, I.A. & Kamelin, R.V., 1992. New Vascular Plants of the Flora of the Mongolian People's Republic. *Bull. of the Moscow Society of Nature Investigators. Div. of Biology.* 2: 119–130 (in Russian).

Gunin, P.D., 1990a. Ecology of desertification processes of arid ecosystems. VASHNIL, Moscow, 354 p. (in Russian).

Gunin, P.D., 1990b. Characteristic features of desertification processes in arid landscapes of the Sahara-Gobi desert area. In: Dorofeyuk, N.I. (ed.) Ecology and nature management in Mongolia, p. 19–22. GUGK MPR, Ulan Bator (in Russian).

Gunin, P.D., 1991. Natural processes of arid ecosystem desertification Dr.Sc. (Biology) Thesis. Moscow, 82 p. (in Russian).

Gunin, P.D., 1992. Diagnostics of desertification processes in arid ecosystems of the Central Asia. In: Dorofeuyk, N.I., Ulziihutag, N. & Gunin, P.D. (eds) Ecology and nature management in Mongolia, p. 271–286. Scientific Center of biological studies of the Russian Academy of Sciences, Pushchino (in Russian).

Gunin, P.D., 1993. Conservation Areas/Wildlife. Mongolia – Biodiversity Project. Ulaan-Baator, 66 p.

Gunin, P.D., Druk, A.Ya., Krasnoshchekov, Yu.N., Prishchepa, A.V., Radziminsky, P.Z. & Chernyshev, E.P., 1991. Anthropogenic disturbances of main natural zones of the Mongolian People's Republic. *Izvestia RAN, Ser. Geography.* 2: 43–55 (in Russian).

Gunin, P.D., Evstifeev, Yu.G., Rachkovskaya, E.I. & Fedorova, I.T., 1984. Oases of the Great Gobi Reserve. Characteristics, significance, and problems of protection. In: Problems of genofond protection and ecosystem management in the reserves of steppe and desert zones, p. 226–229. Nauka, Moscow (in Russian).

Gunin, P.D., Klimanova, O.A. & Miklyaeva, I.M., 1996. Specificity of geoecological monitoring of natural-anthropogenic ecosystems of the Eastern Asian sector of steppes: the Case of Eastern Mongolia. *Arid Ecosystems* 4: 24–37 (in Russian).

Gunin, P.D. & Neronov, V.M., 1986. Ecological principles of genetic fund protection and problems of desertification monitoring organization. In: Results and prospects of reserve organization activity in the USSR, p. 172–193. Nauka, Moscow (in Russian).

Gunin, P.D., Rachkovskaya, E.I. & Fedorova, I.T., 1980. Desert experimental station Ehiin-Gol. Moscow, 11 p. (In Russian and Mongolian).

Gunin, P.D., Samoilova, G.S., Druk, A.Ya., Vostokova, E.A. & Ulziihutag, N., 1992. Ecological assessment of the state of the environment, and principals of map compilation on anthropogenic distortions in the ecosystems of Mongolia. *Vestnik of the Moscow State University. Ser. 5. Geography* 1: 92–99 (in Russian).

Gunin, P.D., Slemnev, N.N., Kazantseva, T.I., Radziminsky, P.Z. & Amarzhargal, B., 1993. On the expansion of *Ephedra sinica* Stapf in mountain ecosystems of Gobi (Mongolia). *Vegetation Resources* 3: 7–21 (in Russian).

Gunin, P.D. & Vostokova, E.A. (eds), 1989. Methodological recommendations on assessment and mapping of the contemporary ecosystems of the MPR. GUGK of the MPR, Moscow, Ulan Bator, 60 p. (in Russian).

Gunin, P.D. & Vostokova, E.A. (eds), 1993. Methodology of assessment of the state and mapping of ecosystems under the extreme conditions. ONTI PNC RAN, Pushchino, 203 p. (in Russian).

Gunin, P.D. & Vostokova, E.A. (eds), 1995a. Ecosystems of Mongolia: spread and modern state. Nauka, Moscow, 223 p. (in Russian).

Gunin, P.D. & Vostokova, E.A. (eds), 1995b. The map «Ecosystems of Mongolia» Scale 1: 1,000,000. 15 sheets. Moscow.

Gunin, P.D., Vostokova, E.A. & Matyushkin, E.N., 1998. Protection of ecosystems of Inner Asia. Nauka, Moscow, 219 p. (in Russian).

Guricheva, N.P., 1983. The influence of life activity of small mammals – shrews – on the structure of vegetation of steppes of the Eastern Hangai of the MPR. In: Sokolov, V.E. & Shagdarsuren, O. (eds) Complex characteristics of desert ecosystems of Transaltai Gobi (The examples of the desert experimental station and the Great Gobi Reserve), p. 87–88. Scientific Center of biological studies of the USSR Academy of Sciences, Pushchino (in Russian).

Guricheva, N.P. & Buevich, Z.G., 1986. Composition, structure and seasonal development of steppe communities. In: Lavrenko, E.M. & Bannikova, I.A. (eds) The Steppes of the Eastern Hangai, p. 53–58. Nauka, Moscow (in Russian).

Guricheva, N.P. & Dmitriev, P.P., 1983. Interrelations between vegetation cover and animals. Main forms of dynamics of mountain steppe structure ruled by fossorial mammals. In: Lavrenko, E.M. & Bannikova, I.A. (eds) Mountain forest-steppe of the Eastern Hangai (MPR). Natural conditions (somon Tevshruleh), p. 172–180. Nauka, Moscow (in Russian).

Harrison, S.P., Yu, G. & Tarasov, P.E., 1996. Late Quaternary lake-level record from northern Eurasia. *Quaternary Research* 45: 138–159.

Hilbig, W., 1984. Die mongoleisammlungim Herbarium der Martin-Luther-Universitaet Halle-Wittenberg. In: Erforschung biologischer ressourcen der MVR. 4: 55–68. Halle (Saale).

Hilbig, W., 1987. Vierter beitrag zur kenntnis der flora Mongolischen Volksrepublik. *Feddes Repertorium* 7–8: 461–467.

Hilbig, W., 1988. On development of ruderal vegetation in the MPR. In: Gubanov, I.A., Dorofeyuk, N.I. & Neronov, V.M. (eds) Natural conditions, vegetation cover and wildlife of Mongolia., p. 256–265. Scientific Center of biological studies of the USSR Academy of Sciences, Pushchino (in Russian).

Hilbig, W., 1991. Bibliographia Phytosociologica: Mongolia. *Excerpta Botanica Sec. B* (*Sociologica*) 28/4: 245–309.

Hilbig, W., 1992. Effect of pasturing on the vegetation cover in Mongolia. Thes. Intern. Sympos. «Erforschung biologischer ressourcen der Mongolei», p. 59. Halle.

Hilbig, W., 1995. The vegetation of Mongolia. SPB Academic Publishing, Amsterdam, The Netherlands, 253 p.

Hilbig, W. & Knapp, H.D., 1983. Vegetationsmosaik und florenelemente an der wald-steppen-grenze in Chentej-Gebirge (Mongolei). *Flora* 174 (1–2): 1–89.

Hilbig, W. & Mirkin, B.M., 1983. Entwicklung und stand der geobotanischen forschung uber die Mongolischen Volksrepublik. In: Erforschung biologischer ressourcen der MVR 3: 33–46. Halle (Saale).

Jolly, D., Prentice, I.C., Bonnefille, R., Ballouche, A., Bengo, M., Brenac, P., Buchet, G., Burney, D., Cazet, J.-P., Cheddadi, R., Edorh, T., Elenga, H., Elmoutaki, S., Guiot, J., Laarif, F., Lamb, H., Lezine, A.-M., Maley, J., Mbenza, M., Peyron, O., Reille, M., Reynaud-Farrera, I., Riollet, G., Ritchie, J.C., Roche, E., Scott, L., Semmanda, I., Straka, H., Umer, M., Van Campo, E., Vilimumbalo, S., Vincens, A. & Waller, M. Biome reconstructed from pollen and plant macrofossil data for Africa and the Arabian peninsula at 0 and 6 ka. *Journal of Biogeography*, in press.

Kalicki, T. & Prokop, P., 1995. Conditions of occurrence and consequences of the extreme floods in the Zaaltaiska Gobi. *Prezeglad Geograficzny*. T. LXVII 1–2: 133–145. (in Polish).

Kalinina, A.W., 1974. Main types of pastures of Mongolia (their structure and productivity). Nauka, Leningrad, 185 p. (in Russian).

Kamelin, R.V. & Gubanov, I.A., 1993. The Mongolian flora and its analysis: composition heterogeneity, geographical elements, biomorphic spectrum. *Journal of General Biology* 1: 59–71 (in Russian).

Kamelin, R.V., Gubanov, I.A., Dariimaa, Sh. & Ganbold, E., 1992. Vascular plants of Eren-Daba Ridge (Mongolian Dauria). *Botanicheskiy Zhurnal* 4: 10–21 (in Russian).

Karamysheva, Z.V., 1988. Latitude and longitude changes in vegetation of Mongolian mountains. In: Vegetable world of the high mountain ecosystems of the USSR, p. 262–273. Vladivostok (in Russian).

Karamysheva, Z.V. & Banzragch, D., 1976. Vegetation of the ridge Han-Huhiiun in the southern part of the Ubsu-Nur depression. In: Lavrenko, E.M. & Rachkovskaya, E.I (eds) The structure and dynamics of main ecosystems of the MPR, p. 99–124. Nauka, Leningrad (in Russian).

Karamysheva, Z.V. & Banzragch, D., 1977. On some botanical – geographical regularities of Hangai in connection with its regionalization. In: Lavrenko, E.M. & Popova, T.A. (eds) The vegetation and wildlife in Mongolia, p. 7–26. Nauka, Leningrad (in Russian).

Kasimov, N.S., Evdokimova, A.K., Lychagin, M.Yu., Pikovski, Yu.I., Dorofeyuk, N.I., Undral, G. & Togoo, G., 1992. Methodology and results of the ecological-geochemical assessment of the Ulan Bator environment state. In: Dorofeyuk, N.I., Gunin, P.D. & Ulziihutag, N. (eds) Ecology and nature management in Mongolia, p. 201–211. Scientific Center of biological studies of the Russian Academy of Sciences, Pushchino (in Russian).

Kazantseva, T.I., 1983. Productivity of some plant communities of the Transaltai Gobi. In: Sokolov, V.E. & Shagdarsuren, O. (eds) The complex characteristics of desert ecosystems of Transaltai Gobi (the examples of desert experimental station and the Great Gobi Reserve), p. 37–40. Scientific Center of biological studies of the USSR Academy of Sciences, Pushchino (in Russian).

Kazantseva, T.I., Yakunin, G.N. & Amarzhargal, B., 1992. Grazing ecosystems of desert zone and assessments of their state. In: Dorofeyuk, N.I., Gunin, P.D. & Ulziihutag, N. (eds) Ecology and nature management in Mongolia, p. 122–133. Scientific Center of biological studies of the Russian Academy of Sciences, Pushchino (in Russian).

Khilko, S.D. & Kurushin, R.A., 1982. Mongolian Altai. In: Florensov, N.A. & Korzhuev, S.S. (eds) Geomorphology of Mongolian People's Republic, p. 40–54. Nauka, Moscow (in Russian).

Khramtsov, V.N., Dmitriev, P.P., Khudiakov, O.I., Lim, V.D., Bayasgalan, D. & Mandahbaiar, B., 1992. Ecological-landscape approach to assessments of anthropogenic distortions of steppe ecosystems of Mongolia. In: Dorofeyuk, N.I., Gunin, P.D. & Ulziihutag, N. (eds) Ecology and nature management in Mongolia, p. 93–103. Scientific Center of biological studies of the Russian Academy of Sciences, Pushchino (in Russian).

Khza, F.K., 1984. Climate, droughts and desertification. *Nature and Resources* 1: 2–10 (in Russian).

Knizhnikov, Yu.F. & Kravtsova, V.I., 1991. Aerospace investigations of dynamics of geographical phenomena. Moscow University Press, Moscow, 205 p. (in Russian).

Komarov, V.L., 1908–1909. Introduction to flora of China and Mongolia, Trans. of St. Petersburg botanical garden, Vol. 29, Iss. 1–2 (in Russian).

Korina, N.A., 1982. Hangai upland. In: Florensov, N.A. & Korzhuev, S.S. (eds) Geomorphology of Mongolian People's Republic, p. 87–105. Nauka, Moscow (in Russian).

Korina, N.A., Kulakov, V.S., Khilko, S.D. & Nazag, Zh., 1982. Orhon-Selenga mountains. In: Florensov, N.A. & Korzhuev, S.S. (eds) Geomorphology of Mongolian People's Republic, p. 122–128. Nauka, Moscow (in Russian).

Korotkov, I.A., 1976. Geographical peculiarities of forest distribution in the Mongolian People's Republic. *Botanicheskiy Zhurnal.* Vol. 61, 2: 145–153 (in Russian).

Korotkov, I.A., 1978. Forest types of the Mongolian People's Republic. In: Zhukov, A.B. (ed.) Forests of the Mongolian People's Republic (geography and typology), p. 36–121. Nauka, Moscow (in Russian).

Korotkov, I.A. & Dorzhsuren, Ch., 1988. Dynamics of phytocenoses of sub-taiga larch forests under the influence of forest clearings and fire. In: Isaiev, A.S. (ed.) Forests of the Mongolian People's Republic. Larch forests of the Eastern Hentei, p. 50–62. Nauka, Moscow (in Russian).

Korzhuev, S.S., 1982. The main features of Mongolia land surfaces. In: Florensov, N.A. & Korzhuev, S.S. (eds) Geomorphology of Mongolian People's Republic, p. 17–34. Nauka, Moscow (in Russian).

Kotlov, F.V., 1978. Changes in geological environment due to human activity. Nedra, Moscow, 260 p. (in Russian).

Kovda, V.A., 1968. Soils of the arid zone. In: Soils of the arid zone as objects of irrigation, p. 5–30. Nauka, Moscow (in Russian).

Kovda, V.A., 1984. Problems of struggle against desertification salinization of irrigated soils. Kolos, Moscow, 304 p. (in Russian).

Kozhemiakin, V.V., 1978. Vegetation changes at the gnawer hills in the desert-steppe band of Mongolia. *Problems of Desert Development* 1: 80–85 (in Russian).

Krasnoshchekov, Yu.N., 1980. On the soil protection role of pseudotaiga larch stands in the Central Hangai. In: Contemporary aspects of studies on erosion processes, p. 166–170. Novosibirsk (in Russian).

Krasnoshchekov, Yu.N., 1992. Protective role of Mongolian forest ecosystems. In: Erforschung biologischer ressourcen der Mongolei, p. 71–72. Halle (in Russian and English).

Krasnoshchekov, Yu.N. & Gombosuren, N., 1988a. Changes in forest growth conditions due to forest cuttings and fires in sub-taiga larch forests. In: Isaiev, A.S. (ed.) Forests of the Mongolian People's Republic. Larch Forests of the Eastern Hentei, p. 16–47. Nauka, Moscow (in Russian).

Krasnoshchekov, Yu.N. & Gombosuren, N., 1988b. Water regulation and soil protection role of sub-taiga larch forests. In: Isaiev, A.S. (ed.) Forests of the Mongolian People's Republic. Larch Forests of the Eastern Hentei, p. 62–74. Nauka, Moscow (in Russian).

Krasnoshchekov, Yu.N. & Korotkov, I.A., 1992. The modern situation of forest ecosystems of the MPR and urgent problems of their protection. In: Erforschung biologischer ressourcen der Mongolei, p. 73–74. Halle (in Russian and English).

Krasnoshchekov, Yu.N., Korotkov, I.A., Cherednikova, Yu.C. & Tsedendash, G., 1990. Methods for assessment and mapping of the contemporary state of the forest ecosystems of the MPR (Methodological recommendations). Ulan-Bator, 30 p. (in Russian).

Krasnoshchekov, Yu.N., Korotkov, I.A., Radziminski, P.Z., Dugarzhav, Ch. & Tsedendash, T., 1992. Principals of the ecological assessment of the state and anthropogenic distortions in the forest ecosystems of the MPR. In: Dorofeyuk, N.I., Gunin, P.D. & Ulziihutag, N. (eds) Ecology and nature management in Mongolia, p. 70–81. Scientific Center of biological studies of the Russian Academy of Sciences, Pushchino (in Russian).

Kravtsova, V.I., 1992. Remote sensing for studies on the environment. Available space images. Moscow University Press, Moscow, 135 p. (in Russian).

Kupriyanova, L.A., 1965. Palynology of Betulaceae. Nauka, Moscow, Leningrad, 213 p. (in Russian).

Kupriyanova, L.A. & Aleshina, L.A., 1972. Pollen and spores of plants from the European part of the USSR. Nauka, Leningrad, 166 p. (in Russian).

Lavrenko, E.M., 1970. Provinces of the Central Asian sub-area of the steppe area of Eurasia. *Botanicheskiy Zhurnal*. Vol. 55, 12: 1734–1747 (in Russian).

Lavrenko, E.M., 1978. On the steppe and desert vegetation of the Mongolian People's Republic. *Problems of Desert Development* 1: 3–19 (in Russian).

Lavrenko, E.M. (ed.), 1979. Vegetation map of Mongolian People's Republic. Scale 1:1,500,000. 4 sheets. GUGK, Moscow (in Russian and Mongolian).

Lavrenko, E.M. (ed.), 1983. The map of forests of the Mongolian People's Republic, scale 1:1,500,000. 2 sheets. GUGK, Moscow (in Russian).

Lavrenko, E.M. & Bannikova, I.A. (eds), 1983. Mountain forest-steppe of the Eastern Hangai (MPR). Natural conditions (Somon Tevshruleh). Nauka, Moscow, 190 p. (in Russian).

Lavrenko, E.M. & Bannikova, I.A. (eds), 1986. Steppes of the Eastern Hangai. Nauka, Moscow, 182 p. (in Russian).

Lavrenko, E.M., Borisova, I.V. & Popova, T.A. (eds), 1981. Desert steppes and northern deserts of the Mongolian People's Republic. Part. 2. Stationary investigations (Bulgan somon). Nauka, Leningrad, 260 p. (in Russian).

Lavrenko, E.M., Karamysheva, Z.V. & Nikulina, R.I., 1991. The steppes of the Eurasia. Nauka, Leningrad, 146 p. (in Russian).

Lavrenko, E.M. & Rachkovskaya, E.I. (eds), 1976. The structure and dynamics of the main ecosystems in MPR. Nauka, Leningrad, 200 p. (in Russian).

Lavrenko, E.M. & Shulzhenko, Ts.F. 1962. Brief survey of activity of the Mongolian Agricultural Expedition of the USSR Academy of Sciences for 1947–1952. *Izvestiya vsesoyuznogo geographicheskogo obshchestva* 94: 168–175 (in Russian).

Lavrenko, E.M. & Yunatov, A.A., 1952. Idle lands regime in steppes as a result of influence of Brandt voles on grass cover and soils. *Botanicheskiy Zhurnal*. Vol. 37, 2: 128–138 (in Russian).

Lavrenko, E.M. & Yunatov, A.A., 1960. Natural oases of deserts of the Transaltai Gobi. In: Problems of evolution biogeography, genetics and selection. p. 125–136. The USSR Academy of Sciences Publishing House, Moscow, Leningrad (in Russian).

Lebedev, A.N. & Kopanev, I.D (eds), 1975. Climates of Foreign Asia. Gidrometeoizdat, Leningrad, 448 p. (in Russian).

Lebedeva, N.V., Kuzikov, I.V., Boldbator, Sh. & Shuktomova, I.I., 1997. Birds and mammals of Mongolia in bioindication of anthropogenic pollution. *Arid ecosystems* 5: 122–131 (in Russian).

Logatchov, N.A. (ed.), 1989. Late Cenozoic of Mongolia. Nauka, Moscow, 213 p. (in Russian).

Lovelius, N.V., 1997. Dendroindication of natural processes and anthropogenic influences. World and Family-95, St.-Petersburg, 320 p.

Lushchekina, A.A., Neronov, V.M. & Shurhal, A.V., 1983. The area structure, interspecific variability, and problems of rational use of the Mongolian dzeren. In: Applied Aspects of the Program «Man and Biosphere», p. 134–146. Moscow (in Russian).

Malaeva, E.M., 1989a. Regional distinctive features of pollen distribution in the present alluvium. Methodical aspects of palaeobotanical reconstructions. In: Logatchov, N.A. (ed.) Late Cenozoic of Mongolia, p. 24–40. Nauka, Moscow.

Malaeva, E.M., 1989b. The history of Pleistocene and Holocene vegetation in Mongolia and palaeoindicative features of fossil pollen floras. In: Logatchov, N.A. (ed.) Late Cenozoic of Mongolia, p. 158–177. Nauka, Moscow (in Russian).

Malaeva, E.M. & Murzaeva, V.E., 1987. Holocene of the Northern Mongolia. *Izvestiya Akademii nauk SSSR, Ser. Geogr.* 2: 69–72 (in Russian).

Malgina, E.A., 1971. Results of surface spore-pollen analysis from Central Mongolia. In: Neistadt, M.I. (ed.) Holocene palynology, p. 239–257. Institute of Geography, Academy of Sciences of the USSR, Moscow (in Russian with English summary).

Malyshev, L.I. & Peshkova, G.A., 1984. Peculiarities and genesis of flora of Siberia (Pre-Baikal and Trans-Baikal regions). Nauka, Novosibirsk, 265 p. (in Russian).

Marinov, N.A., 1967. Geological investigations in the Mongolian People's Republic. Nedra, Moscow, 843 p. (in Russian).

Matyushkin, E.N., 1995. Priorities in developing a net of protected areas in Mongolia as an example. In: Ulziihutag, N. & Gunin, P.D. (eds) Asian ecosystems and their protection, p. 123. Ulaanbaatar.

Maximovicz, C.I., 1859. Index Florae Mongolicae. *Mem. Acad. Sci.* IX: 479–486. St.-Petersbourg.

Maximovicz, C.I., 1889. List of plants of Mongolia and adjacent part of Chinese Turkestan. In: Scientific results of trips of N.M. Przhewalskii to Central Asia. Division of biology. Vol. 2, p. 1–146. St. Petersburg (in Russian).

Medows, D.H, Medows, D.L. & Randers, J., 1994. Out of growth limits. Moscow, 303 p. (in Russian).

Meintzer, O., 1927. Plant as indicators of ground water. In: Water-Supply. Paper, 577 p.

Meteltseva, E.P. (ed.), 1986. Bibliographical index on the results of studies of Joint Soviet-Mongolian Complex Biological Expedition of the Academies of Sciences of the USSR and the MPR (1970–1985). IEMEZh, Academy of Sciences of the USSR, Moscow, 178 p. (in Russian).

Milyutin, L.I., Suntsov, A.V. & Zhamiansuren, S., 1988. Genetics-selection peculiarities of main forest-forming species of the Eastern Hentei. In: Isaev, A.S. (ed.) Forests of the Mongolian People's Republic. Larch forests of the Eastern Hentei, p. 75–120. Nauka, Moscow (in Russian).

Miklyaeva, I.M., 1996. Renewal of steppe vegetation at idle lands of Eastern Mongolia. *Vestnik of the Moscow State University. Ser. 5. Geography* 1: 75–81 (in Russian).

Miklyaeva, I.M. & Lysak, O.A., 1996. Pasture degrading of steppes of Eastern Mongolia. *Bull. of the Moscow Society of Nature Investigators. Div. of Biology* 1: 88–94 (in Russian).

Mongolian environmental laws, 1996. Uyakhan Zambuutiv magazine poublishing hause, Ulaanbaatar, 152 p. (in Mongolian).

Mongolia's Wild Heritage. Biological diversity, protected areas, and conservation in the land of Chingis Khaan, 1996. Avery Press, Boulder, Colorado, U.S.A., 42 p.

Motorina, L.V. & Ovchinnikov, V.A., 1975. Industry and land reclamation, Mysl, Moscow, 240 p. (in Russian).

Murzaev, E.M., 1952. Mongolian People's Republic. Geographical description. Geographgiz, Moscow, 472 p. (in Russian).

Murzaeva, V.E., 1982. Depression of Great Lakes. In: Florensov, N.A. & Korzhuev, S.S. (eds) Geomorphology of Mongolian People's Republic, p. 135–148. Nauka, Moscow (in Russian).

Natural conditions and resources of Hubsugul region (MPR). 1976. Nedra, Moscow, 355 p. (in Russian).

Nechaeva, N.T. (ed.), 1985. Improvement of desert ranges in Soviet Central Asia. Harwood Academic Publishers, Chur, London, Paris, New York, 327 p.

Neronov, V.M. & Lushchekina, A.A., 1980. Regional biogeography of Mongolian People's Republic (the analysis of the published schemes of regionalization). In: Achievements in science and technology. Biogeography. Vol. 2, p. 41–64. VINITI of the Academy of Sciences of the USSR, Moscow (in Russian).

Neronov, V.M., Milanova, E.V. & Zhigzh, S., 1988. Protected areas of the MPR: the state and development prospects. In: Gubanov, I.A., Dorofeyuk, N.I. & Neronov, V.M. (eds) Natural conditions, vegetation cover and wildlife of Mongolia, p. 113–127. Scientific Center of biological studies of the USSR Academy of Sciences, Pushchino (in Russian).

Nogina, N.A. (ed.), 1978. Soil survey of the main natural zones of Mongolia. Nauka, Moscow, 275 p. (in Russian).

Nogina, N.A., 1989. Specific features of soils and soil formation processes in the Central Asian facies (taiga, steppe, desert). *Soil Sciences* 9: 5–14 (in Russian).

Nogina, N.A. & Dorzhgotov, D., 1980. Peculiarities of soils and soil cover of the MPR. In: Bannikova, I.A., Vipper, P.B., Ulziihutag, N. & Tsegmid, G. (eds) Biological resources of the MPR, their use and protection, p. 11–13. Ulan Bator (in Russian).

Nogina, N.A. & Dorzhgotov, D., 1982. Soil-geographical regionalization of Mongolia. *Soil Sciences* 4: 23–32 (in Russian).

Novikova, N.M., 1997. Principles of conservation of botanical diversity of Turan delta plains. Dr.Sc. (Geography) Thesis. Moscow, 104 p. (in Russian).

Pacyna, A., 1984. Vegetation of the Saint Valley in the Khangai Mountains (Mongolia). *Fragmenta Floristica et Geobotanica*, XXX, pars 4: 313–451.

Pankova, E.I., 1986. Salinity of Mongolian soils. *Soil sciences* 10: 81–89 (in Russian).

Pankova, E.I., 1992. Genesis of desert soil salinization (case studies in Mongolia and Central Asia). V.V. Dokuchaev Soil Institute, Moscow, 136 p. (in Russian).

Pavlov, N.V., 1929a. Materials of the flora of Northern and Central Mongolia brought by the botanical expeditions in 1924 and 1926. *Bull. Soc. Nat. Ser. biol.* 1–2: 1–153. Moscow.

Pavlov, N.V., 1929b. Introduction into vegetation cover of Hangai mountain area. In: Materials of the USSR Academy of Sciences Commission on investigations in Mongolian and Tannu-Tuva People's Republics and the Buriat-Mongolian ASSR Iss. 2, p. 3–72. Leningrad (in Russian).

Peterson, G.M., 1993. Vegetation and climate history of the Western Former Soviet Union. In: Wright, H.E., Kutzbach, J.E., Webb III, T., Ruddiman, W.F., Street-Perrott, F.A. & Bartlein, P.J. (eds) Global climates since the last Glacial Maximum, p. 169–193. University of Minnesota Press, Minneapolis.

Petrov, M.P., 1966. Deserts of the Central Asia. Nauka, Moscow, Leningrad, 274 p. (in Russian).

Petukhov, I.A. & Etminus, A.B., 1990. Juniper communities of Gobian and Mongolian Altai. In: Dorofeyuk, N.I. (ed.) Ecology and nature management in Mongolia: Abstr. of Reports, p. 112–114. GUGK MPR, Ulan Bator (in Russian).

Podtyazhkin, O.I., 1992. Zoogeographic analysis of the desert-steppe fauna of rodents in Mongolia. PhD Thesis (Biology). Moscow, 26 p. (in Russian).

Prentice, I.C., Cramer, W., Harrison, S.P., Leemans, R., Monserud, R.A. & Solomon, A.M., 1992. A global biome model based on plant physiology and dominance, soil properties and climate. *Journal of Biogeography* 19: 117–134.

Prentice, I.C., Guiot, J., Huntley, B., Jolly, D. & Cheddadi, R., 1996. Reconstructing biomes from palaeoecological data: a general method and its application to European pollen data at 0 and 6 ka. *Climate Dynamics* 12: 185–194.

Preobrazhensky, V.S., Fadeeva N.V., Kelner, Yu.G., Tsegmid, Sh. & Tulgaa, H., 1984. The experience of landscape-typological zoning of Mongolian territory basing on remote sensing. *Izvestia of the Academy of*

Sciences of the USSR. Ser. Geography 4: 5–14 (in Russian).

Puzachenko, Yu.G., Gunin, P.D., Skulkin, V.S. & Zvenigorodskaya, M.E., 1983. A criterion for allocation of biosphere reserves. In: Applies aspects of the «Man and Biosphere» program, p. 146–167. Moscow (in Russian).

Rachkovskaya, E.I., 1977. The extremely arid desert types of the Transaltai Gobi. In: The problems of ecology, geobotany, botanical geography, and floristics, p. 99–108. Nauka, Leningrad (in Russian).

Rachkovskaya, E.I., 1989. Vegetation of Gobian deserts of the MPR (geography and classification). Dr.Sc. (Biology) Thesis. Tashkent, 41 p. (in Russian).

Rachkovskaya, E.I., 1993. Vegetation of Gobian deserts of Mongolia. Nauka, St.-Petersburg, 134 p. (in Russian).

Rachkovskaya, E.I. & Volkova, E.A., 1977. The vegetation of the Transaltai Gobi. In: Lavrenko, E.M. & Popova, T.A. (eds) The vegetation and wildlife in Mongolia, p. 46–74. Nauka, Leningrad (in Russian).

Reimers, N.F. & Shtilmark, F.R., 1978. Strictly preserved natural territories. Mysl, Moscow, 295 p. (in Russian).

Robinson, T.W., 1957. Phreatophytes. In: U.S. Geol. Surv. Water Supply Paper, 1423 p.

Romanov, G.A., Tsedev, D. & Hudaiberdyiev, G.K., 1988. Soil protective agriculture in Mongolia. Agropromizdat, Moscow, 223 p. (in Russian).

Savin, E.N. & Dugarzhav, Ch., 1980. Renewal processes in forests of the MPR. Pine forests. In: Zhukov, A.B. & Savin, E.N. (eds) Forests of the Mongolian People's Republic (use in economy), p. 6–17. Nauka, Moscow (in Russian).

Savin, E.N., Milyutin, L.I., Krasnoshchekov, Yu.N., Korotkov, I.A., Suntsov, A.V., Dugarzhav, Ch., Dashzeveg, Ts., Tsogoo, Z., Dorzhsuren, Ch., Zhamiasuren, S. & Gombosuren, N., 1988. Forests of the Mongolian People's Republic. Larch forests of the Eastern Hentei. Nauka, Moscow, 176 p. (in Russian).

Savin, E.N., Ogorodnikov, A.V., Semechkin, I.V., Dugarzhav, Ch. & Korotkov, I.A., 1978. Forests of the Mongolian People's Republic (geography and typology). Nauka, Moscow, 127 p. (in Russian).

Savina, L.N. & Burenina, T.A., 1981. Persistence of the larch pollen in forest soils, and reflection of Mongolian larch stands composition in recent spectra. In: Savina, L.N. (ed.) Paleobotanical studies in the forests of the Northern Asia, p. 62–83. Nauka, Novosibirsk (in Russian).

Savina, L.N., Korotkov, I.A., Ogorodnikov, A.V., Savin, E.N. & Burenina, T.A., 1981. Trends in development of forest vegetation of the Mongolian People's Republic. In: Savina, L.N. (ed.) Paleobotanical studies in the forests of Northern Asia, p. 83–158. Nauka, Novosibirsk (in Russian).

Savina, L.N., Savin, E.N. & Korotkov, I.A., 1982. The history of the forest vegetation of the reserve Bogdo-Ula (MPR) according to the data of the spore-pollen analysis of soils. *Izvestiya of the Siberian Division of the Academy of Sciences of the USSR, Ser. biol.* 1: 45–53 (in Russian).

Semechkin, I.V., 1980. Renewal processes in forests of the Mongolian People's Republic. In: Zhukov, A.B. & Savin, E.N. (eds) Forests of the Mongolian People's Republic (use in economy), p. 17–65. Nauka, Moscow, (in Russian).

Sevastyanov, D.V., Seliverstov, Yu.P. & Chernova, G.M., 1993. To the history of landscape development in the Ubsu-Nur depression. *Vestnik St.-Petersburg University. Ser. 7 (Geology)* 28 (4): 71–81 (in Russian).

Sevastyanov, D.V., Shuvalov, V.F. & Neustrueva, I.Yu. (eds), 1994. Limnology and paleolimnology of Mongolia. Nauka, St.-Petersburg, 270 p. (in Russian).

Shilova, G.N., 1973. Development of Mongolian vegetation in Pliocene and Pleistocene. In: Grichuk, V.P. (ed.) Palynology of Pliocene and Pleistocene, p. 167–170. Nauka, Moscow (in Russian).

Shilova, G.N., 1984. To the methodology of interpretation of spore-pollen spectra in mountain regions of Mongolia. In: Khlopova, A.F. (ed.) Problems of the modern palynology, p. 61–65. Nauka, Novosibirsk (in Russian).

Shmidt, G.A., 1974. Main types of Mongolian relief. In: Timofeev, D.A. & Chichagov, V.P. (eds) Geomorphology of foreign countries, p. 92–108. Nauka, Moscow (in Russian).

Shver, A. & Zhadambaa, Sh. (eds), 1971. Climatic gazetteer of the Mongolian People's Republic. Vol. 1. Upravlenie gidrometeorologicheskoi sluzhby pri Sovete Ministrov MNR. Inst. Meteorologii i gidrologii, Ulan-Bator, 339 p. (in Mongolian).

Sinitsyn, V.M., 1948. Some remarks on orographic systematization of Central Asian ridges. *Izvestia of the USSR Academy of Sciences. Ser. geology* 3: 93–98 (in Russian).

Sinitsyn, V.M., 1959. Central Asia. Geografgiz, Moscow, 456 p. (in Russian).

Sinitsyn, V.M., 1962. Palaeogeography of Asia. Nauka, Leningrad, 268 p. (in Russian).

Sirnev, I.P., 1982. East Mongolian plain. In: Florensov, N.A. & Korzhuev, S.S. (eds) Geomorphology of the Mongolian People's Republic, p. 166–176. Nauka, Moscow (in Russian).

Sladkov, A.N., 1967. Introduction in the spore-pollen analysis. Nauka, Moscow, 270 p. (in Russian).

Slemnev, N.N., Gunin, P.D. & Kazantseva, T.I., 1994. To the problem of natural seed renewal of dominating plants in the ecosystems of the desert zone of Mongolia. *Vegetation Resources* 4: 1–14 (in Russian).

Slemnev, N.N., Gunin, P.D., Kazantseva, T.I., Tsoozh, Sh. & Yakunin, G.N., 1997a. Renewal of dominating plants in ecosystems of desert zone of Mongolia:

germination and live ability of regrowth. *Vegetation Resources* 4: 117–138 (in Russian).

Slemnev, N.N., Kazantseva, T.I. & Gunin, P.D., 1997b. Haloxyloneta in Mongolia. Report 1. Rare herbaceous Haloxyloneta, their structure and the peculiarities of development. *Vegetation Resources* 3: 124–139 (in Russian).

Smagin, V.M (ed.), 1983. The forests of the Mongolian People's Republic (Larch forests of the Central Hangai). Nauka, Novosibirsk, 150 p. (in Russian).

Sokolov, V.E., Dash, Ya., Lushchekina, A.A. & Neronov, V.M., 1982. Contemporary distribution and quantitative assessments of dzeren in the MPR. In: Sokolov, V.E. (ed.) Zoological Studies in the MPR, p. 7–20. Nauka, Moscow (in Russian).

Sokolov, V.E. & Gunin, P.D. (eds), 1986. Deserts of the Transaltai Gobi. Natural conditions, ecosystems and zoning. Nauka, Moscow, 208 p. (in Russian).

Sokolov, V.E., Gunin, P.D., Drozdov, A.V. & Puzachenko, Yu.G., 1987. Criteria for allocation of biosphere reserves. In: Nature protection, science and society, p. 117–120. Vneshtorgizdat, Moscow (in Russian).

Sokolov, V.E. & Shagdarsuren, O. (eds), 1983. Complex characteristics of desert ecosystems of Transaltai Gobi (The examples of the desert experimental station, and the Great Gobi Reserve). Scientific Center of biological studies of the USSR Academy of Sciences, Pushchino, 114 p. (in Russian).

Sokolov, V.E., Vostokova, E.A., Gunin, P.D. & Ulziihutag, N., 1990. Experience in ecological mapping of the MPR territory. In: Ecological cooperation. informational bull. on the problem III «Protection of ecosystems (biogeocenoses) and the landscape». Bratislava, 3–4: 12–24.

Summers, R.L., Williams, J.W. & Webb III, T. Applying plant functional types to construct biome maps from eastern North American pollen data: comparison with model results. *Quaternary Science Reviews* (in press).

Tarasov, P.E. & Harrison, S.P., 1997. Lake status records from the Former Soviet Union and Mongolia: a continental-scale synthesis. *Paläoklimaforschung* 25: 115–130.

Tarasov, P.E., Harrison, S.P., Saarse, L., Pushenko, M.Ya., Andreev, A.A., Aleshinskaya, Z.V., Davydova, N.N., Dorofeyuk, N.I., Efremov, Yu.V., Khomutova, V.I., Sevastyanov, D.V., Tamosaitis, J., Uspenskaya, O.N., Yakushko, O.F. & Tarasova, I.V., 1994. Lake Status Records from the Former Soviet Union and Mongolia, Data Base Documentation. NOAA Paleoclimatology Publications Series Report No. 2. Boulder, 274 p.

Tarasov, P.E., Jolly, D. & Kaplan, J.O., 1997. A continuous Late Glacial and Holocene record of vegetation changes in Kazakhstan. *Paleogeography, Paleoclimatology, Paleoecology* 136: 281–292.

Tarasov, P.E., Pushenko, M.Ya., Harrison, S.P., Saarse, L., Andreev, A.A., Aleshinskaya, Z.V., Davydova, N.N., Dorofeyuk, N.I., Efremov, Yu.V., Elina, G.A., Elovicheva, Ya.K., Filimonova, L.V., Gunova, V.S., Khomutova, V.I., Kvavadze, E.V., Neustrueva, I.Yu., Pisareva, V.V., Sevastyanov, D.V., Shelekhova, T.S., Subetto, D.A., Uspenskaya, O.N. & Zernitskaya, V.P., 1996. Lake status records from the Former Soviet Union and Mongolia: Documentation of the second version of the data base. NOAA Paleoclimatology Publications Series Report No. 5. Boulder, 224 p.

Tarasov, P.E., Webb III, T., Andreev, A.A., Afanaseva, N.B., Berezina, N.A., Bezusko, L.G., Blyakharchuk, T.A., Bolikhovskaya, N.S., Cheddadi, R., Chernavskaya, M.M., Chernova, G.M., Dorofeyuk, N.I., Dirksen, V.G., Elina, G.A., Filimonova, L.V., Glebov, F.Z., Guiot, J., Gunova, V.S., Harrison, S.P., Jolly, D., Khomutova, V.I., Kvavadze, E.V., Osipova, I. M., Panova, N.K., Prentice, I.C., Saarse, L., Sevastyanov, D.V., Volkova, V.S. & Zernitskaya, V.P. Present-day and mid-Holocene Biomes reconstructed from pollen and plant macrofossil data from the Former Soviet Union and Mongolia. *Journal of Biogeography*, in press.

Titov, Yu.V., Ogar, N.P., Enhbaiar, B., Baiasgalan, D., Zaslavskaia, T.M., Manibazar, N., Ubugunov, L.L., Baatar, R., Tsoozh, Sh., Sheremetiev, S.N. & Ubugunova, V.I., 1990a. Complex studies on the flood plain ecosystems in the MPR. In: Dorofeyuk, N.I. (ed.) Ecology and nature management in Mongolia: Abstr. of Reports, p. 31–32. Ulan-Bator (in Russian).

Titov, Yu.V., Ogar, N.P., Ubugunov, L.L. & Ubugunova, V.I., 1990b. Methods for assessment of the contemporary state of the river flood plains of the MPR (methodological recommendations). Ulan-Bator, 42 p. (in Russian).

Trautvetter, E.R., 1872. Catalogus plantarum anno 1870 ab Lomonossovio in Mongolia orientali lectarum. *Acta Horti Petrop*. 1: 165–195.

Trofimova, G.Yu. & Novikova, N.M., 1994. The system of ecologically oriented data bases for arid regions of Central Asia. *Problems of the Desert Development* 2: 68–71 (in Russian).

Tsatsenkin, I.A. & Yunatov, A.A., 1951. Natural forage resources of the MPR. USSR Ac. of Sc. Publishing House, Moscow, 350 p. (in Russian).

Tsegmid, Sh., 1962. Physical-geographical zoning of the MPR. *Izvestia of the USSR Academy of Sciences, Ser. Geography* 5: 34–41. (in Russian).

Tsegmid, Sh. & Vorobiev, V.V. (eds), 1990. The National Atlas. Mongolian People's Republic. GUGK SSSR, GUGK MNR, Moscow, Ulan-Bator, 144 p. (in Russian).

Tserensodnom, Zh., 1971. Mongol orny nuur. Ulaanbaatar, 202 p. (in Mongolian).

Turczaninov, N., 1832. Decades tres plantarum Chinae borealis et Mongoliae chinensi incolarum. *Bull. Soc. Nat.* 5: 180–206. Moscou.

Uflyand, A.K., Ilin, A.V. & Spirkin, A.I., 1969. Basins of the Baikal-type of the Northern Mongolia. *Bull. of the Moscow Society of Nature Investigators, Div. Geology* 6: 5–52.

Vasak, V., 1971. *Gnaphalium supinum* L. a new species for Mongolia. *Preslia.* Vol. 43, 3.

Viktorov, S.V., 1973. Aero-landscape indication of consequences of human activity in deserts. Nedra, Moscow, 57 p.(in Russian).

Viktorov, S.V., Vostokova, E.A. & Vyshivkin, D.D., 1962. Introduction to indicational geobotanics. Moscow University Press, Moscow, 228 p. (in Russian).

Vipper, P.B., Dorofeyuk, N.I. & Sokolovskaya, V.T., 1975. The history of formation of the lake Dund-Nur, and the dynamics of the vegetation cover of the Mongolian Altai in Holocene. In: Kvasov, D.D. & Martinson, G.G. (eds) The history of the lakes and inner seas of the arid zone, p. 114–119. Leningrad (in Russian).

Vipper, P.B., Dorofeyuk, N.I., Meteltseva, E.P., Sokolovskaya, V.T. & Shulia, K.S., 1976. The experience of reconstruction of vegetation of Western and Central Mongolia in Holocene on the basis of studies of bottom sediments of the fresh-water lakes. In: Lavrenko, E.M. & Rachkovskaya E.I. (eds) The structure and dynamics of the main ecosystems in the MPR, p. 35–59. Nauka, Leningrad (in Russian).

Vipper, P.B., Dorofeyuk, N.I., Meteltseva, E.P. & Sokolovskaya, V.T., 1989. Landscape-climatic changes in the Central Mongolia in Holocene. In: Khotinckii, N.A. (ed.) Paleoclimates of Late Glacial and Holocene, p. 160–167. Nauka, Moscow (in Russian).

Volkova, E.A., 1976. The influence of the petrographic rock composition on vegetation cover, and the indicating role of vegetation communities at the hummocky terrain of the southern part of the MPR. In: Lavrenko, E.M. & Rachkovskaya, E.I. (eds) The structure and dynamics of the main ecosystems of the MPR, p. 144–157. Nauka, Leningrad, Moscow (in Russian).

Volkova, E.A., 1988. Map of vegetation of the somon Tumen-Tsogt (Mongolian People's Republic). In: Isachenko, T.I. & Gribova, S.A. (eds) Geobotanical mapping, p. 38–45. Nauka, Leningrad.

Volkova, E.A., 1994. Botanical geography of Mongolian and Gobian Altai. Proceedings of Komarov Botanical Institute of Russian Academy of Sciences, St.-Petersburg, 130 p. (in Russian).

Volkova, E.A. & Rachkovskaya, E.I., 1980. Vegetation map of Jungarian Gobi. In: Sochava, V.B & Isachenko, T.I. (eds) Geobotanical mapping, p. 24–39. Nauka, Leningrad (in Russian).

Vostokova, E.A., 1967. Hydrogenic ecological catena of vegetation of desert areas. *Earth Sciences, New series,* Vol. VII (XLVII): 245–256 (in Russian).

Vostokova, E.A., 1980. Peculiarities of interpretation of present-day vegetation at space images. In: Voronov, A.G. & Vyshivkin, D.D. (eds) Ecology of phytocenoses and their dynamics, p. 108–120. MFGO, Moscow (in Russian).

Vostokova, E.A., 1983. Ecological catena of vegetation in closed depressions in the Mongolian People's Republic. In: Ecological-cenotic and geographical peculiarities of vegetation. p. 40–50. Nauka, Moscow (in Russian).

Vostokova, E.A., 1986. Use of the remote information for cartographic monitoring of the Ubsu-Nur lake basin. (Legends for maps «Vegetation», «Landscapes», «Land Use»). In: Bugrovsky, V.V. (ed.) Informational problems of biosphere studies. The Ubsu-Nur Experiment, p. 161–180, 285–292. Scientific Center of biological studies of the USSR Academy of Sciences, Pushchino (in Russian).

Vostokova, E.A., 1995. Use of remote information for cartographic monitoring and nature-protection measures. In: Bugrovsky, V.V. (ed.) Informational problems of biosphere studies. The Ubsu-Nur Experiment, p. 113–122. Scientific Center of biological studies of the Russian Academy of Sciences, Pushchino (in Russian).

Vostokova, E.A., Gunin, P.D., Prishchepa, A.V. & Sarantuiaa, N., 1994. Use of the remote space photo information for water and air migration flux studies (the example of the south of Mongolia). *Studies of the Earth from Space* 1: 93–97 (in Russian).

Vostokova, E.A., Gunin, P.D., Prishchepa, A.V. & Kalitski, T., 1995. The experience of development of the map of natural processes of desertification. *Problems of Desert Development* 2: 20–32 (in Russian).

Vostokova, E.A. & Kazantseva, T.I., 1995. Distribution, ecology and productivity of communities *Achnatherum splendens* (Trin.) Nevski in Mongolia. *Vegetation Resources* 4: 77–87 (in Russian).

Vostokova, E.A. & Kelner, Yu.G., 1986. Space information as the basis for cartographic monitoring of nature. In: Goldman, L.M. & Smekhova, A.E. (eds) Application of aero-space methods for studies and monitoring of land surface features, p. 11–19. Moscow division of Geographical Society, Moscow (in Russian).

Vostokova, E.A. & Khrutsky, V.S., 1996. The set of maps on land resources as the basis for monitoring of Ubsunur depression. In: Bugrovsky, V.V. (ed.) Global monitoring and the Ubsunur depression, p. 89–92. Intellekt, Moscow (in Russian).

Vostokova, E.A., Prishchepa, A.V. & Samoilova, G.S., 1993. Landscapes of the zone of watered fault in

Alashan Gobi (Mongolia). *Problems of Desert Development* 1: 38–45 (in Russian).

Vostokova, E.A., Sushchenya, V.A. & Shevchenko, L.A., 1988a. Methodological recommendations on mapping of dynamics of natural objects basing on remote information. TsNIIGAIK, Moscow, 160 p. (in Russian).

Vostokova, E.A., Sushchenya, V.A. & Shevchenko, L.A., 1988b. Ecological mapping on the basis of space information. Nedra, Moscow, 223 p. (in Russian).

Webb III, T. 1985. A Global Palaeoclimatic Data Base for 6000 B.P. DOE/eV/10097-6, U.S. Department of Energy, Washington, DC., 155 p.

Winkler, M.G. & Wang, P.K., 1993. The Late-Quaternary Vegetation and Climate of China. In: Wright, H.E., Kutzbach, J.E., Webb III, T., Ruddiman, W.F., Street-Perrott, F.A. & Bartlein, P.J. (eds) Global Climates since the Last Glacial Maximum, p. 265–293. University of Minnesota Press, Minneapolis.

Woodward, F.I. 1987. Climate and plant distribution. Cambridge University Press, Cambridge, UK, 174 p.

Wright, H.E., Kutzbach, J.E., Webb III, T., Ruddiman, W.F., Street-Perrott, F.A. & Bartlein, P.J. (eds), 1993. Global Climates since the Last Glacial Maximum. University of Minnesota Press, Minneapolis, 569 p.

Yu, G., Prentice, I.C., Harrison, S.P. & Sun, X. Pollen-based biome reconstructions for China at 0 ka and 6 ka. *Journal of Biogeography*, in press.

Yu, G. & Qin, B., 1997. Holocene temperature and precipitation reconstructions and monsoonal climates in eastern China using pollen data. *Paleoclimates* 2: 1–32.

Yunatov, A.A., 1946. Studies on vegetation of Mongolia during 25 years. In: Trudy Komiteta nauk MPR. Jubilee Series. 2: 1–58.

Yunatov, A.A., 1950. Main features of vegetation cover of the Mongolian People's republic. USSR Academy of Sciences Publishing House, Moscow, Leningrad, 224 p. (in Russian).

Yunatov, A.A., 1954. Forage plants of pastures and meadows of the MPR. USSR Academy of Sciences Publishing House, Moscow, Leningrad, 352 p. (in Russian).

Yunatov, A.A., 1974. Semi-desert steppe of the Northern Gobi in the MPR. Nauka, Leningrad, 132 p. (in Russian).

Zhambazhamts, B. & Bat, B., 1985. The Atlas of the climate and ground water resources in the Mongolian People's Republic. Goskomgidromet SSSR, GUGMS MNR, GUGK SSSR, Ulaanbaatar, 88 p. (in Mongolian).

Zhazvan, S., 1996. Introduction of decorative grass plants of natural flora of Mongolia. PhD Thesis (Biology). Moscow, 19 p. (in Russian).

Zhukov, A.B. & Savin, E.N. (eds), 1980. Forests of the Mongolian People's Republic (use in economy). Nauka, Moscow, 148 p. (in Russian).

Zolotarev, A.G., Kulakov, V.S. & Khilko, S.D., 1982. Khubsugul upland. In: Florensov, N.A. & Korzhuev, S.S. (eds) Geomorphology of Mongolian People's Republic, p. 109–122. Nauka, Moscow (in Russian).

Zolotokrylin, A.N. & Gunin, P.D., 1986. Precipitation Regimes. In: Sokolov, V.E. & Gunin, P.D. (eds) Deserts of the Transaltai Gobi. Natural conditions, ecosystems, and zoning, p. 29–33. Nauka, Moscow (in Russian).

Zonenshain, L.P. & Savostin, L.A., 1979. Introduction into geodynamics. Nedra, Moscow, 311 p. (in Russian).

Appendix 1

List of plant species names used in this study (after Gubanov, 1996)

Abies sibirica Ledeb., 59, 65, 72, 150, 169
Acanthophyllum pungens (Bunge) Boiss., 182
Achnatherum splendens (Trin.) Nevski, 20, 23, 24, 28, 29, 30, 107, 117, 120, 124, 125, 127, 128, 150, 175, 181, 197, 198
Aconitum anthoroideum DC., 182
– *baicalense* Turcz. ex Rapaics, 183
– *barbatum* Pers., 89, 124, 183
– *chasmanthum* Stapf, 183
– *gubanovii* Luferov et Worosch., 168, 201
– *septentrionale* Koelle, 196
Aconogonon alpinum (All.) Schur, 81
– *angustifolium* (Pall.) Hara, 96, 101
– *divaricatum* (L.) Nakai ex Mori, 107, 111, 115, 117, 121, 124, 151
Acorus calamus L., 169
Acroptilon repens (L.) DC., 124, 176
Adenophora changaica Gubanov et R.Kam., 168
– *stenanthina* (Ledeb.) Kitag., 88, 101, 184
Adonis mongolica Simonovicz, 167, 168, 169, 183
– *sibirica* Patr. ex Ledeb., 88, 89, 92, 100, 101, 102, 111, 112, 113, 115, 116, 117, 120, 121, 122, 124
Aegopodium alpestre Ledeb., 36, 88, 89
Agriophyllum pungens (Vahl) Link, 105, 106, 107, 125, 127
Agropyron cristatum (L.) Beauv., 23, 24, 64, 82, 94, 96, 102, 105, 107, 109, 110, 111, 113, 116, 172
Agrostis trinii Turcz., 95, 108
Ajania achilleoides (Turcz.) Poljak. ex Grub., 97, 98, 114
– *fruticulosa* (Ledeb.) Poljak., 97, 98, 102, 106, 116, 141
– *grubovii* Muld., 168
– *trifida* (Turcz.) Tzvel., 199
Alchemilla changaica V.Tichomirov, 168
– *gubanovii* V.Tichomirov, 168, 201
– *pavlovii* Juz., 167, 168
Alisma plantago-aquatica L., 118, 123
Allium altaicum Pall., 167, 169, 201
– *condensatum* Turcz., 169
– *galanthum* Kar. et Kir., 167, 169, 201
– *leucocephalum* Turcz. ex Ledeb., 182
– *macrostemon* Bunge, 169
– *microdictyon* Prokh., 182
– *mongolicum* Turcz. ex Regel, 73, 97, 105, 168, 170
– *obliquum* L., 169
– *polyrrhizum* Turcz. ex Regel, 98, 103, 104, 199
– *ramosum* L., 104
Alnus fruticosa Rupr., 46, 56, 61, 62, 70
– *glutinosa* (L.) Gaertner, 70
– *incana* (L.) Moench, 70
Amblynotus rupestris (Pall. ex Georgi) M.Pop. ex Serg., 82, 138
Ammopiptanthus mongolicus (Maxim. ex Kom.) Cheng fil., 168, 170, 198, 200
Amygdalus mongolica (Maxim.) Ricker, 23, 29, 104, 106, 113, 142, 146, 168, 169, 178, 197, 198, 201
Anabasis aphylla L., 28
– *brevifolia* C.A.Mey., 13, 23, 28, 57, 73, 97, 98, 99, 103, 104, 114, 116, 117, 124, 146, 149, 154, 172
– *eriopoda* (Schrenk) Benth. ex Volkens, 98
– *truncata* (Schrenk) Bunge, 98
Androsace chamaejasme Wulf. subsp. *lehmanniana* (Spreng.) Hult., 113
– *incana* Lam., 101
– *longifolia* Turcz., 170
Anemarrhena asphodeloides Bunge, 169
Anemone crinita Juz., 90, 183
– *sylvestris* L., 48, 61, 62, 64, 65, 69, 72, 92

Arabis mongolica Botsch., 168
Aquilegia glandulosa Fisch. ex Link, 183
– *sibirica* Lam., 140, 150, 151
– *viridiflora* Pall., 183
Arctogeron gramineum (L.) DC., 198
Arenaria capillaris Poir., 101, 102, 107, 109, 113, 176, 198
– *meyeri* Fenzl, 82, 104
Aristida heymannii Regel, 116
Armeniaca sibirica (L.) Lam., 113, 167, 178, 197
Arnebia guttata Bunge, 98
Artemisia adamsii Bess., 112, 113, 125
– *anethifolia* Web. ex Stechm., 124
– *anethoides* Mattf., 114, 115, 117
– *annua* L., 94, 115, 117, 120, 122, 124, 175, 176
– *changaica* Krasch., 101, 108, 176
– *commutata* Bess., 101, 113, 119, 120
– *dracunculus* L., 94, 102, 111, 117, 124
– *frigida* Willd., 20, 23, 24, 25, 64, 93, 94, 96, 99, 100, 107, 108, 109, 110, 112, 113, 115, 116, 117, 153, 172
– *glauca* Pall. ex Willd., 99, 115, 169
– *halodendron* Turcz., 107
– *laciniata* Willd., 120, 121, 124, 176
– *macrocephala* Jacq. ex Bess., 94, 99, 115, 117, 124, 125, 150, 153
– *mongolica* (Bess.) Fisch. ex Nakai, 113
– *palustris* L., 117, 151
– *pamirica* C. Winkl., 98
– *santolinifolia* Turcz. ex Bess., 94
– *scoparia* Waldst. et Kit., 94, 95, 102, 112, 115, 116, 117, 119, 120, 124, 125, 150, 153
– *sericea* Web. ex Stechm., 168
– *sieversiana* Willd., 115
– *sublessingiana* Krasch. ex Poljak., 172
– *tanacetifolia* L., 84, 89, 100, 101, 102, 108, 113, 116
– *terrae-albae* Krasch., 28
– *vulgaris* L., 120
– *xanthochroa* Krasch., 98, 104, 107
– *xerophytica* Krasch., 105, 146
Asparagus gobicus Ivanova ex Grub., 105
– *trichophyllus* Bunge, 182
Asperula saxicola Grub., 168
Aster alpinus L., 84, 184
Asterothamnus centrali-asiaticus Novopokr., 98, 99, 105, 184
– *heteropappoides* Novopokr., 168
– *molluscuius* Novopokr., 98
Astragalus adsurgens Pall., 115, 125
– *alexandri* Ulzij., 168
– *baitagensis* Sancz. ex Ulzij., 168, 201
– *changaicus* Sancz. ex Ulzij., 168, 200
– *galactites* Pall., 198
– *gobicus* Hanelt et Davazamc, 168
– *granitovii* Sancz. ex Ulzij., 168
– *grubovii* Sancz., 168
– *gubanovii* Ulzij., 168, 201
– *klementzii* Ulzij., 13, 168
– *koslovii* B.Fedtsch. et Basil. ex Ulzij., 168
– *mongholicus* Bunge, 88
– *pseudochorinensis* Ulzij., 168
– *tenuis* Turcz., 151
– *variabilis* Bunge ex Maxim., 98
– *viridiflavus* Ulzij., 168
Atragene sibirica L., 183
Atraphaxis frutescens (L.) Ewersm., 106
Atriplex laevis C.A.Mey., 115
– *sibirica* L., 169, 172, 174, 179, 183, 196, 200
– *tatarica* L., 115, 116, 121, 124
Axyris amaranthoides L., 99, 116, 117, 151
A. hybrida L., 95, 112, 115
Bassia dasyphylla (Fisch. ex Mey.) O. Kuntze, 105, 115, 116
Berberis sibirica Pall., 178, 183
Bergenia crassifolia (L.) Fritsch, 196
Betula exilis Sukacz., 82, 86
– *fruticosa* Pall., 82, 84, 86, 98, 101, 118, 123, 182
– *fusca* Pall. ex Georgi, 118
– *humilis* Schrank, 61
– *platyphylla* Sukacz., 72, 118
– *rotundifolia* Spach, 55, 56, 57, 61, 63, 72, 82, 86
Bidens tripartita L., 119, 123
Bistorta alopecuroides (Turcz. ex Meissn.) Kom., 121, 124, 176
– *vivipara* (L.) S.F. Gray, 72, 81, 89, 94, 201
Brachanthemum gobicum Krasch., 170
– *mongolicum* Krasch., 73, 97, 105, 168, 170
Bromopsis inermis (Leys.) Holub, 101, 120
– *korotkiji* (Drob.) Holub, 107
– *sibirica* (Drob.) Peschk., 140, 150, 151
– *ubsunurica* Tzvel., 168

Bromus japonicus Thunb., 115
Bupleurum bicaule Helm, 109
– *scorzonerifolium* Willd., 94, 95, 101, 151
Butomus umbellatus L., 123
Cacalia hastata L., 118
Calamagrostis epigeios (L.) Roth, 95, 104, 107
– *obtusata* Trin., 81, 88
– *purpurea* (Trin.) Trin., 120
Calligonum mongolicum Turcz., 102, 105, 106
Caltha palustris L., 118
Calypso bulbosa (L.) Oakes, 169
Campanula altaica Ledeb., 184
Cannabis sativa L., 121, 124
Caragana alaschanica Grub., 200
– *altaica* (Kom.) Pojark., 200
– *brachypoda* Pojark., 168
– *bungei* Ledeb., 105, 120, 124, 168
– *gobica* Sancz., 13, 168, 200
– *korshinskii* Kom., 199
– *leucophloea* Pojark., 98, 104, 106, 107, 116, 149, 199
– *microphylla* Lam., 13, 64, 95, 99, 100, 101, 102, 104, 107, 112, 113, 197, 198
– *pygmaea* (L.) DC., 13, 107, 149, 199
– *stenophylla* Pojark., 13, 95, 100, 107, 112, 113, 138, 139, 197, 198
– *tibetica* Kom., 200
Cardaria pubescens (C.A. Mey.) Jarm., 115, 176
Carduus crispus L., 120, 121, 124
Carex amgunensis Fr. Schmidt, 89
– *cespitosa* L., 120
– *dichroa* (Freyn) V. Krecz., 118
– *duriuscula* C.A.Mey., 94, 95, 96, 99, 108, 109, 110, 111, 112, 113, 115, 119, 120, 124, 125
– *enervis* C.A. Mey., 120
– *korshinskyi* Kom., 107
– *lanceolata* Boott, 36, 84, 89, 90
– *media* R. Br., 118
– *melanantha* C.A. Mey., 81
– *melanocephala* Turcz., 81
– *obtusata* Liljebl., 81
– *orbicularis* Boott, 81
– *pediformis* C.A.Mey., 101, 108
– *rupestris* All., 55, 81
– *sempervirens* Vill., 81
– *stenophylloides* V. Krecz., 96, 115
Carum carvi L., 116
Caryopteris mongholica Bunge, 97, 170
Ceratocarpus arenarius L., 107, 125
Chamaenerion angustifolium (L.) Scop., 36, 86, 89, 150
Chamaerhodos erecta (L.) Bunge, 124
Chenopodium acuminatum Willd., 94, 96, 115, 121, 124
– *album* L., 94, 95, 99, 115, 121, 124, 125, 138, 151
– *aristatum* L., 95, 115, 121, 124
Chesneya grubovii Yakovlev, 168
– *mongolica* Maxim., 99, 104, 105, 107, 113, 115
Chloris virgata Sw., 114, 115
Cirsium esculentum (Siev.) C.A. Mey., 120
– *setosum* (Willd.) Bess., 120, 121, 124
Clausia trichosepala (Turcz.) Dvorak, 167
Cleistogenes songorica (Roshev.) Ohwi , 141
– *squarrosa* (Trin.) Keng, 20, 23, 26, 30, 94, 95, 96, 99, 107, 109, 110, 111, 112, 113, 116, 197, 198
Clematis aethusifolia Turcz., 107
– *fruticosa* Turcz., 21, 36, 37
– *intricata* Bunge, 183
– *songarica* Bunge, 98, 126
– *tangutica* (Maxim.) Korsch., 183
Convolvulus ammanii Desr., 115, 116, 125
– *chinensis* Ker-Gawl., 99, 110, 111, 116
– *gortschakovii* Schrenk, 97
– *tragacanthoides* Turcz., 98
Corallorhiza trifida Chatel., 169
Corispermum mongolicum Iljin, 105, 115, 116
– *patelliforme* Iljin, 107
Crataegus dahurica Koehne ex Schneid., 118
– *sanguinea* Pall., 118, 120, 124
Cymbaria daurica L., 95, 102, 113, 151
Cynanchum sibiricum Willd., 114, 115, 126, 176
Cynomorium songaricum Rupr., 170
Cypripedium calceolus L., 169
– *guttatum* Sw., 88
– *macranthum* Sw., 169
Cystopteris fragilis (L.) Bernh., 201
Dactylorhiza fuchsii (Druce) Soo, 169
– *salina* (Turcz. ex Lindl.) Soo, 108, 120
Delphinium changaicum Frisen, 168
– *dissectum* Huth, 101
– *grandiflorum* L., 183
– *gubanovii* Frisen, 168

Dendranthema sinuatum (Ledeb.) Tzvel., 170
– *zawadskii* (Herbich)Tzvel., 89, 101, 108, 185
Descurainia sophia (L.) Webb ex Prantl, 124, 153
Dianthus superbus L., 182
– *versicolor* Fisch. ex Link, 182, 185
Diarthron linifolium Turcz., 167
Dictamnus dasycarpus Turcz., 170
Dontostemon integrifolius (L.) C.A. Mey., 114, 115, 120, 138, 151
Draba nemorosa L., 124
Dracocephalum foetidum Bunge, 115
– *junatovii* A.Budantz., 168
– *origanoides* Steph. ex Willd., 82
Dryas oxyodonta Juz., 81
Echinops gmelinii Turcz., 105
– *latifolius* Tausch., 94, 95, 99, 101, 185
Elaeagnus moorcroftii Wall. ex Schlecht., 167, 170
Elymus gmelinii (Ledeb.) Tzvel., 101, 118
Elytrigia repens (L.) Nevski, 120
Empetrum sibiricum V. Vassil., 72
Enneapogon borealis (Griseb.) Honda, 115
Ephedra equisetina Bunge, 104
– *glauca* Regel, 57, 64, 72
– *przewalskii* Stapf, 73, 97, 98, 102, 103, 106, 146, 149
– *sinica* Stapf, 30, 97, 98, 103, 104, 143, 144
Epipogium aphyllum Sw., 169
Equisetum pratense L., 89
Eragrostis minor Host, 114, 115, 116
Eriophorum polystachyon L., 81
Erysimum flavum (Georgi) Bobr., 96
Euonymus maackii Rupr., 170
Euphorbia humifusa Willd., 115, 116, 175, 176
– *kozlovii* Prokh., 168
Fallopia convolvulus (L.) A. Love, 95, 96, 115, 151, 153
Festuca brachyphylla Schult. et Schult. fil., 82
– *kryloviana* Reverd., 81, 101
– *lenensis* Drob., 55, 64, 81, 82, 94, 95, 99, 101, 102, 108, 113, 151, 197
– *ovina* L., 36, 88, 89, 90
– *sibirica* Hack. ex Boiss., 140, 150, 151
– *valesiaca* Gaud., 201
– *venusta* St.-Yves, 200
Filifolium sibiricum (L.) Kitam., 102, 111, 113, 138, 139, 197
Fragaria orientalis Losinsk., 88, 89
Galitzkya macrocarpa (Ik.-Gal.) V.Boczantzeva, 168
Galium boreale L., 89
– *verum* L., 88, 101, 102, 113, 138, 139
Gentiana algida Pall., 167, 170
– *dahurica* Fisch., 184
– *decumbens* L. fil., 184
– *macrophylla* Pall., 184
Gentianopsis barbata (Froel.) Ma, 201
Geranium albiflorum Ledeb., 196
– *eriostemon* Fisch., 36, 89
– *pseudosibiricum* J. Mayer, 89
Glaux maritima L., 120
Glycyrrhiza uralensis Fisch., 126, 127
Gnaphalium baicalense Kirp., 123, 185
Goniolimon speciosum (L.) Boiss., 138, 184, 185
Goodyera repens (L.) R.Br., 169
Gueldenstaedtia monophylla Fisch., 170
Gymnadenia conopsea (L.) R.Br., 169
Gymnocarpos przewalskii Bunge ex Maxim., 169, 198
Gypsophila desertorum (Bunge) Fenzl, 116
– *paniculata* L., 182
Halenia corniculata (L.) Cornaz, 101, 108
Halerpestes salsuginosa (Pall. ex Georgi) Greene, 120
Halimodendron halodendron (Pall.) Voss, 167, 170
Halogeton glomeratus (Bieb.) C.A. Mey., 115
Haloxylon ammodendron (C.A.Mey.) Bunge, 13, 29, 31, 57, 67, 73, 102, 105, 128, 141, 143, 144, 146, 149, 153, 154, 155, 156, 157, 167, 169, 174, 198
Haplophyllum davuricum (L.) G. Don fil., 98, 113
Hedysarum dahuricum Turcz. ex B. Fedtsch., 176
– *fruticosum* Pall., 107
– *kamelinii* Ulzij., 168
– *sangilense* Krasnob. et Timochina, 167
Helictotrichon schellianum (Hack.) Kitag., 95, 108
Hemerocallis lilio-asphodelus L., 167, 169, 182
– *minor* Mill., 58, 69, 82, 96, 99, 152, 165, 182, 192, 193

Heracleum sibiricum L., 124
Heteropappus altaicus (Willd.) Novopokr., 108, 109, 125
– *biennis* (Ledeb.) Tamamsch. ex Grub., 138, 151, 152
– *hispidus* (Thunb.) Less., 94, 95, 115
Hippophaë rhamnoides L., 170
Hordeum brevisubulatum (Trin.) Link, 20, 107, 120, 124
Hyoscyamus niger L., 94, 95
Hypecoum lactiflorum (Kar. et Kir.) Pazij, 176
Hypericum attenuatum Choisy, 170
Iljinia regelii (Bunge) Korov., 29, 98, 99, 144, 149, 153, 154, 156, 157, 169
Incarvillea potaninii Batalin, 170
Inula britannica L., 116, 120
Iris bungei Maxim., 168, 182
– *dichotoma* Pall., 138, 169, 182, 185
– *lactea* Pall., 115, 119, 120, 122, 124, 125
– *sanguinea* Donn, 120
– *tenuifolia* Pall., 96, 105, 115
– *tigridia* Bunge, 182, 185
Juncus gerardii Lois., 120
Juniperus davurica Pall., 169
– *pseudosabina* Fisch. et Mey., 55
– *sabina* L., 16, 182, 201
– *sibirica* Burgsd., 140, 150, 151
Kalidium foliatum (Pall.) Moq., 107, 117, 124, 127, 149, 199
– *gracile* Fenzl, 124, 126, 149
Kobresia humilis (C.A.Mey. ex Trautv.) Serg., 81, 82
– *myosuroides* (Vill.) Fiori, 55, 72, 81, 118
– *sibirica* (Turcz. ex Ledeb.) Boeck., 140, 150, 151
– *smirnovii* Ivanova, 81
Kochia densiflora Turcz. ex Moq., 115, 153
– *prostrata* (L.) Schrad., 100
Koeleria altaica (Domin) Kryl., 82
– *cristata* (L.) Pers., 20, 95, 96, 102, 107, 108, 109, 113, 138, 197, 198
– *cristata* subsp. *mongolica* (Domin) Tzvel., 138, 197, 198
– *glauca* (Spreng.) DC., 64
Krascheninnikovia ceratoides (L.) Gueldenst., 25, 97, 98, 105, 106, 116
Lactuca serriola L., 114, 115, 176
– *sibirica* (L.) Benth ex Maxim., 140, 150, 151
– *tatarica* (L.) C.A. Mey., 176
Lagochilus ilicifolius Bunge, 97
Lagotis integrifolia (Willd.) Schischk., 81, 94
Lancea tibetica Hook. fil. et Thoms., 170
Lappula intermedia (Ledeb.) M.Pop., 138, 139, 151
– *myosotis* Moench, 115, 153, 175, 176
Larix dahurica Laws., 36, 92, 169
– *sibirica* Ledeb., 140, 150, 151
Lathyrus humilis (Ser.) Spreng., 88, 89
– *pratensis* L., 120
Leontopodium leontopodioides (Willd.) Beauverd, 104
– *ochroleucum* Beauverd, 94, 109, 118
Lepidium densiflorum Schrad., 114, 115, 176
– *obtusum* Basiner, 115, 121, 124, 176
Lespedeza davurica (Laxm.) Schindl., 104, 111, 197
– *juncea* (L. fil.) Pers., 138
Leymus chinensis (Trin.) Tzvel., 25, 30, 94, 95, 99, 100, 102, 108, 109, 110, 111, 113, 115, 117, 120, 122, 125, 151, 198
– *paboanus* (Claus) Pilg., 124
– *racemosus* (Lam.) Tzvel., 95, 105, 106, 107, 108, 127
– *secalinus* (Georgi) Tzvel., 122, 124
Lilium buschianum Lodd., 182
– *dauricum* Ker-Gawl., 72, 167, 169, 182
– *martagon* L., 182
– *potaninii* Vrishcz, 182
– *pumilum* Delile, 108
Limonium aureum (L.) Hill, 184
– *bicolor* (Bunge) O. Kuntze, 138, 139, 184
– *flexuosum* (L.) O. Kuntze, 184
– *gobicum* Ik.-Gal., 168
– *tenellum* (Turcz.) O. Kuntze, 97, 98, 116
Linaria acutiloba Fisch. ex Reichenb., 115, 117, 151
Linum baicalense Juz., 108
Lonicera altaica Pall. ex DC., 55
Luzula changaica V. Novikov, 168
Lycium ruthenicum Murr., 128
Malva sylvestris L., 116
Medicago ruthenica (L.) Trautv., 95, 102, 115, 151, 152
Melandrium apricum (Turcz. ex Fisch. et Mey.) Rohrb., 138
– *viscosum* (L.) Celak., 201

Melilotus suaveolens Ledeb., 115
Micropeplis arachnoidea (Moq.) Bunge, 105, 115
Microstigma junatovii Grub., 168
Minuartia verna (L.) Hiern, 94
Moehringia lateriflora (L.) Fenzl, 89
Myricaria longifolia (Willd.) Ehrenb., 118
Nanophyton grubovii Pratov, 12
– *mongolicum* Pratov, 73, 97, 105, 168, 170
Neopallasia pectinata (Pall.) Poljak., 114, 115
Neottia camtschatea (L.) Reichenb. f., 169
Neottianthe cucullata (L.) Schlechter, 169
Nitraria sibirica Pall., 73, 141, 142
– *sphaerocarpa* Maxim., 13, 73, 97, 98, 99, 198, 199
Nuphar pumila (Timm) DC., 169
Nymphaea candida J. Presl, 167, 169
Orchis militaris L., 169
Orostachys fimbriata (Turcz.) Berger, 102
– *malacophylla* (Pall.) Fisch., 99, 102, 151
– *spinosa* (L.) C.A.Mey., 101, 107, 113, 122
Oxycoccus microcarpus Turcz. ex Rupr., 170
Oxytropis aciphylla Ledeb., 96, 97, 105, 115
– *alpina* Bunge, 81
– *bungei* Kom., 168
– *diantha* Bunge ex Maxim., 168
– *filiformis* DC., 197
– *fragilifolia* Ulzij., 168
– *glabra* (Lam.) DC., 115, 124, 176
– *gracillima* Bunge, 107
– *hailarensis* Kitag., 107
– *microphylla* (Pall.) DC.
– *myriophylla* (Pall.) DC., 125
– *nitens* Turcz., 197
– *oligantha* Bunge, 81, 82
– *salina* Vass., 120
– *selengensis* Bunge, 125
– *strobilacea* Bunge, 81, 82
– *trichophysa* Bunge, 82
Paeonia anomala L., 167, 169, 182
– *lactiflora* Pall., 167, 169, 182
Padus avium Mill., 118
Panzerina lanata (L.) Sojak, 115
Papaver baitagense R.Kam. et Gubanov, 168, 201
– *nudicaule* L., 108, 183
– *pseudotenellum* Grub., 168
– *rubro-aurantiacum* subsp. *changaicum* (R. Kam.) R. Kam. et Gubanov, 168
– *rubro-aurantiacum* subsp. *saichanense* (Grub.) R. Kam. et Gubanov, 168
Pedicularis achilleifolia Steph., 94
– *flava* Pall., 176
– *longiflora* J. Rudolph, 176
Peganum harmala L., 115, 167, 170, 176
– *nigellastrum* Bunge, 114, 115, 116, 175, 176
Phlojodicarpus sibiricus (Steph. ex Spreng.) K.-Pol., 109
Phragmites australis (Cav.) Trin. ex Steud., 26, 117, 118, 123, 124, 126, 127, 175, 198
Picea obovata Ledeb., 55, 57, 59, 66, 72, 196
Pinus pumila (Pall.) Regel, 70, 72
– *sibirica* Du Tour, 48, 55, 56, 57, 58, 59, 61, 62, 66, 67, 70, 72
– *sylvestris* L., 20, 92, 140, 174, 183
Plantago depressa Schlecht., 114, 115, 119, 121, 124, 125, 176
– *major* L., 5, 9, 45, 47, 56, 79, 80, 97, 118, 119, 121, 124, 131, 144, 162, 200
– *salsa* Pall., 120
Platanthera bifolia (L.) Rich., 169
Poa altaica Trin., 82
– *attenuata* Trin., 55, 82, 89, 95, 99, 102, 107, 108, 109, 113, 117, 120, 122, 151, 197
– *attenuata* subsp. *botryoides* (Trin. ex Griseb.) Tzvel., 20, 117, 120, 151
– *pratensis* subsp. *angustifolia* (L.) Arcang., 118, 120
– *sibirica* Roshev., 140, 150, 151
– *subfastigiata* Trin., 120
Poacynum pictum (Schrenk) Baill., 170
Polemonium chinense (Brand) Brand, 89
Polygala sibirica L., 138
Polygonatum humile Fisch. ex Maxim., 104
– *odoratum* (Mill.) Druce, 118, 124, 176, 182
Polygonum arenastrum Boreau, 116, 121, 124, 153
– *intramongolicum* A.J. Li, 168
Polypogon monspeliensis (L.) Desf., 114, 115
Populus diversifolia Schrenk, 107, 128, 140, 167, 169, 175, 181, 199
– *laurifolia* Ledeb., 118
– *pilosa* Rehd., 128, 167, 169
– *suaveolens* Fisch., 115, 118
– *tremula* L., 72, 92

Potaninia mongolica Maxim., 98, 99, 167, 169, 198
Potentilla acaulis L., 96, 102, 108, 109, 115, 151
– *anserina* L., 115, 119, 121, 124
– *bifurca* L., 113, 115, 116, 120, 151
– *chenteica* Sojak, 168
– *conferta* Bunge, 94, 95, 120
– *fruticosa* L., 24, 84, 101
– *hilbigii* Sojak, 168
– *ikonnikovii* Juz., 168
– *inopinata* Sojak, 168
– *laevipes* Sojak, 168
– *leucophylla* Pall., 82, 99, 102, 138, 168
– *mongolica* Krasch., 138, 141, 168, 197, 198
– *multifida* L., 89, 94, 95, 101
– *nivea* L., 81, 82
– *sericea* L., 89, 94, 101, 109, 115
– *supina* L., 94, 95
– *tanacetifolia* Willd. ex Schlecht., 36, 108, 151
– *verticillaris* Steph., 99, 102, 138
Prionotrichon kamelinii Botsch., 168
Prunus pedunculata (Pall.) Maxim., 104, 178
Psammochloa villosa (Trin.) Bor, 107, 168, 198
Psathyrostachys lanuginosa (Trin.) Nevski, 107
Psedosophora alopecuroides (L.) Sweet, 127
Ptilagrostis mongholica (Turcz. ex Trin.) Griseb., 81
– *pelliotii* (Danguy) Grub., 97
Ptilotrichum canescens (DC.) C.A. Mey., 105
– *tenuifolium* (Steph.) C.A. Mey., 138, 139
Puccinellia mongolica (Norlindh) Bubnova, 107
– *tenuiflora* (Griseb.) Scribn. et Merr., 96, 120, 122, 124, 125, 150
Pugionium dolabratum Maxim., 168
– *pterocarpum* Kom., 97, 98, 169
Pulsatilla ambigua (Turcz.) Juz., 183
– *bungeana* C.A.Mey., 113, 138, 183
– *multifida* (G. Pritz.) Juz., 89
– *turczaninovii* Kryl. et Serg., 138, 151, 183
Pyrethrum changaicum Krasch. ex Grub., 170
Ranunculus altaicus Laxm., 196
– *propinquus* C.A.Mey., 89, 120
– *repens* L., 120
– *reptans* L., 119
Reaumuria songarica (Pall.) Maxim., 13, 73, 98, 99, 116, 117, 124, 128, 149
Rhamnus parvifolia Bunge, 197
– *ussuriensis* J. Vassil., 170
Rhaponticum carthamoides (Willd.) Iljin, 167, 170
– *uniflorum* (L.) DC., 105, 185
Rheum nanum Siev., 97, 98
– *undulatum* L., 94, 95, 101, 201
Rhodiola quadrifida (Pall.) Fisch. et Mey., 81, 82, 167, 169
– *rosea* L., 167, 169
Rhododendron aureum Georgi, 170, 184
– *dauricum* L., 72
Ribes diacantha Pall., 119
– *nigrum* L., 178
– *pulchellum* Turcz., 197
– *rubrum* L., 105, 118, 178
Rosa acicularis Lindl., 36, 118
– *baitagensis* R.Kam. et Gubanov, 201
– *davurica* Pall., 183
– *laxa* Retz., 183
– *pimpinellifolia* L., 184, 201
Rumex crispus L., 121, 124
Sajanella monstrosa (Willd. ex Spreng.) Sojak, 167
Salicornia perennis Willd., 127
Salix arctica Pall., 81
– *bebbiana* Sarg., 118
– *berberifolia* Pall., 81
– *glauca* L., 57, 72
– *ledebouriana* Trautv., 118
– *microstachya* Turcz. ex Trautv., 107
– *rectijulis* Ledeb., 72
– *rorida* Laksch., 118
– *schwerinii* E. Wolf, 118
– *taraikensis* Kimura, 118
– *triandra* L., 118
Salsola arbuscula Pall., 29, 97, 98, 105
– *collina* Pall., 94, 95, 115, 116, 117, 124, 125, 126
– *laricifolia* Turcz. ex Litv., 97, 98, 117, 198
– *monoptera* Bunge, 117, 138, 139, 151
– *passerina* Bunge, 73, 98, 99, 114, 116, 128, 149, 198
– *paulsenii* Litv., 115
– *tragus* L., 115, 117, 124, 127

Sanguisorba officinalis L., 84, 88, 90, 101, 108
Saussurea amara (L.) DC., 115, 117, 120, 122, 124
– *ceterachifolia* Lipsch., 170
– *elongata* DC., 89
– *grubovii* Lipsch., 168
– *involucrata* (Kar. et Kir.) Sch.Bip., 170, 201
– *leucophylla* Schrenk, 82
– *salicifolia* (L.) DC., 176
Saxifraga hieracifolia Waldst. et Kit., 200
– *hirculus* L., 81
Scabiosa comosa Fisch. ex Roem. et Schult., 101, 108
Schizonepeta multifida (L.) Briq., 101, 138, 139
Schultzia crinita (Pall.) Spreng., 81, 94, 196
Scorzonera austriaca Willd., 138
– *capito* Maxim., 97, 104
– *radiata* Fisch., 90
Senecio nemorensis L., 118
Serratula centauroides L., 94, 95, 100, 102, 113, 138, 151
Setaria glauca (L.) Beauv., 115
– *viridis* (L.) Beauv., 115, 127, 151
Sibbaldianthe adpressa (Bunge) Juz., 198
– *sericea* Grub., 115
Silene altaica Pers., 196
– *jeniseensis* Willd., 101
– *repens* Patr., 94, 120
Smelowskia mongolica Kom., 168, 201
Sonchus arvensis L., 116, 121, 124
– *oleraceus* L., 121, 124
Sophora flavescens Soland., 167, 170
Sorbaria sorbifolia (L.) A.Br., 169
Sparganium glomeratum Laest. ex Beurl., 118
Sphaerophysa salsula (Pall.) DC., 117, 126, 127
Sphallerocarpus gracilis (Bess. ex Trev.) K.-Pol., 96
Spiraea aquilegifolia Pall., 197
– *media* Franz Schmidt, 36
– *pubescens* Turcz., 197
Stellaria dichotoma L., 107, 201
– *petraea* Bunge, 81
– *pulvinata* Grub., 82, 168
Stellera chamaejasme L., 84, 101, 108, 121, 124, 176
Stenosolenium saxatile (Pall.) Turcz., 167
Stevenia sergievskajae (Krasnob.) R.Kam. et Gubanov, 167
Stilpnolepis intricata (Franch.) Shih, 115
Stipa baicalensis Roshev., 94, 95, 101, 108, 109, 113
– *capillata* L., 19, 20, 21, 22, 23, 25
– *glareosa* P. Smirn., 96, 97, 98, 103, 105, 106, 114, 116, 141, 143, 146, 172, 199
– *gobica* Roshev., 143, 149, 154
– *grandis* P. Smirn., 100, 102, 111, 112, 113, 138, 197
– *klementzii* Roshev., 13
– *krylovii* Roshev., 20, 84, 94, 95, 100, 101, 102, 104, 105, 107, 108, 109, 110, 111, 112, 113, 116, 141, 151, 197, 198
– *orientalis* Trin., 172
– *pennata* L., 169
– *sibirica* (L.) Lam., 16, 18, 33, 36
Suaeda corniculata (C.A.Mey.) Bunge, 120, 124
– *heterophylla* (Kar. et Kir.) Bunge, 127
Swertia banzragczii Sancz., 168
– *obtusa* Ledeb., 196
Swida alba (L.) Opiz, 118
Sympegma regelii Bunge, 73, 97, 98, 99, 102, 106, 146, 149, 172
Synurus deltoides (Ait.) Nakai, 170
Tamarix elongata Ledeb., 170, 184
– *gracilis* Willd., 94, 170, 184
– *hispida* Willd., 117, 127, 141, 148, 149, 170, 184
– *karelinii* Bunge, 170
– *leptostachys* Bunge, 170
– *ramosissima* Ledeb., 13, 107, 117, 126, 127, 141, 170, 184, 199
Taraxacum bicorne Dahlst., 124
– *dissectum* (Ledeb.) Ledeb., 101, 125
– *leucanthum* (Ledeb.) Ledeb., 119, 121, 124
– *officinale* Wigg., 124
Thalictrum alpinum L., 81
– *baikalense* Turcz., 183
– *minus* L., 36, 84, 101
– *petaloideum* L., 101
– *simplex* L., 118, 120
– *squarrosum* Steph. ex Willd., 95, 104
Thermopsis lanceolata R.Br., 95, 96, 108, 112, 113, 115, 116, 117, 121, 124
– *mongolica* Czefr., 115

Thesium refractum C.A.Mey., 138, 139
Thymus dahuricus Serg., 115
– *gobicus* Tscherneva, 95, 168, 184, 185
– *pavlovii* Serg., 167, 168
Thypha angustifolia L., 118
– *laxmannii* Lepech., 118, 126
Tribulus terrestris L., 115, 176
Trifolium lupinaster L., 101
Triglochin maritimum L., 120, 122, 124
– *palustre* L., 122, 124
Trisetum sibiricum Rupr., 88, 89
Trollius asiaticus L., 89, 183, 201
Tugarinovia mongolica Iljin, 98
Tulipa uniflora (L.) Bess. ex Baker, 169
Ulmus macrocarpa Hance, 197
– *pumila* L., 46, 64, 70, 72, 87, 95, 104, 107, 119, 140, 175, 197, 199, 201
Urtica cannabina L., 91, 94, 95, 108, 115, 119, 121, 124, 126, 176
Vaccinium myrtillus L., 170
– *vitis-idaea* L., 72, 200
Valeriana alternifolia Ledeb., 36, 101, 120, 167, 170
– *dubia* Bunge, 201
– *saichanensis* Kom., 168
Veronica daurica Stev., 184
– *incana* L., 70, 101, 108, 109, 184, 185
– *laeta* Kar. et Kir., 151
– *pinnata* L., 109, 176
Viburnum mongolicum (Pall.) Rehd., 170
– *sargentii* Koehne, 170
Vicia cracca L., 88, 89
– *magalotropis* Ledeb., 88
– *multicaulis* Ledeb., 101
– *tsydenii* Malyschev, 167
– *unijuga* A.Br., 89
– *venosa* (Willd. ex Link) Maxim., 36, 88
Viola altaica Ker-Gawl., 184
– *biflora* L., 88, 89
– *macroceras* Bunge, 196
– *uniflora* L., 89
Waldheimia tridactylites Kar. et Kir., 81
Woodsia ilvensis (L.) R.Br., 196
Zygophyllum kaschgaricum Boriss., 167
– *neglectum* Grub., 168
– *pterocarpum* Bunge, 97, 98
– *xanthoxylon* (Bunge) Maxim., 13, 73, 98, 99, 106, 141, 142, 144, 146

Appendix 2. Modern surface pollen spectra from Mongolia used in this study. Pollen of individual taxa are expressed as percentage of terrestrial pollen sum, excluding aquatics.

Site NN	Latitude (°N)	Longitude (°E)	Elevation (m)	*Abies*	*Allium*	*Alnus*	Apiaceae	*Artemisia*	Astraceae undif.	*Betula*	Boraginaceae	Brassicaceae	Campanulaceae	Caryophyllaceae	Chenopodiaceae	Convolvulaceae	*Corylus*	Cyperaceae	Dipsacaceae	*Ephedra*	*Epilobium*	Ericales	Euphorbiaceae	Fabaceae	*Galium*	*Gentiana*	Geraniaceae	*Hippophae rhamn.*	*Kochia*	Lamiaceae
1	48.67	88.30	2083	0.5	0.0	0.0	0.0	24.9	0.0	14.6	0.0	6.1	0.5	1.4	13.1	0.0	0.0	0.0	0.0	16.9	0.0	0.0	0.0	0.0	0.0	0.0	0.0	0.0	0.0	0.0
2	48.57	88.60	2072	0.6	0.0	0.0	0.0	14.0	3.5	11.1	0.0	0.6	0.0	5.3	32.2	0.0	0.0	0.0	0.0	0.6	0.0	0.0	0.0	0.0	0.0	0.0	0.0	0.0	0.0	0.0
3	48.60	88.92	2493	0.0	0.4	0.0	0.0	4.9	0.4	8.4	0.0	4.0	0.0	8.9	21.3	0.0	0.0	2.7	0.0	0.0	0.0	0.0	0.0	10.2	0.0	0.0	0.0	0.0	0.0	2.7
4	48.33	88.88	2232	0.0	0.3	0.0	0.0	2.9	0.6	1.7	0.0	1.7	0.0	0.9	71.4	0.0	0.0	1.7	0.0	1.4	0.0	0.0	0.0	2.0	0.0	0.0	0.0	0.0	0.0	0.6
5	48.58	90.10	2079	0.0	0.0	0.0	0.4	28.6	1.4	9.4	0.0	0.0	0.0	0.2	13.5	0.0	0.0	1.4	0.0	1.8	0.0	0.0	0.0	0.0	0.0	0.0	0.0	0.0	0.0	0.0
6	48.28	90.67	2094	0.0	0.5	0.0	0.0	23.6	5.5	3.1	0.0	0.0	0.0	2.8	9.5	0.0	0.0	4.3	0.0	3.4	0.0	0.0	0.0	1.2	0.0	0.0	0.0	0.0	0.0	0.0
7	49.50	89.80	2097	0.0	0.0	0.0	0.0	22.0	3.3	0.5	0.0	0.0	0.0	0.9	31.1	0.0	0.0	9.2	0.0	0.9	0.0	0.0	0.0	0.0	0.0	0.0	0.0	0.0	0.0	0.0
8	49.50	90.60	1435	0.0	0.5	0.0	2.2	8.2	0.5	5.2	0.0	0.0	0.0	3.3	16.4	0.0	0.0	0.3	0.0	1.6	0.0	0.0	0.0	0.0	0.0	0.0	0.0	0.0	2.2	0.0
9	47.87	92.30	1157	0.0	0.0	0.0	0.0	20.5	19.7	1.0	0.0	2.3	0.0	2.6	28.1	0.0	0.0	0.8	0.0	3.3	0.0	0.0	0.0	1.5	0.0	0.0	0.0	0.0	0.0	0.0
10	47.95	93.12	1132	0.0	0.0	0.0	0.0	8.6	0.4	0.4	0.0	0.0	0.0	0.4	45.2	0.0	0.0	0.6	0.0	0.0	0.0	0.0	0.0	0.2	0.0	0.0	0.0	0.0	0.0	0.0
11	48.03	93.28	1132	0.0	6.6	0.0	0.0	33.1	3.3	1.0	0.0	0.3	0.0	6.6	10.6	0.0	0.0	0.7	0.0	0.0	0.0	0.0	0.0	0.0	0.0	0.0	0.0	0.0	0.0	0.0
12	48.92	93.57	1030	0.0	0.0	0.0	0.0	15.4	0.2	1.5	0.0	1.2	0.2	0.0	23.0	0.0	0.0	1.3	0.0	32.6	0.0	0.0	0.0	0.0	0.0	0.0	0.0	0.0	0.0	0.0
13	48.47	95.18	1491	0.0	0.0	0.0	0.5	36.0	0.9	0.0	0.0	0.0	0.0	9.0	18.0	0.0	0.0	9.0	0.0	9.0	0.0	0.0	0.0	0.0	0.0	0.0	0.0	0.0	0.0	0.0
14	49.98	94.03	931	0.0	0.0	0.0	0.4	4.7	0.4	3.4	0.0	0.9	0.0	0.0	62.1	0.0	0.0	1.7	0.0	1.7	0.0	0.0	0.0	0.9	0.0	0.0	0.0	0.0	1.3	0.0
15	51.27	99.40	1538	0.0	0.3	0.0	0.0	5.7	1.6	3.5	0.0	1.9	0.0	1.6	2.2	0.0	0.0	4.5	0.0	0.3	0.0	0.0	0.0	8.3	0.0	0.0	0.0	0.0	0.0	0.6
16	51.20	99.45	1539	0.0	0.0	0.0	0.0	2.1	2.1	2.1	0.0	2.1	0.0	0.0	4.2	0.0	0.0	2.1	0.0	0.0	0.0	0.0	0.0	3.5	0.0	0.0	0.0	0.0	0.0	0.0
17	51.38	99.37	1538	0.0	0.0	0.0	0.0	1.2	3.3	8.3	0.0	2.5	0.0	0.0	3.7	0.0	0.0	11.2	0.0	0.0	0.0	0.0	0.0	0.8	0.0	0.0	0.0	0.0	0.0	0.0
18	50.40	100.15	1650	0.0	0.0	0.0	1.0	7.2	1.9	7.7	0.0	2.4	0.0	1.4	2.9	0.0	0.0	0.0	0.0	0.0	0.0	1.0	0.0	0.0	0.0	0.0	0.0	0.0	0.0	0.0
19	50.40	100.15	1662	0.0	0.0	0.0	0.0	15.4	0.0	0.0	0.0	0.0	0.0	5.1	2.6	0.0	0.0	0.0	0.0	0.0	0.0	0.0	0.0	0.0	0.0	0.0	0.0	0.0	0.0	0.0
20	49.88	102.57	1000	0.0	0.0	0.0	0.0	4.5	3.0	14.9	0.0	11.2	0.0	0.0	1.5	0.0	0.0	0.7	0.0	0.0	0.0	0.7	0.0	3.7	0.0	0.0	0.0	0.0	0.0	0.0
21	48.82	102.93	1440	0.0	0.0	0.0	0.0	3.1	4.9	5.6	0.0	4.3	0.0	4.3	3.1	0.0	0.0	4.9	0.0	0.0	0.0	0.0	0.0	1.9	0.0	0.0	0.0	0.0	0.0	0.0
22	48.80	103.32	1540	0.0	0.0	0.0	1.9	17.0	0.0	18.9	0.0	0.0	0.0	5.7	0.0	0.0	0.0	3.8	0.0	0.0	0.0	0.0	0.0	0.0	0.0	0.0	0.0	0.0	0.0	0.0
23	48.18	99.55	2060	0.0	0.0	0.0	0.0	55.1	2.1	1.3	0.0	0.0	0.0	2.5	4.2	0.0	0.0	4.2	0.0	1.7	0.0	0.8	0.0	0.0	0.0	0.0	0.0	0.0	0.0	0.0
24	48.13	99.83	2060	0.0	0.0	0.0	0.0	5.2	6.7	1.1	0.0	0.0	0.0	8.1	10.7	0.0	0.0	3.3	0.0	29.3	0.0	0.0	0.0	0.0	0.0	0.0	0.0	0.0	0.0	0.0
25	48.18	99.62	2060	0.0	0.0	0.0	0.2	29.2	17.2	0.5	0.0	2.0	0.0	10.0	21.5	0.0	0.0	1.1	0.0	0.8	0.0	0.0	0.0	0.5	0.0	0.0	0.0	0.0	0.0	0.0
26	47.38	101.85	1700	0.0	0.0	0.0	0.3	27.8	5.0	7.1	0.0	0.0	0.0	6.2	3.8	0.0	0.0	1.5	0.0	0.6	0.0	0.0	0.0	0.0	0.0	0.0	0.0	0.0	0.0	0.0
27	47.33	101.93	1640	0.0	0.0	0.0	4.1	14.4	2.1	2.1	0.0	1.0	0.0	3.1	3.1	0.0	0.0	0.0	0.0	0.0	0.0	0.0	0.0	6.2	0.0	0.0	0.0	0.0	0.0	0.0
28	47.45	101.92	1500	0.0	0.0	0.0	0.0	31.3	9.2	5.0	0.0	8.4	0.0	1.5	10.3	0.0	0.0	0.4	0.0	0.8	0.0	0.0	0.0	2.7	0.0	0.0	0.0	0.0	0.0	0.0
29	47.42	101.92	1600	0.0	4.1	0.0	1.0	14.4	1.0	2.1	0.0	13.8	0.0	0.0	3.6	0.0	0.0	0.5	0.0	2.6	0.0	0.0	0.0	6.7	0.0	0.0	0.5	0.0	0.0	2.6
30	47.35	101.87	1650	0.0	3.5	0.0	2.1	9.8	0.0	2.8	0.0	0.0	0.0	1.4	0.0	0.0	0.0	9.1	0.0	2.1	2.8	0.0	0.0	0.0	0.0	0.0	0.0	0.0	0.0	0.0
31	46.58	105.67	1500	0.0	0.0	0.0	0.0	58.0	1.8	1.8	0.0	0.0	0.0	1.8	19.5	0.0	0.0	3.5	0.0	0.0	0.0	0.0	0.0	0.0	0.0	0.0	0.0	0.0	0.4	0.0
32	46.75	105.95	1316	0.0	0.7	0.0	2.4	31.3	1.0	2.4	0.0	1.9	0.0	1.7	10.0	0.0	0.0	4.8	0.0	0.2	0.0	0.0	0.0	0.0	0.0	0.0	0.0	0.0	0.0	0.0
33	46.90	105.92	1500	0.0	0.9	0.0	0.0	27.4	0.0	7.5	0.0	0.0	0.0	2.8	11.3	0.0	0.0	0.0	0.0	0.0	0.0	0.0	0.0	29.2	0.0	0.0	0.0	0.0	0.0	0.0
34	46.70	105.47	1550	0.0	0.0	0.0	0.0	31.6	2.3	2.7	0.0	16.6	0.0	0.7	2.3	0.0	0.0	9.3	0.0	0.0	0.0	0.0	0.0	1.0	0.0	0.0	0.0	0.0	0.0	7.0

Appendix 2. Continued.

Site NN	*Larix*	Liliaceae	*Linaria*	*Lonicera*	*Pedicularis*	*Picea*	*Pinus sylvestris*	*Pinus Haploxylon*	*Pinus undif.*	*Plantago*	Plumbaginaceae	Poaceae	Polygonaceae	*Potentilla*	Primulaceae	Ranunculaceae	Rosaceae	Rubiaceae	Rutaceae	*Salix*	*Salsola*	Saxifragaceae	*Scabiosa*	Scrophulariaceae	*Stellera*	*Ulmus*	Valerianaceae	Zygophyllaceae
1	0.0	0.0	0.0	0.0	0.0	4.7	0.0	8.9	0.0	0.0	0.0	4.7	3.3	0.0	0.0	0.0	0.5	0.0	0.0	0.0	0.0	0.0	0.0	0.0	0.0	0.0	0.0	0.0
2	2.9	0.0	0.0	0.0	0.0	4.7	0.0	9.4	0.0	0.0	0.0	11.7	0.6	0.0	0.0	2.9	0.0	0.0	0.0	0.0	0.0	0.0	0.0	0.0	0.0	0.0	0.0	0.0
3	0.9	0.0	0.0	0.0	0.0	2.7	1.3	2.2	0.0	0.0	0.0	19.1	0.0	0.0	0.0	4.0	5.8	0.0	0.0	0.0	0.0	0.0	0.0	0.0	0.0	0.0	0.0	0.0
4	0.0	0.0	0.0	0.0	0.0	0.3	0.6	0.0	0.0	0.0	0.0	8.0	1.1	0.0	0.0	4.9	0.0	0.0	0.0	0.0	0.0	0.0	0.0	0.0	0.0	0.0	0.0	0.0
5	0.0	0.0	0.0	0.0	0.0	2.9	0.0	0.0	28.6	0.0	0.0	7.0	0.0	0.0	0.0	3.9	0.0	0.0	0.8	0.0	0.0	0.0	0.0	0.0	0.0	0.0	0.0	0.0
6	1.1	0.0	0.0	0.0	0.0	3.5	6.1	2.1	0.0	0.0	0.0	31.3	1.1	0.0	0.0	0.9	0.0	0.0	0.0	0.0	0.0	0.0	0.0	0.0	0.0	0.0	0.0	0.0
7	0.5	0.0	0.0	0.0	0.0	1.5	0.0	0.0	22.7	0.0	0.0	7.3	0.0	0.0	0.0	0.0	0.0	0.0	0.0	0.0	0.0	0.0	0.0	0.0	0.0	0.0	0.0	0.0
8	0.5	0.0	0.0	0.0	0.0	0.8	0.0	0.0	39.5	0.0	0.0	2.2	0.0	0.0	0.0	16.4	0.0	0.0	0.0	0.0	0.0	0.0	0.0	0.0	0.0	0.0	0.0	0.0
9	0.3	0.0	0.0	0.0	0.0	0.5	6.4	0.5	0.0	0.0	0.0	4.6	4.3	0.0	0.0	3.6	0.0	0.0	0.0	0.0	0.0	0.0	0.0	0.0	0.0	0.0	0.0	0.0
10	0.0	0.0	0.0	0.0	0.0	0.0	0.0	0.0	3.9	0.0	0.0	0.4	31.4	0.0	0.0	8.3	0.2	0.0	0.0	0.0	0.0	0.0	0.0	0.0	0.0	0.0	0.0	0.0
11	3.0	0.0	0.0	0.0	0.0	0.0	0.0	0.0	1.3	0.0	0.0	18.2	0.0	0.0	0.0	14.6	0.7	0.0	0.0	0.0	0.0	0.0	0.0	0.0	0.0	0.0	0.0	0.0
12	0.0	0.0	0.0	0.0	0.0	0.2	0.0	0.0	19.8	0.0	0.0	0.6	0.6	0.0	0.0	2.7	0.0	0.0	0.8	0.0	0.0	0.0	0.0	0.0	0.0	0.0	0.0	0.0
13	0.0	0.0	0.0	0.0	0.0	0.0	0.0	0.0	1.8	0.0	0.0	13.5	0.0	0.0	0.0	1.8	0.0	0.5	0.0	0.0	0.0	0.0	0.0	0.0	0.0	0.0	0.0	0.0
14	0.0	0.0	0.0	0.0	0.0	1.3	5.2	2.6	0.0	0.0	0.0	13.4	0.0	0.0	0.0	0.0	0.0	0.0	0.0	0.0	0.0	0.0	0.0	0.0	0.0	0.0	0.0	0.0
15	13.4	0.0	0.0	0.0	0.0	1.9	7.0	26.8	0.0	0.0	0.0	16.2	1.3	0.0	0.0	0.6	2.2	0.0	0.0	0.0	0.0	0.0	0.0	0.0	0.0	0.0	0.0	0.0
16	2.8	0.0	0.0	0.0	0.0	4.9	9.2	35.2	0.0	0.0	0.0	28.9	0.7	0.0	0.0	0.0	0.0	0.0	0.0	0.0	0.0	0.0	0.0	0.0	0.0	0.0	0.0	0.0
17	5.4	0.0	0.0	0.0	0.0	6.2	27.3	19.8	0.0	0.0	0.0	9.5	0.0	0.0	0.0	0.0	0.8	0.0	0.0	0.0	0.0	0.0	0.0	0.0	0.0	0.0	0.0	0.0
18	19.2	0.0	0.0	0.0	0.0	1.9	13.9	19.7	0.0	0.0	0.0	12.0	4.8	0.0	1.4	1.4	0.0	0.0	0.0	0.0	0.0	0.0	0.0	0.0	0.0	0.0	0.0	0.0
19	10.3	0.0	0.0	0.0	0.0	0.0	20.5	33.3	0.0	0.0	0.0	12.8	0.0	0.0	0.0	0.0	0.0	0.0	0.0	0.0	0.0	0.0	0.0	0.0	0.0	0.0	0.0	0.0
20	6.0	0.0	0.0	0.0	0.0	3.0	13.4	18.7	0.0	0.0	0.0	13.4	2.2	0.0	0.0	0.0	3.0	0.0	0.0	0.0	0.0	0.0	0.0	0.0	0.0	0.0	0.0	0.0
21	6.8	0.0	0.0	0.0	0.0	3.7	24.7	14.8	0.0	0.0	0.0	17.9	0.0	0.0	0.0	0.0	0.0	0.0	0.0	0.0	0.0	0.0	0.0	0.0	0.0	0.0	0.0	0.0
22	26.4	0.0	0.0	0.0	0.0	0.0	13.2	0.0	0.0	0.0	0.0	9.4	0.0	0.0	0.0	3.8	0.0	0.0	0.0	0.0	0.0	0.0	0.0	0.0	0.0	0.0	0.0	0.0
23	4.7	0.0	0.0	0.0	0.0	0.4	0.8	4.7	0.0	0.0	0.0	12.7	0.0	0.0	0.0	4.2	0.4	0.0	0.0	0.0	0.0	0.0	0.0	0.0	0.0	0.0	0.0	0.0
24	4.4	0.0	0.0	0.0	0.0	0.4	11.5	18.5	0.0	0.0	0.0	0.0	0.0	0.0	0.0	0.0	0.0	0.0	0.0	0.7	0.0	0.0	0.0	0.0	0.0	0.0	0.0	0.0
25	0.5	0.0	0.0	0.0	0.0	0.0	6.3	9.1	0.0	0.0	0.0	1.1	0.0	0.0	0.0	0.0	0.3	0.0	0.0	0.0	0.0	0.0	0.0	0.0	0.0	0.0	0.0	0.0
26	0.6	0.0	0.0	0.0	0.0	0.0	3.3	14.8	0.0	0.0	0.0	20.1	0.3	0.0	0.3	0.9	0.0	0.0	0.0	0.0	0.0	0.0	0.0	0.0	7.4	0.0	0.0	0.0
27	21.6	0.0	0.0	0.0	0.0	0.0	0.0	13.4	0.0	0.0	0.0	18.6	0.0	0.0	3.1	4.1	1.0	0.0	0.0	0.0	0.0	0.0	0.0	2.1	0.0	0.0	0.0	0.0
28	0.8	0.0	0.0	0.0	0.0	0.0	8.4	1.9	0.0	0.0	0.0	1.9	3.1	0.0	0.0	4.6	0.8	0.0	0.0	0.0	0.0	0.0	9.2	0.0	0.0	0.0	0.0	0.0
29	0.5	0.0	0.0	0.0	0.0	0.0	0.5	26.7	0.0	0.0	0.0	8.2	0.0	0.0	0.0	0.0	2.6	0.0	0.0	0.0	0.0	0.0	0.0	0.0	8.7	0.0	0.0	0.0
30	0.0	0.0	0.0	0.0	0.0	0.0	6.3	16.8	0.0	0.0	0.0	30.1	0.7	0.0	0.0	3.5	0.0	0.0	0.0	0.0	0.0	0.0	0.0	0.0	9.1	0.0	0.0	0.0
31	0.0	0.0	0.0	0.0	0.0	0.0	3.5	1.9	0.0	0.0	0.0	7.0	0.0	0.0	0.0	0.7	0.0	0.0	0.0	0.0	0.0	0.0	0.0	0.0	0.2	0.0	0.0	0.0
32	0.0	0.0	0.0	0.0	0.0	0.0	1.4	1.7	0.0	0.0	0.0	39.4	0.0	0.0	0.0	1.0	0.2	0.0	0.0	0.0	0.0	0.0	0.0	0.0	0.0	0.0	0.0	0.0
33	0.0	0.0	0.0	0.0	0.0	0.0	2.8	0.0	0.0	0.0	0.0	9.4	0.0	0.0	0.0	8.5	0.0	0.0	0.0	0.0	0.0	0.0	0.0	0.0	0.0	0.0	0.0	0.0
34	0.0	0.0	0.0	0.0	0.0	0.0	0.0	8.0	0.0	0.0	0.0	12.6	0.3	0.0	0.0	4.3	1.3	0.0	0.0	0.0	0.0	0.0	0.0	0.0	0.0	0.0	0.0	0.0

Appendix 2. Continued

Site NN	Latitude (°N)	Longitude (°E)	Elevation (m)	*Abies*	*Allium*	*Alnus*	Apiaceae	*Artemisia*	Astraceae undif.	*Betula*	Boraginaceae	Brassicaceae	Campanulaceae	Caryophyllaceae	Chenopodiaceae	Convolvulaceae	*Corylus*	Cyperaceae	Dipsacaceae	*Ephedra*	*Epilobium*	Ericales	Euphorbiaceae	Fabaceae	*Galium*	*Gentiana*	Geraniaceae	*Hippophae rhamn.*	*Kochia*	Lamiaceae
35	45.95	106.28	1200	0.0	0.5	0.0	0.0	47.4	0.5	0.5	0.0	12.9	0.0	2.1	14.9	0.0	0.0	0.0	0.0	0.5	0.0	0.0	0.0	1.0	0.0	0.0	0.0	0.0	0.0	2.1
36	45.77	106.48	1350	0.0	0.7	0.0	0.0	20.7	2.7	12.0	0.0	8.0	0.0	2.0	24.0	0.0	0.0	1.3	0.0	2.0	0.0	0.0	0.0	5.3	0.0	0.0	0.0	0.0	0.0	4.7
37	45.77	106.73	1450	0.0	3.1	0.0	0.0	12.2	0.3	3.1	0.0	0.8	0.0	0.8	50.8	0.0	0.0	0.6	0.0	1.1	0.0	0.0	0.0	20.0	0.0	0.0	0.0	0.0	0.0	0.8
38	47.68	102.05	1400	0.0	1.0	0.0	0.3	29.4	11.2	2.9	0.0	9.9	0.0	1.0	6.1	0.0	0.0	0.3	0.0	1.0	0.0	0.0	0.0	0.0	0.0	0.0	0.0	0.0	0.0	8.6
39	47.72	101.85	1400	0.0	0.0	0.0	0.0	21.7	0.0	4.5	0.0	0.0	0.0	0.9	29.4	0.0	0.0	1.4	0.0	0.5	0.0	0.0	0.0	3.6	0.0	0.0	0.0	0.0	0.0	7.2
40	47.80	104.75	1050	0.0	1.2	0.0	0.4	24.7	1.6	6.8	0.0	16.7	0.0	1.2	10.8	0.0	0.0	0.0	0.0	0.4	0.0	0.0	0.0	0.0	0.0	0.0	0.0	0.0	0.0	1.2
41	47.87	105.43	1000	0.0	0.0	0.0	0.0	19.0	3.0	5.7	0.0	20.7	0.0	0.0	28.0	0.0	0.0	0.7	0.0	0.0	0.0	0.0	0.0	0.0	0.0	0.0	0.0	0.0	0.0	4.7
42	47.95	106.08	1195	0.0	0.3	0.0	0.3	18.0	3.1	10.2	0.0	10.8	0.0	1.7	5.1	0.0	0.0	0.0	0.0	0.3	0.0	0.0	0.0	0.7	0.0	0.0	0.0	0.0	0.0	1.0
43	48.35	106.57	1200	0.0	0.7	0.0	0.0	28.4	2.8	11.7	0.0	9.9	0.0	0.0	3.9	0.0	0.0	1.4	0.0	0.0	0.0	0.0	0.0	4.6	0.0	0.0	0.0	0.0	0.0	1.1
44	48.45	106.62	1400	0.0	0.0	0.0	0.0	14.6	0.5	61.7	0.0	2.9	0.0	0.0	1.0	0.0	0.0	0.0	0.0	0.0	0.0	0.0	0.0	1.9	0.0	0.0	0.0	0.0	0.0	0.0
45	48.90	106.28	780	0.0	0.0	0.0	0.8	16.7	6.8	9.9	0.0	5.7	0.0	0.0	3.4	0.0	0.0	0.0	0.0	0.4	0.0	0.0	0.0	1.5	0.0	0.0	0.0	0.0	0.0	4.6
46	49.02	106.13	800	0.0	1.4	0.0	0.0	34.4	2.1	8.1	0.0	2.5	0.0	0.0	22.1	0.0	0.0	0.0	0.0	0.0	0.0	0.0	0.0	1.1	0.0	0.0	0.0	0.0	0.0	1.1
47	49.37	105.72	750	0.0	0.3	0.0	0.0	17.2	0.8	9.0	0.0	16.3	0.0	0.0	14.7	0.0	0.0	0.3	0.0	0.0	0.0	0.0	0.0	0.0	0.0	0.0	0.0	0.0	0.0	0.3
48	49.72	106.27	600	0.0	0.0	0.0	0.0	28.9	1.2	6.3	0.0	19.8	0.0	0.0	7.1	0.0	0.0	0.0	0.0	0.0	0.0	0.0	0.0	0.0	0.0	0.0	0.0	0.0	0.0	2.0
49	50.20	106.08	560	0.0	0.0	0.0	0.0	13.6	1.9	4.5	0.0	3.0	0.0	1.1	5.0	0.0	0.0	0.9	0.0	0.0	0.0	0.4	0.0	0.2	0.0	0.0	0.0	0.0	0.0	1.9
50	50.27	106.17	550	0.0	0.0	0.0	0.0	2.7	1.6	3.8	0.0	9.9	0.0	0.0	3.3	0.0	0.0	0.5	0.0	0.0	0.0	0.0	0.0	0.0	0.0	0.0	0.0	0.0	0.0	0.5
51	50.03	105.95	600	0.0	0.0	0.0	0.0	15.1	0.5	3.4	0.0	8.3	0.0	0.0	2.0	0.0	0.0	0.0	0.0	0.0	0.0	0.0	0.0	0.0	0.0	0.0	0.0	0.0	0.0	0.0
52	49.52	106.92	570	0.0	2.9	0.0	0.0	6.4	0.2	8.8	0.0	1.7	0.0	0.0	1.7	0.0	0.0	1.5	0.0	0.0	0.0	0.0	0.0	0.0	0.0	0.0	0.0	0.0	0.0	0.2
53	50.25	106.62	600	0.0	1.2	0.0	0.0	4.1	4.7	3.5	0.0	4.7	0.0	0.6	4.1	0.0	0.0	0.6	0.0	0.0	0.0	0.0	0.0	0.0	0.0	0.0	0.0	0.0	0.0	0.0
54	47.80	106.87	1850	0.0	1.0	0.0	0.0	13.0	1.0	2.9	0.0	1.9	0.0	1.0	1.0	0.0	0.0	0.0	0.0	0.0	0.0	0.0	0.0	0.0	0.0	0.0	0.0	0.0	0.0	0.0
55	48.07	106.35	1800	0.0	0.0	0.0	0.0	1.6	0.0	3.7	0.0	6.9	0.0	0.0	0.5	0.0	0.0	0.0	0.0	0.0	0.0	0.0	0.0	0.0	0.0	0.0	0.0	0.0	0.0	0.0
56	48.13	109.40	1326	0.0	0.3	0.0	0.0	18.3	1.3	7.8	0.0	3.3	0.0	2.9	4.6	0.0	0.0	0.7	0.0	0.3	0.0	0.0	0.0	0.7	0.0	0.0	0.0	0.0	0.0	0.0
57	48.35	108.58	1600	0.0	0.0	0.0	0.0	2.9	0.0	10.6	0.0	6.3	0.0	0.5	1.9	0.0	0.0	9.7	0.0	0.0	0.0	0.0	0.0	1.0	0.5	0.0	0.0	0.0	0.0	0.0
58	48.32	108.67	1400	0.0	1.7	0.0	0.3	7.0	3.4	3.9	0.0	3.4	0.0	1.7	1.4	0.0	0.0	1.1	0.0	0.3	0.0	0.0	0.0	1.1	0.0	0.0	0.0	0.0	0.0	0.3
59	47.40	109.92	1100	0.0	0.8	0.0	0.0	25.3	0.8	9.4	0.0	15.1	0.0	0.0	10.2	0.0	0.0	1.9	0.0	0.4	0.0	0.0	0.0	0.4	0.0	0.0	0.0	0.0	0.0	0.0
60	48.22	114.12	800	0.0	11.1	0.0	0.3	16.9	3.0	8.8	0.0	4.7	0.0	7.8	12.2	0.0	0.0	0.7	0.0	1.7	0.0	0.0	0.0	0.3	0.0	0.0	0.0	0.0	0.0	2.0
61	47.22	116.88	660	0.0	1.8	0.0	0.0	32.4	2.1	7.4	0.0	4.4	0.0	0.0	19.2	0.0	0.0	1.2	0.0	0.9	0.0	0.0	0.0	0.3	0.0	0.0	0.0	0.0	0.0	0.0
62	47.67	117.75	600	0.0	2.9	0.0	0.2	5.7	1.2	7.2	0.0	5.3	0.0	0.6	4.3	0.0	0.0	9.0	0.0	0.2	0.0	0.0	0.0	0.0	0.0	0.0	0.0	0.0	0.0	0.2
63	46.60	109.33	1300	0.0	0.0	0.0	0.0	36.9	0.0	1.5	0.0	15.0	0.0	1.8	22.3	0.0	0.0	0.4	0.0	0.0	0.0	0.0	0.0	1.1	0.0	0.0	0.0	0.0	0.0	0.4
64	47.53	118.88	700	0.0	0.3	0.0	0.0	28.7	4.8	6.6	0.0	9.0	0.0	4.8	5.5	0.0	0.0	0.0	0.0	0.3	0.0	0.0	0.0	0.7	0.0	1.7	0.0	0.0	0.0	5.2
65	46.72	119.77	1050	0.0	1.9	0.0	0.0	24.4	3.8	3.8	0.0	9.1	1.3	3.1	4.4	0.0	0.0	4.7	0.0	0.6	0.0	0.0	0.0	3.4	0.0	0.0	0.0	0.0	0.0	0.6
66	47.48	118.70	710	0.0	2.9	0.0	0.0	18.7	3.9	5.3	0.0	7.1	0.0	2.4	22.1	0.0	0.0	2.4	0.0	2.9	0.0	0.0	0.0	2.1	0.0	0.0	0.0	0.0	0.0	3.4
67	46.33	108.35	1160	0.0	1.3	0.0	0.0	29.6	4.6	5.9	0.0	3.9	0.0	0.7	32.9	0.0	0.0	0.0	0.0	2.0	0.0	0.0	0.0	2.0	0.0	0.0	0.0	0.0	0.0	0.0
68	45.00	110.22	900	0.0	1.7	0.0	0.0	37.5	0.0	2.0	0.0	4.0	0.0	0.9	33.5	0.0	0.0	0.0	0.0	0.3	0.0	0.0	0.0	0.0	0.0	0.0	0.0	0.0	0.0	0.6

Appendix 2. Continued.

Site NN	*Larix*	Liliaceae	*Linaria*	*Lonicera*	*Pedicularis*	*Picea*	*Pinus sylvestris*	*Pinus Haploxylon*	*Pinus undif.*	*Plantago*	Plumbaginaceae	Poaceae	Polygonaceae	*Potentilla*	Primulaceae	Ranunculaceae	Rosaceae	Rubiaceae	Rutaceae	*Salix*	*Salsola*	Saxifragaceae	*Scabiosa*	Scrophulariaceae	*Stellera*	*Ulmus*	Valerianaceae	Zygophyllaceae
35	2.1	0.0	0.0	0.0	0.0	0.5	3.1	5.7	0.0	0.0	0.0	3.6	0.0	0.0	0.0	0.0	2.6	0.0	0.0	0.0	0.0	0.0	0.0	0.0	0.0	0.0	0.0	0.0
36	0.0	0.0	0.0	0.0	0.0	0.0	1.3	4.0	0.0	0.0	0.0	9.3	0.0	0.0	0.0	2.0	0.0	0.0	0.0	0.0	0.0	0.0	0.0	0.0	0.0	0.0	0.0	0.0
37	0.0	0.0	0.0	0.0	0.0	0.0	0.0	2.5	0.0	0.0	0.0	3.3	0.0	0.0	0.0	0.0	0.6	0.0	0.0	0.0	0.0	0.0	0.0	0.0	0.0	0.0	0.0	0.0
38	0.0	0.0	0.0	0.0	0.0	0.0	7.0	19.8	0.0	0.0	0.0	0.6	0.0	0.0	0.0	0.0	1.0	0.0	0.0	0.0	0.0	0.0	0.0	0.0	0.0	0.0	0.0	0.0
39	0.0	0.0	0.0	0.0	0.0	0.0	0.9	9.0	0.0	0.0	0.0	0.9	0.0	0.0	0.0	0.0	19.9	0.0	0.0	0.0	0.0	0.0	0.0	0.0	0.0	0.0	0.0	0.0
40	0.0	0.0	0.0	0.0	0.0	0.0	11.2	19.9	0.0	0.0	0.0	2.4	0.0	0.0	0.0	0.4	1.2	0.0	0.0	0.0	0.0	0.0	0.0	0.0	0.0	0.0	0.0	0.0
41	0.0	0.0	0.0	0.0	0.0	0.0	1.0	12.3	0.0	0.0	0.0	2.7	0.0	0.0	0.0	1.0	1.3	0.0	0.0	0.0	0.0	0.0	0.0	0.0	0.0	0.0	0.0	0.0
42	0.0	0.0	0.0	0.0	0.0	0.0	8.5	31.5	0.0	0.0	0.0	4.1	0.0	2.7	0.0	1.7	0.0	0.0	0.0	0.0	0.0	0.0	0.0	0.0	0.0	0.0	0.0	0.0
43	3.2	0.0	0.0	0.0	0.0	0.0	1.8	17.0	0.0	0.0	0.0	13.1	0.0	0.0	0.0	0.0	0.0	0.0	0.0	0.0	0.0	0.0	0.0	0.0	0.4	0.0	0.0	0.0
44	1.0	0.0	0.0	0.0	0.0	0.0	0.0	9.7	0.0	0.0	0.0	3.9	0.5	0.0	0.0	0.5	1.5	0.0	0.0	0.0	0.0	0.0	0.0	0.0	0.5	0.0	0.0	0.0
45	1.1	0.0	0.0	0.0	0.0	0.0	0.8	23.6	0.0	0.0	0.0	12.5	0.4	11.0	0.0	0.8	0.0	0.0	0.0	0.0	0.0	0.0	0.0	0.0	0.0	0.0	0.0	0.0
46	0.7	0.0	0.0	0.0	0.0	0.0	0.4	14.0	0.0	0.0	0.0	9.5	0.4	0.0	0.0	0.0	2.1	0.0	0.0	0.0	0.0	0.0	0.0	0.0	0.4	0.0	0.0	0.0
47	0.8	0.0	0.0	0.0	0.0	0.0	0.8	28.6	0.0	0.0	0.0	9.0	0.0	0.0	0.0	1.6	0.0	0.0	0.0	0.0	0.0	0.0	0.3	0.0	0.0	0.0	0.0	0.0
48	0.8	0.0	0.0	0.0	0.0	0.0	2.0	27.7	0.0	0.0	0.0	4.0	0.0	0.0	0.0	0.4	0.0	0.0	0.0	0.0	0.0	0.0	0.0	0.0	0.0	0.0	0.0	0.0
49	0.0	0.0	0.0	0.0	0.0	0.0	50.0	0.0	0.0	0.0	0.0	10.6	6.1	0.0	0.0	0.2	0.4	0.0	0.0	0.0	0.0	0.0	0.0	0.0	0.0	0.0	0.0	0.0
50	0.5	0.0	0.0	0.0	0.0	0.0	73.1	0.5	0.0	0.0	0.0	3.3	0.0	0.0	0.0	0.0	0.0	0.0	0.0	0.0	0.0	0.0	0.0	0.0	0.0	0.0	0.0	0.0
51	1.0	0.0	0.0	0.0	0.0	0.0	67.3	0.5	0.0	0.0	0.0	1.0	0.0	0.0	0.0	1.0	0.0	0.0	0.0	0.0	0.0	0.0	0.0	0.0	0.0	0.0	0.0	0.0
52	1.2	0.0	0.0	0.0	0.0	0.7	69.5	0.0	0.0	0.0	0.0	3.9	0.2	0.0	0.0	0.0	0.2	0.0	0.0	0.2	0.0	0.0	0.0	0.0	0.0	0.2	0.0	0.0
53	1.7	0.0	0.0	0.0	0.0	0.0	65.1	0.0	0.0	0.0	0.0	9.9	0.0	0.0	0.0	0.0	0.0	0.0	0.0	0.0	0.0	0.0	0.0	0.0	0.0	0.0	0.0	0.0
54	1.0	0.0	0.0	0.0	0.0	58.0	1.9	8.7	0.0	0.0	0.0	4.3	0.0	0.0	0.0	2.9	1.4	0.0	0.0	0.0	0.0	0.0	0.0	0.0	0.0	0.0	0.0	0.0
55	1.1	0.0	0.0	0.0	0.0	0.0	1.1	85.1	0.0	0.0	0.0	0.0	0.0	0.0	0.0	0.0	0.0	0.0	0.0	0.0	0.0	0.0	0.0	0.0	0.0	0.0	0.0	0.0
56	2.6	0.0	0.0	0.0	0.0	1.0	1.3	45.8	0.0	0.0	0.0	4.2	0.0	0.0	0.0	1.0	1.3	0.0	0.0	0.0	0.0	0.0	0.0	0.0	2.6	0.0	0.0	0.0
57	3.4	0.0	0.0	0.0	0.0	0.0	30.0	15.9	0.0	0.0	0.0	11.1	0.0	0.0	0.0	1.9	1.0	0.0	0.0	0.0	0.0	0.0	0.0	0.0	3.4	0.0	0.0	0.0
58	0.0	0.0	0.0	0.0	0.0	0.0	47.9	0.0	0.0	0.0	0.0	5.6	0.3	5.9	0.0	1.1	0.0	0.0	0.0	0.0	0.0	0.0	6.4	0.0	7.3	0.0	0.0	0.0
59	0.0	0.0	0.0	0.0	0.4	0.0	23.0	3.4	0.0	0.0	0.0	7.5	0.0	0.0	0.0	1.5	0.0	0.0	0.0	0.0	0.0	0.0	0.0	0.0	0.0	0.0	0.0	0.0
60	0.0	0.0	0.0	0.0	0.0	0.0	11.5	0.0	0.0	0.0	0.0	12.2	0.3	4.1	0.0	2.4	0.0	0.0	0.0	0.0	0.0	0.0	0.0	0.0	0.0	0.0	0.0	0.0
61	0.0	0.0	0.0	0.0	0.0	0.0	4.1	9.4	0.0	0.0	0.0	5.9	0.0	1.5	0.0	9.4	0.0	0.0	0.0	0.0	0.0	0.0	0.0	0.0	0.0	0.0	0.0	0.0
62	0.0	0.0	0.0	0.0	0.0	0.0	4.3	0.0	0.0	0.0	0.0	58.4	0.0	0.0	0.0	0.2	0.0	0.0	0.0	0.0	0.0	0.0	0.0	0.0	0.0	0.0	0.2	0.0
63	0.0	0.0	0.0	0.0	0.0	0.0	5.8	1.1	0.0	0.0	0.0	9.1	0.0	0.0	0.0	4.0	0.7	0.0	0.0	0.0	0.0	0.0	0.0	0.0	0.0	0.0	0.0	0.0
64	0.0	0.0	0.0	0.0	0.0	0.0	13.8	1.4	0.0	0.0	0.0	10.0	0.0	0.0	0.0	4.2	1.4	0.0	0.0	0.0	0.0	0.0	0.0	0.0	1.4	0.0	0.0	0.0
65	0.0	0.0	0.9	0.0	0.0	0.0	15.6	0.6	0.0	0.0	0.0	7.8	1.6	0.0	0.0	3.8	6.3	0.0	0.0	0.0	0.0	0.0	1.9	0.0	0.0	0.0	0.6	0.0
66	0.0	0.0	0.0	0.0	0.0	0.0	7.4	0.0	0.0	0.0	0.0	8.9	1.8	0.0	0.0	1.8	5.8	0.0	0.0	0.0	0.0	0.0	0.5	0.0	0.5	0.0	0.0	0.0
67	0.0	0.0	0.0	0.0	0.0	2.0	0.0	11.2	0.0	0.0	0.0	2.0	0.0	0.0	0.0	2.0	0.0	0.0	0.0	0.0	0.0	0.0	0.0	0.0	0.0	0.0	0.0	0.0
68	0.0	0.0	0.0	0.0	0.0	0.0	2.8	0.0	0.0	0.0	0.0	0.9	0.6	0.9	0.0	0.0	0.0	0.0	0.0	0.0	14.5	0.0	0.0	0.0	0.0	0.0	0.0	0.0

Appendix 2. Continued.

Site NN	Latitude (°N)	Longitude (°E)	Elevation (m)	*Abies*	*Allium*	*Alnus*	Apiaceae	*Artemisia*	Astraceae undif.	*Betula*	Boraginaceae	Brassicaceae	Campanulaceae	Caryophyllaceae	Chenopodiaceae	Convolvulaceae	*Corylus*	Cyperaceae	Dipsacaceae	*Ephedra*	*Epilobium*	Ericales	Euphorbiaceae	Fabaceae	*Galium*	*Gentiana*	Geraniaceae	*Hippophae rhamn.*	*Kochia*	Lamiaceae
69	44.70	109.55	1000	0.0	2.7	0.0	0.0	38.4	1.6	1.4	0.0	5.9	0.0	0.9	27.8	0.0	0.0	0.2	0.0	0.7	0.0	0.0	0.0	0.0	0.0	0.0	0.0	0.0	0.0	2.3
70	44.75	111.25	1000	0.0	2.0	0.0	0.0	31.7	0.6	0.6	0.0	3.2	0.0	0.9	38.9	0.0	0.0	0.0	0.0	11.5	0.0	0.0	0.0	0.3	0.0	0.0	0.0	0.0	0.0	0.6
71	48.42	96.03	2100	0.0	0.0	0.0	0.0	48.6	3.9	5.1	0.0	1.6	0.0	2.3	12.1	0.0	0.0	0.0	0.0	3.1	0.0	0.0	0.0	2.3	0.0	0.0	0.0	0.0	0.0	0.0
72	47.53	98.52	2455	0.0	0.0	0.0	0.0	11.0	0.0	21.3	0.0	0.0	1.6	3.1	18.9	0.0	0.0	4.7	0.0	3.1	0.0	0.0	0.0	0.8	0.0	0.0	0.0	0.0	0.0	0.0
73	47.58	97.52	2200	0.0	0.3	0.0	0.0	33.6	2.7	3.9	0.0	3.3	0.0	2.1	12.3	0.0	0.0	3.9	0.0	2.1	0.0	0.0	0.0	7.5	0.0	0.0	0.0	0.0	0.0	2.4
74	48.60	95.43	1600	0.0	0.0	0.0	0.0	15.4	0.3	0.9	0.0	0.0	0.0	0.3	67.7	0.0	0.0	0.6	0.0	2.8	0.0	0.0	0.0	1.8	0.0	0.0	0.0	0.0	0.0	0.0
75	46.52	100.10	2029	0.0	0.0	0.0	0.0	6.5	0.0	5.3	0.0	3.6	0.0	1.2	65.1	0.0	0.0	2.4	0.0	0.6	0.0	0.0	0.0	0.6	0.0	0.0	0.0	0.0	0.0	0.0
76	46.52	101.85	2500	0.0	0.0	0.0	0.0	15.0	0.9	2.8	0.0	17.4	0.0	3.3	16.0	0.0	0.0	6.1	0.0	1.4	0.0	0.0	0.0	3.3	0.0	0.0	0.0	0.0	0.0	1.9
77	46.13	94.98	1013	0.0	5.9	0.0	0.0	29.8	2.5	3.4	0.0	16.1	0.0	0.6	12.7	0.0	0.0	0.3	0.0	0.9	0.0	0.0	0.0	0.0	0.0	0.0	0.0	0.0	0.0	0.0
78	46.25	95.37	1020	0.0	2.2	0.0	0.0	7.0	0.9	4.8	0.0	11.8	0.0	1.3	21.8	0.0	0.0	0.0	0.0	3.9	0.0	0.0	0.0	11.8	0.0	0.0	0.0	0.0	0.0	1.7
79	47.42	94.58	1400	0.0	0.7	0.0	0.0	6.0	0.4	6.4	0.0	6.4	0.0	0.0	53.2	0.0	0.0	0.4	0.0	0.7	0.0	0.0	0.0	4.1	0.0	0.0	0.0	0.0	0.0	0.0
80	51.42	99.33	1543	0.6	0.0	0.3	0.2	16.7	1.2	33.1	0.0	0.3	0.0	0.2	6.4	0.0	0.0	4.3	0.0	2.1	0.0	0.3	0.0	0.0	0.0	0.0	0.0	0.0	0.0	0.0
81	51.43	99.33	1543	0.0	0.0	0.6	0.3	34.1	3.7	32.9	0.0	0.0	0.0	0.0	7.3	0.0	0.0	1.2	0.0	0.6	0.0	0.3	0.0	0.0	0.0	0.0	0.0	0.0	0.0	0.0
82	50.17	99.02	1750	0.4	0.0	2.6	0.0	2.6	1.7	5.7	0.4	0.0	0.0	0.4	1.3	0.0	0.4	1.3	0.0	0.0	0.0	3.1	0.0	0.0	0.0	0.0	0.0	0.0	0.0	0.4
83	49.27	98.25	1700	0.0	0.0	0.0	0.0	21.6	1.7	5.8	0.0	0.0	0.0	0.4	4.6	0.0	0.0	15.0	0.0	0.0	0.0	0.0	0.0	0.0	0.0	0.0	0.0	0.0	0.0	0.0
84	48.73	98.02	1760	0.0	0.0	0.0	0.0	8.1	0.5	0.5	0.0	0.2	0.0	53.1	8.5	0.0	0.0	2.8	0.0	0.2	0.0	0.0	0.0	0.0	0.0	0.0	0.0	0.0	0.0	10.0
85	48.48	100.57	1500	0.0	0.0	0.0	0.2	19.9	2.5	1.7	0.0	0.2	0.0	1.3	2.1	0.2	0.0	24.2	0.2	0.8	0.0	0.0	0.0	0.8	0.0	0.0	0.0	0.0	0.0	0.0
86	48.45	100.50	1790	0.0	0.0	0.0	0.0	29.3	1.3	4.3	0.0	0.0	0.0	1.0	3.3	0.0	0.0	18.0	0.0	0.3	0.0	0.7	0.0	0.0	0.0	0.0	0.0	0.0	0.0	0.0
87	48.45	100.57	1500	0.0	0.0	0.0	0.3	42.0	6.5	2.0	0.5	0.0	0.0	1.0	6.0	0.0	0.0	2.5	1.5	0.5	0.0	0.0	0.0	4.5	0.0	0.0	0.0	0.0	0.0	0.3
88	48.35	100.45	1780	0.0	0.0	0.0	0.0	33.2	3.4	0.6	0.0	0.0	0.0	3.4	26.7	0.0	0.0	9.1	0.4	0.2	0.0	0.0	0.0	0.0	0.0	0.0	0.0	0.0	0.0	0.2
89	47.72	100.18	1900	0.2	0.0	0.0	0.0	23.0	0.8	2.1	0.0	0.2	0.0	7.4	2.5	0.0	0.0	24.7	0.2	1.2	0.0	0.0	8.6	0.2	0.0	0.0	0.0	0.0	0.0	0.4
90	47.70	100.32	1900	0.0	0.0	0.0	0.0	25.7	2.1	0.2	0.0	0.0	0.0	4.7	1.6	0.0	0.0	43.5	0.0	4.2	0.0	0.0	0.0	0.0	0.0	0.0	0.0	0.0	0.0	0.0
91	47.62	100.30	1900	0.0	0.0	0.0	0.0	21.0	0.9	0.2	0.0	0.4	0.0	5.8	3.6	0.0	0.0	47.8	1.1	2.2	0.0	0.0	0.0	0.0	0.0	0.0	0.0	0.0	0.0	0.0
92	47.27	99.92	2250	0.0	0.0	0.0	0.0	42.3	0.3	1.1	0.0	0.0	0.0	12.6	3.3	0.0	0.0	9.3	0.0	3.3	0.0	0.0	0.3	0.0	0.0	0.0	0.0	0.0	0.0	0.3
93	47.18	99.83	2300	0.0	0.0	0.0	0.3	20.1	0.7	0.7	0.0	0.7	0.0	2.1	8.3	0.0	0.0	24.9	0.0	4.2	0.0	0.0	0.0	0.0	0.0	0.0	0.0	0.0	0.0	3.1
94	47.17	99.92	2500	0.0	0.0	0.0	0.0	23.1	2.0	0.0	0.0	6.0	0.0	0.5	2.0	0.0	0.0	28.1	0.0	3.0	0.0	0.0	0.0	0.0	0.0	0.0	0.0	0.0	0.0	0.0
95	47.10	100.53	2300	0.0	0.0	0.0	0.3	11.5	2.6	0.5	0.0	0.0	0.0	9.4	4.7	0.0	0.0	46.0	0.0	0.3	0.0	0.0	0.0	0.0	0.0	0.0	0.0	0.0	0.0	0.3
96	46.48	102.62	1800	0.0	0.0	0.2	0.0	57.7	1.0	0.7	0.0	0.0	0.0	11.2	7.3	0.0	0.0	9.3	0.2	1.0	0.0	0.0	0.0	0.0	0.0	0.0	0.0	0.0	0.0	0.0
97	46.32	101.20	2100	0.0	0.0	0.0	0.0	51.9	6.2	0.1	0.0	0.3	0.0	0.3	24.9	0.0	0.0	6.2	0.0	0.3	0.0	0.0	0.0	0.0	0.0	0.0	0.0	0.0	0.0	0.0
98	45.45	100.63	1450	0.0	0.0	0.4	0.0	75.8	3.0	0.9	0.0	1.7	0.0	1.3	4.3	0.0	0.0	3.5	0.0	1.7	0.0	0.0	0.0	0.0	0.0	0.0	0.0	0.0	0.0	0.0
99	48.13	93.88	1200	0.0	0.0	0.0	0.0	39.4	0.1	0.5	0.0	0.0	0.0	0.1	46.6	0.0	0.1	0.9	0.0	0.9	0.0	0.0	0.0	0.0	0.0	0.0	0.0	0.0	0.0	0.0
100	47.53	95.25	1400	0.0	0.0	0.8	0.0	30.4	0.8	4.8	0.0	0.8	0.0	1.6	35.2	0.0	0.0	0.8	0.0	0.0	0.0	0.0	0.0	0.0	0.0	0.0	0.0	0.0	0.0	0.8
101	47.55	95.30	1350	0.0	0.0	0.0	0.0	12.0	1.8	1.4	0.0	0.0	0.0	0.0	4.1	0.0	0.0	2.8	0.0	0.5	0.0	0.0	0.0	0.0	0.0	0.0	0.0	24.0	0.0	0.0
102	47.40	95.95	1450	0.1	0.0	0.4	0.0	33.9	3.9	0.6	0.0	0.5	0.0	0.5	32.0	0.5	0.2	7.7	0.0	1.9	0.0	0.0	0.0	0.0	0.0	0.0	0.0	0.0	0.0	0.5

Appendix 2. Continued.

Site NN	*Larix*	Liliaceae	*Linaria*	*Lonicera*	*Pedicularis*	*Picea*	*Pinus sylvestris*	*Pinus Haploxylon*	*Pinus undif.*	*Plantago*	Plumbaginaceae	Poaceae	Polygonaceae	*Potentilla*	Primulaceae	Ranunculaceae	Rosaceae	Rubiaceae	Rutaceae	*Salix*	*Salsola*	Saxifragaceae	*Scabiosa*	Scrophulariaceae	*Stellaria*	*Ulmus*	Valerianaceae	Zygophyllaceae
69	0.0	0.0	0.0	0.0	0.0	0.0	3.4	0.0	0.0	0.0	0.0	4.1	1.4	0.0	0.0	1.4	7.4	0.0	0.0	0.0	0.7	0.0	0.0	0.0	0.0	0.0	0.0	0.0
70	0.0	0.0	0.0	0.0	0.0	0.0	0.9	0.0	0.0	0.0	0.0	2.0	0.9	0.3	0.0	0.6	4.9	0.0	0.0	0.0	0.3	0.0	0.0	0.0	0.0	0.0	0.0	0.0
71	0.0	0.0	0.0	0.0	0.0	0.0	3.1	5.1	0.0	0.0	0.0	10.5	0.0	0.0	0.0	2.3	0.0	0.0	0.0	0.0	0.0	0.0	0.0	0.0	0.0	0.0	0.0	0.0
72	0.0	0.0	0.0	0.0	0.0	0.0	0.0	0.8	0.0	0.0	0.0	20.5	6.3	0.0	0.0	7.9	0.0	0.0	0.0	0.0	0.0	0.0	0.0	0.0	0.0	0.0	0.0	0.0
73	0.6	0.0	0.0	0.0	0.0	0.6	0.6	1.8	0.0	0.0	0.0	15.9	0.3	2.1	0.0	3.9	0.0	0.0	0.0	0.0	0.0	0.0	0.0	0.0	0.0	0.0	0.0	0.0
74	0.6	0.0	0.0	0.0	0.0	1.2	0.9	6.2	0.0	0.0	0.0	0.6	0.0	0.0	0.0	0.0	0.6	0.0	0.0	0.0	0.0	0.0	0.0	0.0	0.0	0.0	0.0	0.0
75	1.2	0.0	0.0	0.0	0.0	0.0	0.0	0.6	0.0	0.0	0.0	10.1	0.6	0.0	0.0	1.2	0.0	0.0	0.0	0.0	1.2	0.0	0.0	0.0	0.0	0.0	0.0	0.0
76	0.0	0.0	0.0	0.0	0.0	0.0	0.0	13.6	0.0	0.0	0.0	8.0	7.0	0.0	0.0	3.3	0.0	0.0	0.0	0.0	0.0	0.0	0.0	0.0	0.0	0.0	0.0	0.0
77	0.0	0.0	0.0	0.0	0.0	0.6	0.0	4.3	0.0	0.0	0.0	9.0	0.0	0.0	0.0	0.0	0.0	0.0	0.0	0.0	13.7	0.0	0.0	0.0	0.0	0.0	0.0	0.0
78	0.0	0.0	0.0	0.0	0.0	0.9	22.7	0.0	0.0	0.0	0.0	5.7	0.4	0.0	0.0	2.2	0.9	0.0	0.0	0.0	0.0	0.0	0.0	0.0	0.0	0.0	0.0	0.0
79	0.0	0.0	0.0	0.0	0.0	0.0	0.0	0.7	0.0	0.0	0.0	6.4	0.0	0.0	0.0	0.4	10.1	0.0	0.0	0.0	4.1	0.0	0.0	0.0	0.0	0.0	0.0	0.0
80	0.3	0.0	0.0	0.0	0.0	1.2	0.0	0.0	23.4	0.0	0.0	8.5	0.0	0.0	0.0	0.0	0.0	0.0	0.0	0.6	0.0	0.3	0.0	0.0	0.0	0.0	0.0	0.0
81	0.0	0.0	0.0	0.0	0.0	3.0	0.0	0.0	7.3	0.0	0.0	7.3	0.0	0.0	0.0	0.0	0.0	0.0	0.0	0.6	0.0	0.0	0.0	0.0	0.0	0.6	0.0	0.0
82	0.9	0.4	0.0	0.0	0.0	29.3	0.0	0.0	46.7	0.0	0.0	0.9	0.4	0.0	0.0	0.4	0.0	0.0	0.0	0.4	0.0	0.0	0.0	0.0	0.0	0.0	0.0	0.0
83	2.9	0.0	0.0	0.0	0.0	0.4	0.0	0.0	31.6	0.0	0.0	15.0	0.2	0.0	0.0	0.4	0.0	0.0	0.0	0.4	0.0	0.0	0.0	0.0	0.0	0.0	0.0	0.0
84	4.7	0.0	0.0	0.0	0.0	1.9	0.0	0.0	8.1	0.0	0.0	0.9	0.2	0.0	0.0	0.0	0.0	0.0	0.0	0.2	0.0	0.0	0.0	0.0	0.0	0.0	0.0	0.0
85	4.2	0.2	0.0	0.2	0.0	0.2	0.0	0.0	22.5	0.0	0.0	17.4	0.0	0.0	0.0	0.0	0.0	0.0	0.0	0.8	0.0	0.0	0.0	0.0	0.0	0.1	0.0	0.0
86	2.0	0.3	0.0	0.0	0.0	0.0	0.0	0.0	32.3	0.0	0.0	6.7	0.0	0.0	0.0	0.3	0.0	0.0	0.0	0.0	0.0	0.0	0.0	0.0	0.0	0.0	0.0	0.0
87	2.5	0.5	0.0	0.0	0.0	0.0	0.0	0.0	14.5	0.0	1.0	12.0	1.5	0.0	0.0	0.0	0.0	0.0	0.0	0.5	0.0	0.0	0.0	0.0	0.0	0.0	0.0	0.0
88	2.4	0.4	0.0	0.0	0.0	0.2	0.0	0.0	11.6	0.0	0.0	7.3	0.0	0.0	0.0	0.0	0.0	0.4	0.0	0.2	0.0	0.0	0.0	0.0	0.0	0.0	0.0	0.0
89	1.6	0.0	0.0	0.0	0.0	0.2	0.0	0.0	21.0	0.0	0.0	4.9	0.2	0.0	0.0	0.4	0.0	0.0	0.0	0.0	0.0	0.0	0.0	0.0	0.0	0.0	0.0	0.0
90	2.1	0.0	0.0	0.0	0.0	0.0	0.0	0.0	3.1	0.0	0.0	12.6	0.0	0.0	0.0	0.0	0.0	0.0	0.0	0.3	0.0	0.0	0.0	0.0	0.0	0.0	0.0	0.0
91	2.7	0.0	0.0	0.0	0.0	0.7	0.0	0.0	6.7	0.0	0.0	5.6	0.2	0.0	0.0	0.2	0.0	0.0	0.0	0.9	0.0	0.0	0.0	0.0	0.0	0.0	0.0	0.0
92	0.0	0.3	0.0	0.0	0.0	0.0	0.0	0.0	4.9	0.0	0.0	21.4	0.3	0.0	0.0	0.3	0.0	0.0	0.0	0.0	0.0	0.0	0.0	0.0	0.0	0.0	0.0	0.0
93	1.4	0.7	0.0	0.0	0.0	0.0	0.0	0.0	12.1	0.0	0.0	20.1	0.3	0.0	0.0	0.3	0.0	0.0	0.0	0.0	0.0	0.0	0.0	0.0	0.0	0.0	0.0	0.0
94	6.3	1.0	0.0	0.0	0.0	0.0	0.0	0.0	3.8	0.0	0.0	18.1	0.5	0.0	0.0	5.0	0.0	0.0	0.0	0.0	0.0	0.0	0.0	0.0	0.0	0.0	0.5	0.0
95	1.8	0.3	0.0	0.0	0.0	0.3	0.0	0.0	8.4	0.0	0.0	13.1	0.5	0.0	0.0	0.3	0.0	0.0	0.0	0.0	0.0	0.0	0.0	0.0	0.0	0.0	0.0	0.0
96	0.0	0.2	0.0	0.0	0.0	0.1	0.0	0.0	6.4	0.0	0.0	3.9	0.0	0.0	0.0	0.2	0.0	0.0	0.0	0.4	0.0	0.0	0.0	0.0	0.0	0.0	0.0	0.0
97	0.0	0.0	0.0	0.0	0.0	0.1	0.0	0.0	1.8	0.0	0.3	6.2	0.3	0.0	0.0	0.3	0.0	0.0	0.0	0.1	0.0	0.0	0.0	0.0	0.0	0.0	0.0	0.3
98	0.0	0.9	0.0	0.0	0.0	0.0	0.0	0.0	1.3	0.0	0.0	4.8	0.0	0.0	0.0	0.0	0.0	0.0	0.0	0.4	0.0	0.0	0.0	0.0	0.0	0.0	0.0	0.0
99	0.1	0.0	0.0	0.0	0.0	0.2	0.0	0.0	9.1	0.9	0.1	0.9	0.1	0.0	0.0	0.0	0.0	0.0	0.0	0.0	0.0	0.0	0.0	0.0	0.0	0.0	0.0	0.0
100	1.6	0.8	0.0	0.0	0.0	0.8	0.0	0.0	9.6	4.0	0.0	7.2	0.0	0.0	0.0	0.0	0.0	0.0	0.0	0.0	0.0	0.0	0.0	0.0	0.0	0.0	0.0	0.0
101	0.0	0.2	0.0	0.0	0.0	0.0	0.0	0.0	2.8	0.0	0.2	8.3	0.0	0.0	0.0	0.0	0.0	0.0	0.0	41.9	0.0	0.0	0.0	0.0	0.0	0.0	0.0	0.0
102	0.2	0.7	0.0	0.0	0.0	0.4	0.0	0.0	4.3	0.0	0.5	10.7	0.0	0.0	0.0	0.5	0.0	0.0	0.0	0.1	0.0	0.0	0.0	0.0	0.0	0.0	0.0	0.0

INDEX

Aboveground biomass, 110, 144
Achit-Nur, 25, 50, 51, 52, 57, 58, 59, 74
Adgiin-Tsagan-Nur, 126, 199
Aesthetic resources, 178
After-fire successions, 86
Aggressive plant species, 94
Aimak, 40, 132, 175
Ak-Su, 55
Alag Khairkhan (Nature reserve), 190
Alagnur, 23
Alashan desert, 29
Alashan Gobi, 15, 29, 97, 98, 116, 125, 127, 149, 150, 169, 170, 182, 184, 202
Altai, 166, 168, 169, 170, 178, 182, 183, 184, 186, 190, 194, 195, 196
Altai, 9, 11, 12, 14, 15, 16, 23, 24, 25, 28, 42, 47, 54, 55, 57, 66, 67, 68, 69, 76, 81, 82, 94, 96, 158
Altai-Sayan region, 15
Altai Tavan Bogd (National park), 190, 196
Altan-Bulgii, 18
Altan Shira, 199
Annual atmospheric precipitation, 10, 11
Annuals plants, 152, 153
Anthropogenic factors, 5, 29, 33, 34, 43, 80, 81, 85, 93
Anthropogenic influence, 33, 35, 113, 152, 177
Arable farming, 116
Arable land, 40, 123, 152, 177
Arkhangai, 175, 190
Arig-Gol, 16
Arvaikhere, 42, 153, 186
Aspen forests, 19, 93
Asgat-Tsagan-Tolgoi, 195, 198
Asralt-Khairkhan, 18
Assessing, 5, 188

Baga-Gazaryn-Nuruu, 197
Baganur, 42, 186
Baidrag-Gol, 27, 199
Baikal Lake, 10
Baitag-Bogdo, 168, 169, 170, 172, 201
Barun-Khurai, 28, 172
Barun-Sukhain river, 150
Batkhan Uul (Nature reserve), 190
Bayandelger, 187
Bayan-Gobi, 11
Bayankhongor, 132
Bayan-Sair, 50, 53, 66, 67, 68
Bayan-Tsagan, 11
Bayan-Tukhemiin Gobi, 194, 195, 198
Bayan-Ulgii, 175, 190
Bayan-Under, 11
Biennials plants, 150, 152, 153
Biological resources, 173
Biosphere reserve, 188, 192
Birch forests, 19, 86, 87, 88, 93, 136, 140, 150, 183, 196
Black Irtysh river, 16, 196
Bogd Khan Uul (Biosphere reserve), 188, 190, 192, 193, 197
Bogging, 134
Bon-Tsagan-Nur, 27, 132, 194, 195, 199
Borig-Del-Els, 25
Borkhar-Els, 26
Borzongiin Gobi, 29, 195, 198, 199, 200, 201
Botanical Institute of the USSR Academy of Sciences, 3, 4
Botanical-geographical investigations, 4, 5
Botanical-geographical zoning, 3
Botanical successions, 173
Buir-Nur lake, 19
Bulgan, 116, 133, 175, 190
Bulgan Uul (Natural monument), 190
Bulgan-Gol (Nature reserve), 178
Burenglin Nuruu, 18
Burengiin ridges, 18
Burrower mammals, 99
Burkhan Buudai (Nature reserve), 190

Burunnur, 23
Busseiin-Gol, 16
Buteiliin-Nuruu, 18
Buteiliin ridges, 15, 18

Channeling processes, 118
Chingiz Khan memorial, 197
Choibalsan, 40, 42, 121, 124, 186, 187
Chulut, 121
Cis-Khyangan, 172
Classification of forests, 5
Climate fluctuations, 123
Climatic changes, 29, 33, 43, 132
Climatic cycles, 131, 132
Cluster reserve, 189, 191
Cold aridization, 94
Community diversity, 172, 198
Community structure, 110, 139, 151
Conifer species, 93, 140
Conservation, 93, 158, 164, 165, 173, 178, 186, 188, 189, 193, 194, 195, 198, 199, 201, 202, 204
Cover percentage, 95
Cryogenic process, 134
Cryo-hydrogenic process, 9, 10, 134
Cryomorphic taiga soils, 85
Cryophite forb sedge communities, 18, 21, 82

Daba-Nur, 50, 52, 60, 61, 62
Dagurian, 19, 167, 191, 175, 190
Dalanzadgad, 15, 42, 133
Dariganga, 15, 19, 20, 107, 117, 182, 184, 194, 195, 198
Darkhat depression, 15, 18
Davst, 106, 145, 162
Decorative grasses and bushes, 182
Deeper water, 118
Deflation, 8
Degradation successions, 158, 166
Degrading successions, 134
Delger-Muren, 16
Deris, 24
Desertifed steppes, 12, 13, 159
Desertification, 9, 10, 140
Disturbing factors, 35
Dolongyn Gobi, 15, 19, 20
Dood-Nur, 50, 53, 59, 60, 73
Dornod (aimak), 97, 98, 175, 190, 193
Dornod Gobi, 97, 98
Dornod Mongol (Strictly protected area), 190, 193
Dornogobi (aimak), 175, 190
Dund, 59
Dundgobi, 175, 190
Dynamics ecosystems, 81
Dzakhain-Us, 172
Dzakhoi-Tooroi, 11
Dzungaria, 170, 178
Dzungarian Gobi, 14, 15, 28, 29, 170, 172, 182, 183, 184, 185, 192, 194, 201

East Mongolian Plain, 14, 19
Ecological factors, 29
Ecological potential, 186
Ecological profile, 79, 81
Ecosystems of Mongolia, 5, 42, 158, 163
Eej Khairkhan (Natural monument), 190
Egiin-Gol, 18
Ekhiin-Gol, 11, 127, 149, 153, 154, 156
Endangered species, 167, 173
Endemic plants, 167, 168, 178
Erdene, 11
Erdenet, 40, 42, 186
Eren-Daba, 169, 172, 195, 196
Ergeliin Zoo (Nature reserve), 190
Ero-Gol, 19, 121
Evergreens species, 167
Expert assessment, 159, 165
Extra-arid deserts, 11, 13, 103, 114, 144, 147, 153, 156, 176
Extreme arid deserts, 12, 41, 154, 156, 172
Extreme ecological conditions, 8

Feathergrass steppe, 101
Flood plain plant communities, 122
Flora of Mongolia, 4
Floristic and systematic investigation, 4
Forb-features-bistort community, 18, 82, 94
Forb-grass meadows, 108
Forest canopy, 39, 84, 88, 93
Forest fells, 38
Forest Institute of the Sibirian Section of the USSR Academy of Sciences, 4
Forest hydrology, 5
Forest islands, 196
Forest productivity, 93

Forest-growing properties, 93
Fossorial animals, 33, 100, 101, 102
Galbyn Gobi, 194, 195, 198, 199
Ganga Nuur (Natural monument), 190
Gashun Gobi, 15, 29, 30
Geobotanical surveys, 4
Gobi Altai, 12, 23, 27, 28, 29, 47, 66, 67, 68, 77, 81, 82, 94, 168, 172
Gobi Deserts, 114
Gobi Gurvansaikhan Uul (National park), 190, 191, 193
Gobi Tyan-Shan, 12, 23, 24, 28, 29, 82, 103, 104, 143, 172
Gobian Altai, 168, 169, 170, 182, 183, 184
Gorkhi Terelj (National park), 190
Grass-forb communities, 102, 120
Grassing of livestock, 34
Great Gobi (Biosphere reserve), 189, 190, 192, 198, 200
Great Khyangan, 7, 14, 166
Great Lakes depression, 24, 25, 27, 182, 184
Gun-Nur, 50, 53, 64, 65, 76
Gurvansaikhan, 23, 24, 193, 199
Gurvan-Tes, 11, 24, 29, 143

Halohydromorphic communities, 117
Halophytization, 33
Haloxeromorphic species, 30, 120
Hara-Nur, 11, 25
Hara-Us-Nur, 11, 25, 26, 50, 52
Health resources, 178
Heavy forest fires, 84
Hirgis-Nur, 11, 25
Hoit-Gol, 50, 52, 75, 76
Hoton-Nur, 50, 51, 54, 55, 56, 57, 68
Hubsugul area, 9, 16
Hubsugul Lake, 5, 18, 172
Hubsugul Region, 5, 16
Human-associated disturbances, 42, 43, 161, 163, 192, 193
Human-associated factors, 43, 79, 118, 128, 140, 163, 203
Human-induced successios, 41, 42
Human-stimulated natural factors, 81
Hydromorphic plant communities, 118

Ider River, 10
Ikh Nart (Nature reserve), 190
Indicator species for grazing pressure, 114, 123
Ingenii Khovraiin Kholoi, 148, 149
Institute of Botany of the Academy of Sciences of Mongolia, 4
Irrigated agriculture, 34, 41, 186
Irrigated croplands, 116, 117, 123

Joint Soviet (Russian)-Mongolian Complex Biological Expedition, 3, 166

Karatyr, 55
Kerulen, 10, 15, 19, 84, 94, 109, 110, 120, 121, 186, 187
Khalkha, 14, 22, 30, 123, 124, 132, 169, 176, 182, 183, 184, 185, 194, 195, 197
Khalkhin-Gol, 20
Khangai, 2, 4, 9, 11, 12, 14, 15, 20, 21, 24, 27, 33, 38, 42, 47, 82, 83, 84, 86, 88, 93, 95, 96, 100, 101, 105, 112, 118, 166, 168, 169, 170, 172, 176, 178, 180, 182, 183, 184, 185, 186, 190, 194, 195, 196, 197
Khan Khentii (Strictly protected area), 190
Khangai Nuruu (National park), 190
Khankh, 193
Khankhukhiin ridge, 20, 21, 24
Khara-Gol, 10, 19, 116, 121, 123
Khara Us Nuur (National park), 190
Kharkhorin, 121, 123, 186
Kharmai, 59
Khassagt-Khaiirkhan (Strictly protected area), 27
Khatgal, 193
Khentii, 4, 9, 11, 12, 14, 15, 18, 19, 36, 38, 61, 66, 70, 82, 84, 85, 86, 88, 91, 94, 112, 118, 166, 169, 170, 171, 172, 176, 178, 182, 183, 184, 185, 186, 190, 196
Khiidiin-Sardag, 18
Khokh Serkh (Strictly protected area), 190
Khorgo Tsagaan Nuur Terkh (National park), 190
Khovd, 25, 26, 55, 57, 175, 186, 190
Khovsgol Nuur (National park), 190, 193, 200
Khudzhirt, 178
Khugne Khan Uul, 197
Khuisiin Naiman Nuur (Natural monument), 190
Khungii-Gol, 25
Khuren-Khana, 194, 199

Khurshut, 116
Khustain Nuruu (Nature reserve), 190
Khuvsgul, 175
Khuv-Usny-Gol, 57

Lagnuur, 194, 195, 198
Lake Valley, 27, 30, 115, 123, 126, 132, 148, 169, 170, 176, 182, 183, 184, 194, 195, 199
Lake water level, 118, 131, 132
Landscape-ecological regions, 14
Landscape factors, 14
Larch forest, 13, 48, 56, 73, 82, 84, 85, 86, 87, 88, 90, 92, 93, 118, 140, 150, 167, 172, 183, 196
Lhachinvandad Uul (Nature reserve), 190
Linear erosion, 8, 41, 87
Liquid surface runoff, 93
Little Gobi (Strictly protected area), 190, 196, 199
Low-grass steppes, 108
Lyme grass-deris stands, 124

Manchurian, 20
Mandal Gobi, 15, 22, 42
Map of Mongolian Vegetation, 3
Map of the Forest of Mongolia, 5
Map of Vegetation of Mongolia, 5
Mapping vegetation dynamics, 157
Matad, 19
Mesophyte types, 112
Middle Gobi province, 27
Middle Khalkha province, 19
Middle Selenge province, 18
Minning idustry, 40, 162
Mixed pine-birch forests, 87, 91
Modern changes, 131
Mogen-Buren River, 57
Moisture, 9
Moltsog-Els, 20
Mongol Dagurian (Transborder reserve), 190, 191, 193, 201
Mongol-Els, 26, 105
Mongolian Agricultural Expedition, 3
Mongolian Altai, 9, 14, 16, 24, 195, 196
Mongolian Protected Areas, 189, 190
Monitoring, 131, 163, 188
Moscow State University, 4
Mt. Kuiten-Uul, 9
Mt. Munkhe-Khairkhan, 9
Munku Sardyk, 200
Muren, 62, 121

Nagalkhan Uul (Nature reserve), 190
Nairyin-Gol, 172
National concervation park, 188, 189
National park, 173
Natural and Historical monument, 190
Natural oasis, 14, 41
Natural resources, 178
Nature reserve, 190
Negative degradation processes, 173
Nomrog (Strictly protected area), 190, 193
Northern deserts, 12

Obon-Els, 107
Obot-Khural, 116, 125, 149, 181
Oldrakh-Khinu, 199
Ombrophyte plants, 141
Ongiin-Gol, 28
Onon, 10, 19, 121, 196
Open woodlands under bald mountains, 92
Orkhon river, 179, 186, 194, 200
Orkhon-Tuul province, 22
Orkhon valley, 22
Orog-Nur, 27, 132, 145, 195
Otgon-Tenger (Strictly protected area), 9, 20
Overgrazing, 36, 82, 110, 114, 145

Pastor degradation, 94
Pastoral succession, 109, 110
Pasture degrading, 108, 109, 124
Pasture overload, 35
Perennial plants, 151, 153
Phreatophyte plants, 142
Pine forests, 18, 64, 72, 84, 85, 86, 87, 90, 91, 104, 106, 136, 140, 182, 184, 197
Plant community, 79, 82, 96, 98, 104, 106, 107, 108, 112, 114, 118, 123, 124, 126, 127, 132, 134, 136, 139, 144, 149, 152, 156, 158, 163, 166, 167, 172, 173, 175, 196, 198, 203, 204
Plants of Central Asia, 3
Present-day plant cover, 79
Preserves, 173, 188, 189, 196
Primary coniferous species, 93
Processes of denudation, 32
Progressive succession, 134, 149

Rainfed agriculture, 41, 174
Rare species, 167

Recreational sites, 178
Recultivation, 173, 178, 185, 186, 204
Recultivation measures, 204
Red Data Book of Mongolia, 167
Regeneration processes, 4, 153, 165
Regenerative successions, 86, 93, 163, 178
Regressive successions, 96, 135, 136, 139, 149
Reserves, 173, 174, 188, 189, 190, 191, 192, 193, 194, 195, 198, 201, 202, 204
Restoration, 152, 173, 188, 204
River floodplains, 11, 14, 118
River valleys, 8, 14, 16, 19, 21, 22, 25, 40, 46, 55, 66, 76, 105, 118, 119, 123, 134, 194, 196, 197
Rock eluvium, 98, 99
Rock streams, 84, 85, 93, 134, 136
Rocky-sandy soils, 95
Root system, 11, 13, 30, 36, 97, 134, 140, 143, 144, 154, 203
Ruderal plants, 35

Saikhan Dulan, 199
Saksai Depression, 96
Sand-loving plants, 14, 106, 107, 114
Sandy soils, 112, 139, 145
Sandy-rubble alluvium, 118
Satellite photos, 80, 148, 158
Selenga river, 5, 18, 85, 86, 172, 188
Selenge (aimak), 175, 190
Semi-fixed sand dunes, 148, 172
Sengilen ridge, 18, 24
Sharga-Mankhan (Nature reserve), 190
Shargain Gobi, 27
Shargain Gobi depression, 27
Shargain-Gol, 27
Shinedzhinst, 11
Shishkhid-Gol, 16, 17, 59, 200
Siberian pine forest, 84, 85, 86
Silty crust, 148
Slope erosion, 39
Soil moisture, 140
Solar activity, 131, 132, 133
Solonchak deris stands, 124
Somon, 40
Southern deserts, 46, 114
Spring, 116
Stages in succession series, 118
Steppe-like communities, 81
Steppized deserts, 4, 30, 114
Strategies for natural management, 165
Strategies for vegetation conservation, 173
Stream communities, 118
Strictly Protected Areas, 173, 202
Sudzhiin-Khuduk, 50, 54, 67
Suikhent Uul (Natural monument), 190
Sukhbaatar (aimak), 145, 175, 190
Synopsis of Flora of Mongolia, 3

Tarbagan marmot, 33
Tarbagatai Ridge, 60
Tannu-Ola, 12
Tatsyn-Tsagan-Nur, 27, 30, 132, 145
Tesiin-Gol, 24, 172
Tevshrulekh, 108
Tov (aimak), 175, 190
Transaltai Gobi, 11, 14, 15, 29, 30, 31, 41, 97, 98, 99, 103, 106, 127, 128, 144, 146, 147, 148, 153, 154, 156, 168, 169, 170, 181
Transbaikal region, 15, 18
Transboundary reserves, 191, 193, 201
True deserts, 14, 27, 29, 97, 99, 103, 114, 144, 172
Tsagan-Bogdo, 156
Tsagangol, 23
Tumentsogt, 100, 111, 112, 113, 117
Tunkin National Park, 200
Turan-Dzungarian, 14
Turgen mountains, 191
Tuul-Gol, 22, 186, 197

Ubsu-Nur, 11, 24, 25, 172
Ugtam Uul (Nature reserve), 190
Ulaan-Taiga, 16
Ulan Gobi, 194, 198, 199
Ulan-Bator, 40, 41, 42, 153, 179, 186, 188, 193
Ulangom, 42, 57
Ulan-Nur, 132
Uldza, 19, 196
Uliastai, 61
Umnugobi, 175
Underkhan, 42, 186, 187
Undisturbed plots, 112
Undzhul, 197

Unfixed sands, 93, 108, 114
Unpalatable species, 119
Uran-Togoo Tulga Uul (Natural monument), 190
Ure-Gol, 16
Ureg-Nur lake, 15, 16
Urumqi River Valley, 68
Usun-Kholoi River, 57
Uvs, 175, 189, 190, 191, 195, 196, 201
Uvs Nuur, 189, 190, 191, 195, 196, 201
Uvs Nuur Basin (Strictly protected area), 190, 191, 196
Uvurkhangai, 175

Vegetation distribution, 13, 21
Vegetation successions, 32, 86, 92, 96, 97, 100, 102, 108, 118, 119, 120, 123, 129, 135, 153, 157, 158, 162, 163, 173, 174, 197, 203

Water erosion, 30, 131, 134, 136, 148, 163, 165, 177
Water protection zones, 187
Water-accumulating ruts, 176
Weed community, 112
Weed groups, 109
Wetland-forb communities, 118
Winter grazing, 120
Wolf number, 132
Wormwood-hard-sedge communities, 94
Wormwood-saltwort deris stands, 124

Xeromorphic species, 96, 112, 119, 140
Xerophytozation, 33

Yamant-Nur, 50, 53, 62, 63

Zagiin Us (Nature reserve), 190
Zavkhan, 25, 26, 175, 190
Zoogenic microcomplexes, 32, 33, 100, 101
Zoogenic successions, 100, 101

Geobotany

1. J.B. Hall and M.D. Swaine (eds.): *Distribution and Ecology of Vascular Plants in a Tropical Rain Forest.* Forest Vegetation in Ghana. 1981 ISBN 90-6193-681-0
2. W. Holzner and M. Numata (eds.): *Biology and Ecology of Weeds.* 1982 ISBN 90-6193-682-9
3. N.J.M. Gremmen: *The Vegetation of the Subantarctic Islands Marion and Prince Edward.* 1982 ISBN 90-6193-683-7
4. R.C. Buckley (ed.): *Ant-Plant Interactions in Australia.* 1982 ISBN 90-6193-684-5
5. W. Holzner, M.J.A. Werger and I. Ikusima (eds.): *Man's Impact on Vegetation.* 1983 ISBN 90-6193-685-3
6. P. Denny (ed.): *The Ecology and Management of African Wetland Vegetation.* 1985 ISBN 90-6193-509-1
7. C. Gómez-Campo (ed.): *Plant Conservation in the Mediterranean Area.* 1985 ISBN 90-6193-523-7
8. J.B. Faliński: *Ecological Studies in Białowieza Forest.* 1986 ISBN 90-6193-534-2
9. G.A. Ellenbroek: *Ecology and Productivity of an African Wetland System.* The Kafue Flats, Zambia. 1987 ISBN 90-6193-638-1
10. J. van Andel, J.P. Bakker and R.W. Snaydon (eds.): *Disturbance in Grasslands.* Causes, Effects and Processes. 1987 ISBN 90-6193-640-3
11. A.H.L. Huiskes, C.W.P.M. Blom and J. Rozema (eds.): *Vegetation Between Land and Sea.* Structure and Processes. 1987 ISBN 90-6193-649-7
12. G. Orshan (ed.): *Plant Pheno-morphological Studies in Mediterranean Type Ecosystems.* 1988 ISBN 90-6193-656-X
13. B. Dell, J.J. Havel and N. Malajczuk (eds.): *The Jarrah Forest.* A Complex Mediterranean Ecosystem. 1988 ISBN 90-6193-658-6
14. J.P. Bakker: *Nature Management by Grazing and Cutting.* 1989 ISBN 0-7923-0068-8
15. J. Osbornová, M. Kovářová, J. Lepš and K. Prach (eds.): *Succession in Abandoned Fields.* Studies in Central Bohemia, Czechoslovakia. 1990 ISBN 0-7923-0401-2
16. B. Gopal (ed.): *Ecology and Management of Aquatic Vegetation in the Indian Subcontinent.* 1990 ISBN 0-7923-0666-X
17. B.A. Roberts and J. Proctor (eds.): *The Ecology of Areas with Serpentinized Rocks.* A World View. 1991. ISBN 0-7923-0922-7
18. J.T.A. Verhoeven (ed.): *Fens and Bogs in the Netherlands.* Vegetation, History, Nutrient Dynamics and Conservation. 1992 ISBN 0-7923-1387-9
19. Woo-seok Kong and D. Watts: *The Plant Geography of Korea.* With an Emphasis on the Alpine Zones. 1993 ISBN 0-7923-2068-9

Geobotany

20. R. Aerts and G.W. Heil (eds.): *Heathlands*. Patterns and Processes in a Changing Environment. 1993 ISBN 0-7923-2094-8
21. W. van Vierssen, M. Hootsmans and J. Vermaat (eds.): *Lake Veluwe, a Macrophyte-dominated System under Eutrophication Stress*. 1994 ISBN 0-7923-2320-3
22. Y. Laumonier: *The Vegetation and Physiography of Sumatra*. 1997 ISBN 0-7923-3761-1
23. C.M. Finlayson and I. von Oertzen (eds.): *Landscape and Vegetation Ecology of the Kakadu Region, Northern Australia*. 1996 ISBN 0-7923-3770-0
24. R. Peters: *Beech Forests*. 1997 ISBN 0-7923-4485-5
25. S.A. Ghazanfar and M. Fisher (eds.): *Vegetation of the Arabian Peninsula*. 1999 ISBN 0-7923-5015-4
26. P.D. Gunin, E.A. Vostokova, N.I. Dorofeyuk, P.E. Tarasov and C.C. Black (eds.): *Vegetation Dynamics of Mongolia*. 1999 ISBN 7923-5582-2

KLUWER ACADEMIC PUBLISHERS – DORDRECHT / BOSTON / LONDON